深度学习训练营

21天实战
TensorFlow+Keras+scikit-learn

张强／编著

图书在版编目（CIP）数据

深度学习训练营：21天实战TensorFlow+Keras+scikit-learn / 张强编著. -- 北京：人民邮电出版社，2020.4
 ISBN 978-7-115-44615-2

Ⅰ. ①深… Ⅱ. ①张… Ⅲ. ①机器学习 Ⅳ. ①P181

中国版本图书馆CIP数据核字(2019)第268449号

内容提要

本书基于TensorFlow、Keras和scikit-learn，介绍了21个典型的人工智能应用场景。全书共3篇，分别是预测类项目实战篇、识别类项目实战篇和生成类项目实战篇。其中预测类项目包括房价预测、泰坦尼克号生还预测、共享单车使用情况预测、福彩3D中奖预测、股票走势预测等8个项目；识别类项目包括数字识别、人脸识别、表情识别、人体姿态识别等7个项目；生成类项目包括看图写话、生成电视剧剧本、风格迁移、生成人脸等6个项目。

本书代码丰富，注释详尽，适合有一定Python基础的读者，包括计算机相关专业的学生、程序员和人工智能神经网络的技术爱好者。

◆ 编　著　张　强
　　责任编辑　任芮池
　　责任印制　王　郁　马振武
◆ 人民邮电出版社出版发行　北京市丰台区成寿寺路 11 号
　　邮编　100164　电子邮件　315@ptpress.com.cn
　　网址　http://www.ptpress.com.cn
　　北京鑫正大印刷有限公司印刷
◆ 开本：800×1000 1/16
　　印张：22.25
　　字数：602千字　　2020年4月第 1 版
　　印数：1 – 2 500 册　2020年4月北京第 1 次印刷

定价：69.00 元
读者服务热线：(010)81055410　印装质量热线：(010)81055316
反盗版热线：(010)81055315
广告经营许可证：京东工商广登字 20170147 号

前　言

当下我们的生活和工作已经被人工智能和深度学习的技术所包围，例如人机对话、人脸识别过闸机或付款、各种语言间的自动翻译、自动驾驶汽车、淘宝的个性化推荐等，各大科技公司都在努力发展的"千人千面"精细化产品也离不开人工智能（它背后的技术就是深度学习）。

坦白来讲，深度学习就是对深度神经网络架构的学习，基于神经网络设计出各种架构的模型，如卷积神经网络（CNN）、循环神经网络（RNN）、长短期记忆网络（LSTM）、生成对抗网络（GAN）等。每一种神经网络在不同的领域起着不同的作用，例如在图像识别领域的人脸识别一般使用 CNN；语言翻译或一些需要有记忆状态的领域一般用 RNN 或者 LSTM；需要使用神经网络模型生成各种图像和视频的领域就可以用 GAN。

自从 Google 公司开源了 TensorFlow 深度学习框架后，深度学习这门技术成为广大开发者、科研人员和企业最实用的技术；之后不久，基于 TensorFlow 深度学习框架编写而成的 Keras 深度学习框架就诞生了，Keras 使得编写神经网络模型更简单、更易理解、更高效，只需几行代码就可以完成一个 CNN 模型。Keras 有着快速编写神经网络模型的原型的称号，其后端之一就是 TensorFlow。对这两个框架的掌握，现在基本已成为各大公司对深度学习工程师的招聘要求的标配。

本书特色

本书内容以实践、应用为导向，以实际编写代码和运行通过为准则，无须求解方程，无须理解复杂的公式。本书以项目编码实践为主，无方程公式的理论知识，只求读者能又快又准地运行代码并看到模型最终效果，在仅配置普通 CPU 的计算机上也能训练模型和完成所有的项目实践。

本书介绍的 21 个项目选取了目前比较流行的应用案例，为读者呈现生动有趣的实践应用。当初学者掌握了一定的统计学知识后，亟需的就是神经网络模型的项目编写练习，此时学习本书再适合不过了。

本书内容分为三大类，分别是预测类、识别类和生成类，各部分内容中均有从浅入深的项目案例。每个项目的每一行代码，都是笔者实际运行通过的。有的地方代码过多，我已将之写到了 Python 脚本文件中，读者需要下载相关脚本文件，引入后再运行。每个神经网络模型的每一层都有激活函数和损失函数，在隐藏层时用以调节权重和误差，在输出层时用以计算概率。

本书内容及体系结构

第1章 房价预测

本章通过对波士顿和北京的房价进行分析和预测，讲解如何使用scikit-learn（sklearn）进行网格搜索来训练和预测，以及如何使用Keras构建神经网络模型来训练、评估和预测。

第2章 泰坦尼克号生还预测

本章通过对泰坦尼克号的船上乘员的人数情况进行数据清洗与分割，讲解如何使用决策树、逻辑回归、梯度提升分类器、神经网络模型来构建、训练、评估和预测模型。

第3章 共享单车使用情况预测

本章将根据Capital Bikeshare共享单车公司的用户骑行数据绘制多种图表来进行呈现、分析和总结，最终编写一个基于TensorFlow的长短期记忆网络模型来讲解如何进行预测。

第4章 福彩3D中奖预测

本章通过分析从福彩开奖的网站上获取的从2004年10月到2018年7月的3D中奖数据（读者也可以自行获取更多的数据），构建基于Keras的时间序列和多层感知器模型来讲解如何进行中奖预测。

第5章 股票走势预测

本章主要是通过从美国纳斯达克股票交易市场获取的百度和微软的股票价格走势数据，对从上市开始到2018年的股票收盘价格数据进行观察和分析，假定购买策略，构建Prophet模型来讲解如何进行预测。

第6章 垃圾邮件预测

本章通过判定给定的邮件是否是垃圾邮件来进行分析，然后构建多项式朴素贝叶斯模型和多层感知器模型来讲解如何预测邮件是否是垃圾邮件。

第7章 影评的情感分析

本章通过分析和预测给定的电影评论是正面的还是负面的，讲解如何基于TensorFlow和Keras构建长短期记忆网络模型来分析影评是正面的或者负面的。

第8章 语言翻译

本章通过从网上下载的英文和法文的平行语料库来进行机器翻译，讲解如何基于Keras的长短期记忆网络构建从法文到英文的语言神经机器翻译引擎。

第9章 MNIST手写数字识别

本章将介绍如何识别开源的MNIST手写数字数据库的图像，并基于TensorFlow和Keras构建多层感知器和卷积神经网络模型，讲解如何进行训练、评估和识别。

第 10 章　狗的品种识别

本章将介绍通过由 120 个品种的狗的图像数据库构建的数据集进行图像基本分析，然后构建 Keras 的卷积神经网络识别模型，再构建基于 InceptionV3 预训练模型的迁移学习技术来讲解如何进行训练、评估和识别。

第 11 章　人脸识别

本章通过提供的 LFW 人脸图像数据集，进行图像中的人脸对齐，然后训练模型、预测模型，还讲解了如何使用流行的开源库 FaceNet 和 FaceRecognition 来进行实时的人脸识别。

第 12 章　人脸面部表情识别

本章将讲解如何使用 fer2013 的数据集构建卷积神经网络模型来进行人脸面部表情识别。首先是对单张图片中的人脸面部表情进行识别，然后再对视频中的人脸面部表情进行识别，最后通过开启摄像头实时识别拍摄到的人脸面部表情。

第 13 章　人体姿态识别

本章将介绍 OpenPose 开源库，它是由 CMU 计算机感知实验室研发并开源的。我们将依据 OpenPose 开源库基于 TensorFlow 和 Keras 的实现版本，对单张图片、视频和摄像头拍摄到的画面进行人体姿态识别。

第 14 章　皮肤癌分类

本章将讲解如何通过 ISIC 2017: Skin Lesion Analysis 提供的 3 种皮肤癌症的图像数据集进行构建卷积神经网络模型的分类和预测，基于 TensorFlow 的迁移学习技术和 Keras 的 CNN 模型进行识别。

第 15 章　对象检测

本章将讲解如何使用 Mask R-CNN Inception COCO、Faster R-CNN Inception COCO 和 SSD MobileNet COCO，对 COCO 图像数据集里实时拍摄的画面中的对象进行识别。

第 16 章　看图说话

本章将讲解如何通过 MSCOCO 图像数据集，使每张图像都有对应的图像文本描述，使用 Show and Tell 模型来实现模型的构建，生成 TFRecords 图像格式数据，训练、评估和测试模型。

第 17 章　生成电视剧剧本

本章讲解如何使用 *The Simpsons* 剧本的某一个片段作为训练数据集，通过 TensorFlow 构建循环神经网络和 Textgenrnn 来实现电视剧剧本的生成。

第 18 章　风格迁移

本章将讲解如何通过给定的大师画作和自己提供的图片来构建 TensorFlow 和 Keras 的神经网络模型，把大师的画作风格迁移到指定的图片上。

第 19 章　生成人脸

本章将讲解如何使用 TensorFlow 构建生成对抗网络模型，先用于简单的 MNIST 手写数字图像识别，然后实现基于 LFW 人脸图像数据集的人脸生成。

第 20 章　图像超分辨率

本章将讲解如何使用 LFW 人脸图像数据集和 srez 开源库构建深度对抗生成网络模型，训练和生成高度清晰的图像（放大图像 4 倍也不模糊）。

第 21 章　移花接木

本章将讲解如何通过提供的 3 种图片数据集，基于 CycleGAN 技术来实现根据橘子生成苹果、根据马生成斑马，以及将男性人脸面貌和女性人脸面貌进行互换。

读者对象

- 统计学从业者。
- 人工智能研究人员。
- 深度学习开发者。
- 机器学习开发者。
- 人工智能工程师。
- 互联网行业的创业者。

阅读本书的前提

本书介绍的项目主要运行环境配置信息和安装流程分别如下。

① Linux 系统的分支 Ubuntu，版本 16.04.5。
② 虚拟环境 virtualenv，版本 16.0.0。
③ Python 脚本编程语言，版本 Python3.6.6。
④ 试验代码运行环境 Jupyter Notebook，版本 5.4.0。
⑤ 深度学习框架 TensorFlow，版本 1.10.1；Keras，版本 2.1.6。

各个项目不同的安装包版本信息，在各个项目开始的时候会提及。

在写这本书的时间里，各种技术框架和工具不断地更新和迭代，各自的版本也升级了，所以如果读者在本书的项目中会看到所提及的一些技术框架和工具的版本高于这里提及的版本，请以具体项目里所提及的版本为主；这么做的目的是让初学者适应和习惯持续更新的技术框架和工具。

本书对应的每一章节的资源或代码的 Github 仓库下载地址如下。

- https://github.com/21-projects-for-deep-learning/MyBook

致谢

感谢互联网时代，感恩深度学习领域的先驱者们为神经网络技术的基础架构做出的贡献，包括但不限于 Geoffrey Hinton（杰弗里·希尔顿）、Yann LeCun（杨立昂）、Yoshua Bengio（约书亚·本吉奥）、Andrew Ng（吴恩达）、Francois Chollet（弗朗索瓦·肖莱）等资深专家。Google 作为全球科技公司对开源做出的贡献是巨大的，让无数的开发者和研究者可以沿着 Google 前沿的技术一起前进，也让我们可以为此做出微薄的贡献。

还要真诚地感谢北京源智天下科技有限公司的许亚楠女士，她在知名博客网站上看到了我的博客，和我进行了详细的沟通，经过她的认可和热情的推动后，我决定编写和出版本书。也感谢王蕾编辑，因为她的重视和诚恳的建议，本书终于完美收官。

最后，由于作者水平和成书时间所限，本书难免存有疏漏和不当之处，敬请专家和读者给出批评和指正。

张强

2019 年 5 月

目　录

第一篇　预测类项目实战

第1章　房价预测 ·· 2
 1.1　数据准备 ·· 2
  1.1.1　环境准备 ································ 2
  1.1.2　预处理数据 ······························ 3
  1.1.3　数据可视化分析 ························ 5
 1.2　基于 scikit-learn 实现房价预测 ····· 7
  1.2.1　衡量 R2 值 ······························· 7
  1.2.2　模型性能对比 ···························· 7
  1.2.3　网格搜索模型 ·························· 12
  1.2.4　波士顿房价预测 ······················ 13
  1.2.5　北京房价预测 ·························· 16
 1.3　基于 Keras 实现房价预测 ············ 19
  1.3.1　数据准备 ································ 19
  1.3.2　创建神经网络模型 ··················· 20
  1.3.3　训练网络模型 ·························· 21
  1.3.4　可视化模型的结果 ··················· 22
  1.3.5　评估和预测模型 ······················ 23
  1.3.6　预测可视化显示 ······················ 24
 1.4　小结 ·· 26

第2章　泰坦尼克号生还预测 ······················ 27
 2.1　数据准备 ······································ 27
  2.1.1　环境准备 ································ 27
  2.1.2　预处理数据 ···························· 28
  2.1.3　缺失值处理 ···························· 29
  2.1.4　数据清洗与分割 ······················ 32
 2.2　基于决策树模型预测 ···················· 33
  2.2.1　训练 ·· 33
  2.2.2　预测 ·· 33
 2.3　基于逻辑回归模型预测 ················ 34
  2.3.1　训练 ·· 34
  2.3.2　预测 ·· 35
 2.4　基于梯度提升分类器模型预测 ······ 35
  2.4.1　训练 ·· 35
  2.4.2　预测 ·· 35
 2.5　基于神经网络模型预测 ················ 36
  2.5.1　训练 ·· 36
  2.5.2　预测 ·· 36
  2.5.3　绘制曲线图 ···························· 37
 2.6　基于 Keras 的神经网络模型
   预测 ·· 38
  2.6.1　训练 ·· 38
  2.6.2　预测 ·· 39
 2.7　小结 ·· 40

第3章　共享单车使用情况预测 ··················· 41
 3.1　数据准备 ······································ 41
  3.1.1　环境准备 ································ 41
  3.1.2　数据可视化 ···························· 41
  3.1.3　预处理数据 ···························· 51
  3.1.4　数据清洗与分割 ······················ 52
 3.2　基于 TensorFlow 的长短期记忆
   网络模型预测 ······························ 52
  3.2.1　处理序列 ································ 52
  3.2.2　参数准备 ································ 53

3.2.3 创建 LSTM 模型 …… 53
3.2.4 训练模型 …… 54
3.2.5 模型预览与测试 …… 56
3.2.6 对比预测值模型预览 …… 58
3.3 小结 …… 59

第 4 章 福彩 3D 中奖预测 …… 60
4.1 数据准备 …… 60
4.1.1 环境准备 …… 60
4.1.2 数据准备 …… 60
4.1.3 数据预处理 …… 61
4.1.4 数据可视化 …… 62
4.2 基于神经网络模型预测 …… 68
4.2.1 决策树 …… 69
4.2.2 多层感知器 …… 69
4.2.3 时间序列基础 …… 70
4.2.4 时间序列预测 …… 73
4.2.5 根据中奖号码单变量单个位数预测 …… 74
4.3 小结 …… 76

第 5 章 股票走势预测 …… 77
5.1 数据准备 …… 77
5.1.1 环境准备 …… 77
5.1.2 数据集说明 …… 77
5.2 百度股票预测 …… 78
5.2.1 数据准备 …… 78
5.2.2 数据可视化 …… 80
5.2.3 计算购买的股票收益 …… 82
5.2.4 训练和评估模型 …… 84
5.2.5 股票预测 …… 87
5.2.6 股票买入策略 …… 88
5.3 微软股票预测 …… 89
5.3.1 数据准备 …… 89
5.3.2 数据可视化 …… 89
5.3.3 计算购买的股票收益 …… 90

5.3.4 训练和评估模型 …… 91
5.3.5 股票预测 …… 92
5.3.6 股票买入策略 …… 92
5.4 小结 …… 93

第 6 章 垃圾邮件预测 …… 94
6.1 数据准备 …… 94
6.1.1 环境准备 …… 94
6.1.2 数据准备 …… 95
6.1.3 数据预处理 …… 95
6.2 基于多项式朴素贝叶斯分类器的邮件分类 …… 101
6.2.1 数据处理 …… 101
6.2.2 创建和训练模型 …… 102
6.2.3 测试模型 …… 102
6.3 基于 TensorFlow 的神经网络模型的邮件分类 …… 103
6.3.1 构建 N-Gram 向量化数据 …… 103
6.3.2 创建模型 …… 104
6.3.3 训练模型 …… 104
6.3.4 可视化训练结果 …… 106
6.4 小结 …… 107

第 7 章 影评的情感分析 …… 108
7.1 数据准备 …… 108
7.1.1 环境准备 …… 108
7.1.2 预处理数据 …… 108
7.1.3 数据集编码 …… 110
7.1.4 数据集分割 …… 114
7.2 基于 TensorFlow 的长短期记忆网络实现影评的情感分析 …… 115
7.2.1 参数准备 …… 115
7.2.2 创建 LSTM 模型 …… 116
7.2.3 训练模型 …… 117
7.2.4 模型测试 …… 119

7.3　基于 Keras 的长短期记忆网络
　　实现影评的情感分析 ·············· 119
　　7.3.1　数据预处理 ················ 119
　　7.3.2　创建模型 ··················· 120
　　7.3.3　预览模型架构 ············ 120
　　7.3.4　训练模型 ··················· 120
　　7.3.5　模型评估 ··················· 121
7.4　小结 ··································· 121

第 8 章　语言翻译 ···························· 122
8.1　数据准备 ··························· 122
8.1.1　环境准备 ··················· 122
8.1.2　数据准备 ··················· 122
8.1.3　数据预处理 ··············· 123
8.2　基于 Keras 的长短期记忆网络
　　实现语言翻译 ······················ 126
　　8.2.1　Tokenize 文本数据 ····· 126
　　8.2.2　数据编码和填充 ······· 127
　　8.2.3　创建模型 ··················· 128
　　8.2.4　训练模型 ··················· 129
　　8.2.5　测试模型 ··················· 130
8.3　小结 ··································· 132

第二篇　识别类项目实战

第 9 章　MNIST 手写数字识别 ········ 134
9.1　MNIST 数据集 ·················· 134
　　9.1.1　简介 ·························· 134
　　9.1.2　数据下载 ··················· 135
　　9.1.3　可视化数据 ··············· 136
9.2　基于多层感知器的 TensorFlow
　　实现 MNIST 识别 ················ 138
　　9.2.1　参数准备 ··················· 138
　　9.2.2　创建模型 ··················· 138
　　9.2.3　训练模型 ··················· 139
　　9.2.4　模型预测 ··················· 140
9.3　基于多层感知器的 Keras
　　实现 MNIST 识别 ················ 141
　　9.3.1　数据准备 ··················· 141
　　9.3.2　创建模型 ··················· 142
　　9.3.3　训练模型 ··················· 143
　　9.3.4　模型预测 ··················· 144
　　9.3.5　单个图像预测 ··········· 144
9.4　基于卷积神经网络的 TensorFlow
　　实现 MNIST 识别 ················ 144
　　9.4.1　参数准备 ··················· 145
　　9.4.2　创建模型 ··················· 145
　　9.4.3　训练模型 ··················· 146
　　9.4.4　模型预测 ··················· 147
9.5　基于卷积神经网络的 Keras
　　实现 MNIST 识别 ················ 148
　　9.5.1　数据准备 ··················· 148
　　9.5.2　创建模型 ··················· 149
　　9.5.3　训练模型 ··················· 150
　　9.5.4　模型预测 ··················· 150
　　9.5.5　单个图像预测 ··········· 151
9.6　小结 ··································· 151

第 10 章　狗的品种识别 ···················· 152
10.1　数据准备 ··························· 152
　　10.1.1　环境准备 ················· 153
　　10.1.2　数据可视化 ············· 153
　　10.1.3　预处理数据 ············· 159
10.2　基于 Keras 的卷积神经网络
　　　模型预测 ··························· 160
　　10.2.1　创建模型 ················· 160
　　10.2.2　训练模型 ················· 162
　　10.2.3　模型评估 ················· 162
10.3　基于 Keras 的 InceptionV3 预训
　　　练模型实现预测 ··············· 163

- 10.3.1 模型函数声明 …… 163
- 10.3.2 预测单张图片 …… 164
- 10.4 基于 TFHUB 的 Keras 的迁移学习实现预测 …… 166
  - 10.4.1 数据集下载和准备 …… 166
  - 10.4.2 预训练模型下载 …… 167
  - 10.4.3 创建模型 …… 168
  - 10.4.4 训练模型 …… 169
  - 10.4.5 测试模型 …… 171
  - 10.4.6 模型预测单张图片 …… 173
- 10.5 小结 …… 175

第 11 章 人脸识别 …… 176
- 11.1 数据准备 …… 176
  - 11.1.1 环境准备 …… 177
  - 11.1.2 数据下载和分析 …… 177
  - 11.1.3 人脸图片数据预览 …… 178
- 11.2 基于 FaceNet 的人脸对齐和验证 …… 181
  - 11.2.1 下载和对齐图片 …… 181
  - 11.2.2 在 LFW 上验证 …… 182
- 11.3 训练自己的人脸识别模型 …… 183
  - 11.3.1 图片数据准备和对齐 …… 183
  - 11.3.2 训练模型 …… 184
  - 11.3.3 验证模型 …… 184
  - 11.3.4 再训练模型 …… 185
  - 11.3.5 再评估模型 …… 187
  - 11.3.6 将模型 CheckPoints 文件转换成 pb 文件 …… 188
- 11.4 基于 FaceRecognition 的人脸识别 …… 188
  - 11.4.1 配置环境 …… 189
  - 11.4.2 人脸检测 …… 189
  - 11.4.3 实时人脸识别 …… 190
- 11.5 小结 …… 193

第 12 章 人脸面部表情识别 …… 194
- 12.1 基于 Keras 的卷积神经网络实现人脸面部表情识别 …… 194
  - 12.1.1 环境准备 …… 194
  - 12.1.2 数据准备 …… 195
  - 12.1.3 数据集分割 …… 196
  - 12.1.4 数据集预处理 …… 196
  - 12.1.5 构建 CNN 模型 …… 197
  - 12.1.6 图片增强与训练模型 …… 199
  - 12.1.7 评估模型 …… 200
  - 12.1.8 保存与读取模型 …… 201
  - 12.1.9 单张图片测试模型 …… 202
- 12.2 视频中的人脸面部表情识别 …… 205
  - 12.2.1 读取模型 …… 206
  - 12.2.2 模型参数定义 …… 206
  - 12.2.3 视频的帧处理函数定义 …… 206
  - 12.2.4 识别与转换视频 …… 207
- 12.3 实时人脸面部表情识别 …… 208
  - 12.3.1 模型参数定义 …… 208
  - 12.3.2 启动摄像头和识别处理 …… 209
- 12.4 小结 …… 210

第 13 章 人体姿态识别 …… 211
- 13.1 基于 TensorFlow 实现人体姿态识别 …… 211
  - 13.1.1 环境准备 …… 211
  - 13.1.2 下载与安装 …… 212
  - 13.1.3 单张图片识别 …… 212
  - 13.1.4 视频中的人体姿态识别 …… 215
  - 13.1.5 实时摄像识别 …… 217
- 13.2 基于 Keras 实现人体姿态识别 …… 218

	13.2.1	环境准备 219		15.2.2	Faster R-CNN 的介绍与分析 237
	13.2.2	下载仓库 219		15.2.3	Mask R-CNN 的介绍与分析 238
	13.2.3	单张图片识别 219	15.3	基于 Mask R-CNN Inception COCO 的图片对象检测 239
	13.2.4	视频中的人体姿态识别 220		15.3.1	环境准备 239
	13.2.5	实时摄像识别 221		15.3.2	导入 Packages 240
13.3	小结 221		15.3.3	下载 Mask R-CNN Inception 2018 预训练模型 242

第 14 章 皮肤癌分类 222

- 14.1 数据准备 222
  - 14.1.1 环境准备 223
  - 14.1.2 数据下载 223
  - 14.1.3 数据可视化 224
- 14.2 基于 Keras 的卷积神经网络实现分类 226
  - 14.2.1 数据预处理 226
  - 14.2.2 创建 CNN 模型 227
  - 14.2.3 编译模型 229
  - 14.2.4 训练模型 229
  - 14.2.5 评估模型和图像测试 230
- 14.3 基于 TensorFlow 的迁移学习实现分类 232
  - 14.3.1 数据准备 232
  - 14.3.2 训练模型 232
  - 14.3.3 验证模型 233
  - 14.3.4 Tensorboard 可视化 233
- 14.4 小结 234

第 15 章 对象检测 235

- 15.1 对象检测的应用领域 236
  - 15.1.1 无人机应用领域 236
  - 15.1.2 自动驾驶汽车应用领域 236
  - 15.1.3 无人超市应用领域 236
- 15.2 原理分析 236
  - 15.2.1 R-CNN 的介绍与分析 237
- 15.3 基于 Mask R-CNN Inception COCO 的图片对象检测 239
  - 15.3.1 环境准备 239
  - 15.3.2 导入 Packages 240
  - 15.3.3 下载 Mask R-CNN Inception 2018 预训练模型 242
  - 15.3.4 加载模型到内存中 242
  - 15.3.5 加载类别映射 242
  - 15.3.6 定义函数将图片转为 NumPy 数组 243
  - 15.3.7 定义图片对象检测函数 243
  - 15.3.8 检测图片中的对象 244
  - 15.3.9 效果预览 245
- 15.4 基于 Faster R-CNN Inception COCO 的视频实时对象检测 246
  - 15.4.1 环境准备 246
  - 15.4.2 导入 Packages 246
  - 15.4.3 下载 Faster R-CNN Inception 2018 预训练模型 247
  - 15.4.4 加载模型到内存中 247
  - 15.4.5 加载类别映射 247
  - 15.4.6 定义视频中的图像对象检测函数 248
  - 15.4.7 定义视频中的图像处理函数 249
  - 15.4.8 视频中的图像对象检测 249
  - 15.4.9 效果预览 250
- 15.5 基于 SSD MobileNet COCO 的实时对象检测 250
  - 15.5.1 环境准备 250

15.5.2 导入 Packages ·········250
15.5.3 下载 SSD MobileNet 2018 预训练模型 ·········251
15.5.4 加载模型 ·········251
15.5.5 加载类别映射 ·········252
15.5.6 开启实时对象检测 ·········252
15.5.7 效果预览 ·········253
15.6 小结 ·········254

第三篇 生成类项目实战

第16章 看图写话 ·········256
16.1 数据准备 ·········256
16.1.1 环境准备 ·········257
16.1.2 数据下载 ·········257
16.1.3 数据预处理 ·········258
16.2 基于 TensorFlow 的 Show and Tell 实现看图写话 ·········264
16.2.1 介绍 ·········265
16.2.2 数据统计 ·········265
16.2.3 构建 TFRecords 格式数据 ·········266
16.2.4 训练模型 ·········268
16.2.5 评估模型 ·········268
16.2.6 测试模型 ·········269
16.3 小结 ·········270

第17章 生成电视剧剧本 ·········271
17.1 数据准备 ·········271
17.1.1 环境准备 ·········271
17.1.2 数据预处理 ·········272
17.1.3 数据可视化分析 ·········274
17.2 基于 TensorFlow 的循环神经网络实现电视剧剧本生成 ·········277
17.2.1 创建检查表 ·········278
17.2.2 数据 token 化预处理 ·········278
17.2.3 创建 Tensor 占位符和学习率 ·········279
17.2.4 初始化 RNN Cell ·········279
17.2.5 创建 Embedding ·········280
17.2.6 创建神经网络 ·········280
17.2.7 创建超参数和优化器 ·········280
17.2.8 训练神经网络模型 ·········281
17.2.9 生成电视剧剧本 ·········283
17.3 基于 textgenrnn 来实现电视剧剧本生成 ·········285
17.3.1 介绍 ·········285
17.3.2 训练模型 ·········285
17.3.3 生成剧本文本 ·········286
17.4 小结 ·········286

第18章 风格迁移 ·········287
18.1 基于 TensorFlow 实现神经风格迁移 ·········287
18.1.1 环境准备 ·········287
18.1.2 图像预览 ·········287
18.1.3 处理图像 ·········289
18.1.4 模型获取 ·········289
18.1.5 损失函数计算 ·········290
18.1.6 训练模型与图像生成 ·········292
18.2 基于 Keras 实现神经风格迁移 ·········295
18.2.1 图像预览 ·········295
18.2.2 图像处理 ·········296
18.2.3 获取模型 ·········297
18.2.4 损失函数计算 ·········298
18.2.5 迭代与生成风格图像 ·········298
18.3 小结 ·········300

第19章 生成人脸 ······················· 301
19.1 基于TensorFlow的GAN实现MNIST数字图像生成 ·············· 301
19.1.1 环境准备 ····················· 302
19.1.2 MNIST数字图像数据准备 ···················· 302
19.1.3 随机查看25张图片 ····· 303
19.1.4 构建模型输入 ············· 304
19.1.5 构建鉴别器 ················· 305
19.1.6 构建生成器 ················· 306
19.1.7 计算模型损失 ············· 307
19.1.8 构建优化器 ················· 307
19.1.9 构建训练模型时的图像输出 ······················ 308
19.1.10 构建训练模型函数 ····· 309
19.1.11 训练MNIST数据集的GAN模型 ···················· 310
19.2 基于TensorFlow的GAN实现LFW人脸图像生成 ················ 313
19.2.1 人脸图像数据准备 ······ 314
19.2.2 训练LFW数据集的GAN模型 ························ 314
19.3 小结 ······························· 315

第20章 图像超分辨率 ··············· 316
20.1 效果预览与数据准备 ········ 316
20.1.1 效果预览 ····················· 316
20.1.2 环境准备 ····················· 317
20.1.3 数据准备 ····················· 317
20.2 基于TensorFlow的DCGAN实现超分辨率 ······················· 318
20.2.1 下载srez代码库 ··········· 318
20.2.2 训练模型根据模糊图像生成清晰图像 ················ 318
20.2.3 输出效果预览 ············· 320
20.2.4 生成效果图视频 ········· 321
20.2.5 图片放大高清化 ········· 321
20.3 srez库的代码分析 ············· 322
20.3.1 主入口函数代码分析 ··· 322
20.3.2 创建模型代码分析 ····· 323
20.3.3 训练模型代码分析 ····· 323
20.4 小结 ······························· 324

第21章 移花接木 ······················· 325
21.1 基本信息 ·························· 325
21.1.1 3种模型效果预览 ······· 325
21.1.2 环境准备 ····················· 326
21.1.3 图片数据集准备 ········· 326
21.1.4 CycleGAN网络模型架构 ·························· 327
21.2 基于CycleGAN根据苹果生成橘子 ···························· 327
21.2.1 下载代码库 ················· 327
21.2.2 图片数据处理 ············· 328
21.2.3 训练模型 ····················· 328
21.2.4 导出模型 ····················· 329
21.2.5 测试图片 ····················· 330
21.3 基于CycleGAN根据马生成斑马 ···························· 332
21.3.1 图片数据处理 ············· 332
21.3.2 训练模型 ····················· 332
21.3.3 导出模型 ····················· 333
21.3.4 测试图片 ····················· 334
21.4 男性和女性的人脸面貌互换 ···························· 335
21.4.1 环境准备 ····················· 335
21.4.2 计算和生成模型 ········· 336
21.4.3 代码分析 ····················· 337
21.5 小结 ······························· 338

第一篇

预测类项目实战

- 第 1 章　房价预测
- 第 2 章　泰坦尼克号生还预测
- 第 3 章　共享单车使用情况预测
- 第 4 章　福彩 3D 中奖预测
- 第 5 章　股票走势预测
- 第 6 章　垃圾邮件预测
- 第 7 章　影评的情感分析
- 第 8 章　语言翻译

第 1 章 房价预测

大多数编程入门教程都从 "Hello,World!" 开始,深度学习领域也有入门教程。在预测类项目实战篇中,我们就把房价预测作为入门教程。

房价预测是一个很现实,也很接地气的问题。也就是说,读者学完本章关于房价预测的流程和代码的用法后,可以将其用到房产公司的房屋价格统计,买房时的价格预测和定位,或者做房屋数据分析时的统计等方面。这里的分析和预测不会讲解看房买房的方方面面,但是可以给读者一个启发,通过这些代码和读者自己扩展的数据,达到解决现实问题的目的。

本章将会讲解著名的波士顿房屋价格预测(Boston Housing Prediction)和北京房屋价格预测,读者可以在前言中找到所提供的数据集并自行下载。

1.1 数据准备

数据集是公开的,可以在前言所提供的链接下找到。本章使用的数据集是早期的房屋统计数据集,作为项目数据集来讲解也较为经典。波士顿的房屋数据集文件是 housing.csv,北京的房屋数据集文件是 bj_housing.csv。开始的时候,我们会用 scikit-learn 库来分析和预测,然后会用 TensorFlow 下的 Keras 的神经网络模型来预测,在预测结束后,都会使用一张对照图来直观地查看。

1.1.1 环境准备

- numpy = 1.14.6。
- pandas = 0.22.0。
- sklearn = 0.19.2。
- seaborn = 0.7.1。

- matplotlib = 2.1.2。
- tensorflow = 1.10.0。

1.1.2 预处理数据

在训练人工智能模型前，我们需要先观察和分析要训练的数据集，然后进行必要的预处理。这里我们先通过 pandas 的 read_csv() 函数加载 housing.csv 数据，获得 489 行乘以 4 列的数据，返回的是一个 pandas 的 DataFrame 对象（DataFrame 可以简单地理解成类似 Excel 表格的数据对象，但它并不是 Excel 表格），然后查看前 5 行数据以观察。

```
import pandas as pd
# 加载数据集
data = pd.read_csv('housing.csv')
print("波士顿房屋数据有{}行 x {}列。".format(data.shape[0], data.shape[1]))
# 查看前 5 行数据
data.head()
```

输出如下。

波士顿房屋数据有 489 行 x 4 列。

输出前 5 行数据，如图 1.1 所示。

	RM	LSTAT	PTRATIO	MEDV
0	6.575	4.98	15.3	504000.0
1	6.421	9.14	17.8	453600.0
2	7.185	4.03	17.8	728700.0
3	6.998	2.94	18.7	701400.0
4	7.147	5.33	18.7	760200.0

图 1.1 波士顿房屋数据集的前 5 行数据

图中该数据集的 4 列数据的含义分别如下。
- RM：该地区每座房屋的平均房间数量。
- LSTAT：该地区的房主中低收入阶层（有工作但收入微薄）所占百分比。
- PTRATIO：该地区的中小学里，学生和老师的人数之比（学生数/老师数）。
- MEDV：该地区业主自住房屋的价格中位数。

输出结果还有其他的字段（列），此处附图中只显示了主要的字段。然后，我们通过 describe() 函数来观察数据的基本统计信息，代码如下。

```
data.describe()
```

输出如图 1.2 所示。

	RM	LSTAT	PTRATIO	MEDV
count	489.000000	489.000000	489.000000	4.890000e+02
mean	6.240288	12.939632	18.516564	4.543429e+05
std	0.643650	7.081990	2.111268	1.653403e+05
min	3.561000	1.980000	12.600000	1.050000e+05
25%	5.880000	7.370000	17.400000	3.507000e+05
50%	6.185000	11.690000	19.100000	4.389000e+05
75%	6.575000	17.120000	20.200000	5.187000e+05
max	8.398000	37.970000	22.000000	1.024800e+06

图 1.2　波士顿房屋数据基本统计信息

如果我们想自己写代码统计房屋价格的信息，可以这样写。

```
import numpy as np
# 计算最低价格
minimum_price = np.min(data["MEDV"])
# 计算最高价格
maximum_price = np.max(data["MEDV"])
# 计算均值
mean_price = np.mean(data["MEDV"])
# 计算中间价格
median_price = np.median(data["MEDV"])
# 计算标准差价格
std_price = np.std(data["MEDV"])
# 打印输出所有计算的价格统计
print("波士顿房价数据的统计：")
print("最低价格： ${}".format(minimum_price))
print("最高价格： ${}".format(maximum_price))
print("均价： ${}".format(mean_price))
print("中间价格： ${}".format(median_price))
print("标准差价格： ${}".format(std_price))
```

输出如下。

```
波士顿房价数据的统计：
最低价格： $105000.0
最高价格： $1024800.0
均价： $454342.9447852761
中间价格： $438900.0
标准差价格： $165171.13154429474
```

然后，我们来分割一下训练集的训练数据 X 和 y，也就是房屋特征数据 features 和房屋价格 prices。变量 data 是一个 DataFrame 对象，它可以通过调用 drop() 函数来丢弃某个列。所以我们先取出 MEDV 列作为训练数据 y，而训练数据 X 就是将 MEDV 列丢弃后的数据。

```
# 获取房屋价格
prices = data['MEDV']
# 获取房屋特征
features = data.drop('MEDV', axis=1)
```

1.1.3 数据可视化分析

我们接下来分别分析 RM 列、LSTAT 列和 PTRATIO 列与 MEDV 列的相关性，会使用到 seaborn 库。seaborn 库是一个用于统计数据可视化的库，它是基于 Matplotlib 库编写而来的，提供了更为方便、高级的封装和调用。这里就用 seaborn 来绘图显示。

首先，我们来分析 RM 列和 MEDV 列的相关性。先导入 Matplotlib 库，再导入 seaborn 库，最后调用 regplot()函数，它会根据数据和线性回归的模型进行拟合。

```
import matplotlib.pyplot as plt
import seaborn as sns
# regplot()函数根据数据绘制线性回归（Linear Regression）模型图
# 参数1：X轴，表示每座房屋的平均房间数量
# 参数2：y轴，表示房屋价格
# 参数3：绘图时使用的颜色，这里是红色
sns.regplot(data['RM'], prices, color='red')
# 显示绘图
plt.show()
```

输出结果如图 1.3 所示。

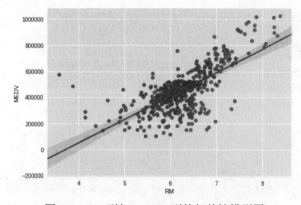

图 1.3 RM 列与 MEDV 列的相关性模型图

通过这个线性回归模型图，我们可以观察到，房间的数量越多，房屋价格就越高，也就是空间越大越舒适，还可以看到大部分房屋的房间数量在 6 个左右。

然后，我们分析 LSTAT 列和 MEDV 列的相关性。

```
# regplot()函数根据数据绘制线性回归（Linear Regression）模型图
# 参数1：X轴，表示该地区的房主中低收入阶层所占百分比
# 参数2：y轴，表示房屋价格
# 参数3：表示绘制的图是用什么标记绘制出来的，这里是"+"
# 参数4：绘图时使用的颜色，这里是绿色
```

```
sns.regplot(data['LSTAT'], prices, marker='+', color='green')
# 显示绘图
plt.show()
```

输出如图 1.4 所示。

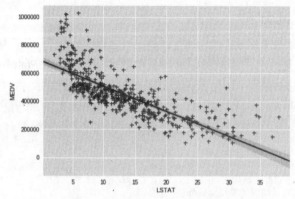

图 1.4　LSTAT 列和 MEDV 列的相关性模型图

通过这个线性回归模型图，我们可以看到 LSTAT 列的数值越高，房屋价格就越低。LSTAT 列表示该地区的房主中低收入阶层所占百分比，低收入阶层所占百分比的值越高，就意味着，周边的消费水平越低，房屋价格自然也会越低。

最后，我们来分析一下 PTRATIO 列和 MEDV 列的相关性。

```
# regplot()函数根据数据绘制线性回归（Linear Regression）模型图
# 参数 1：x 轴，表示该地区的中小学里，学生和老师的人数之比
# 参数 2：y 轴，表示房屋价格
# 参数 3：表示绘制的图是用什么标记绘制出来的，这里是"^"
# 参数 4：绘图时使用的颜色，这里是蓝色
sns.regplot(data['PTRATIO'], prices, marker='^', color='blue')
# 显示绘图
plt.show()
```

输出如图 1.5 所示。

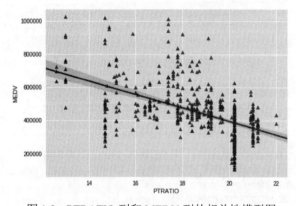

图 1.5　PTRATIO 列和 MEDV 列的相关性模型图

通过这个线性回归模型图，我们得知 PTRATIO 列的值的增加会降低房屋价格。这是因为拥有学生和老师的人数之比较高表明该地区缺乏教学设施或资源；相反的是，学生和老师的人数之比较低的地方，教学设施或资源较好，该地区的房屋价格也就高一些。

1.2 基于 scikit-learn 实现房价预测

本节将使用 scikit-learn 来实现房价预测。首先使用决定系数 R2 Score 来做一个小测验；然后使用决策树回归器配合不同的深度值来创建和对比模型；最后使用网格搜索（grid search）来查找最佳模型，以便进行房价预测。

1.2.1 衡量 R2 值

通过 sklearn 提供的 r2_score()函数来计算拟合优度（Goodness of Fit），R2 值越接近 1 表明拟合的程度越高；反之，R2 值越小，则说明回归直线的拟合程度越低。

```
# 导入 r2_score
from sklearn.metrics import r2_score
# 定义函数，计算目标值和预测值之间的分值
def performance_metric(y_true, y_predict):
    score = r2_score(y_true, y_predict)
    return score
```

这里有一个小测验，假设有 5 个真实数值和 5 个预测数值，分别是[3,–0.5,2,7,4.2]和[2.5,0.0,2.1,7.8,5.3]，然后通过 r2_score()函数来看它的拟合优度的程度值。

```
# 真实数值
test_y_true    = [3, -0.5, 2, 7, 4.2]
# 预测数值
test_y_predict = [2.5, 0.0, 2.1, 7.8, 5.3]
# 通过 r2_score 函数来计算
score = performance_metric(test_y_true, test_y_predict)
print("决定系数，R^2=: {}。".format(score))
```

输出如下。

决定系数，R^2=: 0.9228556485355649。

可以看到，R2 值非常接近 1，说明相关程度高。

1.2.2 模型性能对比

我们将通过 4 个不同的深度值创建 4 个不同的模型，然后来对比这 4 个模型的性能表现，最后通过图表直观显示。现在，我们把房屋特征和价格的数据进行清洗和分割。

```
# 从 sklearn 库里导入 train_test_split 方法
from sklearn.model_selection import train_test_split
# train_test_split()方法用来清洗和分割数据
# 参数 1：特征样本，就是房屋特征数据
```

```
# 参数 2：特征样本对应的目标（房屋）价格
# 参数 3：分配给测试集的大小为 0.1，也就是 10% 的数据用于测试、90% 的数据用于训练
# 参数 4：random_state 表示随机数生成器的种子，如果希望第二次调用 train_test_split() 方法
#         的结果和第一次调用的结果一致，那么就可以设置一个值，多少都可以，生产环境不要设值
X_train, X_test, y_train, y_test = train_test_split(features, prices, test_size=0.1, random_state=50)
print("X_train.shape={}, y_train.shape={}.".format(X_train.shape, y_train.shape))
print("X_test.shape={}, y_test.shape={}.".format(X_test.shape, y_test.shape))
```

输出如下。

```
X_train.shape=(440, 3), y_train.shape=(440,).
X_test.shape=(49, 3), y_test.shape=(49,).
```

得到 440 行的数据用于训练，49 行的数据用于测试。接下来，我们定义一个函数来衡量模型的性能，深度值分别是 1、3、6、10，然后输出这些模型的图，便于更直观地对比。

```
# 导入绘图库
import matplotlib.pyplot as plt
# 使用 matplotlib 在 Jupyter Notebook 中绘图时，需要使用这个
%matplotlib inline
# 导入 sklearn 的清洗分割、学习曲线和决策树回归器的对象和函数
from sklearn.model_selection import ShuffleSplit, learning_curve
from sklearn.tree import DecisionTreeRegressor
# 定义模型的性能对比函数
# 通过不同大小的深度值来创建 for 循环里的模型，然后以图的形式展现
def ModelLearningGraphMetrics(X, y):
    # 清洗和分割数据对象定义
    # 参数 1：n_splits 表示重新清洗和分割数据的迭代次数，默认值为 10
    # 参数 2：test_size=0.2 表示有 0.2 的数据用于测试，也就是 20% 的数据用于测试，80% 的数据用于训练
    # 参数 3：random_state 表示随机数生成器的种子，如果希望第二次调用 ShuffleSplit() 方法
    #         的结果和第一次调用的结果一致，那么就可以设置一个值，多少都可以，生产环境不要设值
    cv = ShuffleSplit(n_splits=10, test_size=0.2, random_state=0)
    # 生成训练集大小
    # 函数 np.rint() 是计算数组各元素的四舍五入的值到最近的整数
    # 函数 np.linspace(start_i, stop_i, num) 表示从起始值到结束值之间，以均匀的间隔返回指定个数的值。
    # 那么这里就是从 1 开始、以 X 结束的总行数的 80% 的数据，数据间隔为 9，最后将数据元素都转换成整型
    train_sizes = np.rint(np.linspace(1, X.shape[0]*0.8 - 1, 9)).astype(int)
    # 创建一个绘图窗口，大小 10×7，单位是英寸（inch）
    fig = plt.figure(figsize=(10, 7))
    # 根据深度值创建不同的模型
    # 这里的深度值就是 1、3、6、10 这 4 个
    for k, depth in enumerate([1,3,6,10]):
        # 根据深度(max_depth)值来创建决策树回归器
        regressor = DecisionTreeRegressor(max_depth=depth)
        # 通过学习曲线函数计算训练集和测试集的分值
        # 参数 1：评估器，这里就是决策树回归器
        # 参数 2：特征样本，房屋特征
        # 参数 3：目标标签，房屋价格
        # 参数 4：训练样本的个数，这是用来生成学习曲线的
        # 参数 5：交叉验证生成器，或可迭代对象
        # 参数 6：评分器，是一个可调用对象
```

```python
        sizes, train_scores, test_scores = learning_curve(
            regressor, X, y, train_sizes=train_sizes, cv=cv, scoring='r2')
        # 计算训练集分值和测试集分值的标准差
        train_std = np.std(train_scores, axis=1)
        test_std = np.std(test_scores, axis=1)
        # 计算训练集分值和测试集分值的均值
        train_mean = np.mean(train_scores, axis=1)
        test_mean = np.mean(test_scores, axis=1)
        # 根据学习曲线值来绘制图,四个图的位置通过 k+1 来控制
        ax = fig.add_subplot(2, 2, k+1)
        # 绘制训练得分线,plot()方法
        # 参数 1:x 轴方向的值
        # 参数 2:y 轴方向的值
        # 参数 3:绘制出来的线的样式风格,比如这里的 "o" 表示一个圆点标记,而 "-" 表示实线
        # 参数 4:绘制的线的颜色
        # 参数 5:图例上的标题
        ax.plot(sizes, train_mean, 'o-', color='r', label='Training Score')
        # 绘制测试得分线
        ax.plot(sizes, test_mean, 'o-', color='g', label='Testing Score')
        # fill_between()方法表示为训练得分线描边
        # 参数 1:x 轴方向的值
        # 参数 2:y 轴方向的覆盖下限
        # 参数 3:y 轴方向的覆盖上限
        # 参数 4:设置覆盖区域的透明度
        # 参数 5:设置覆盖区域的颜色
        ax.fill_between(sizes, train_mean - train_std,
            train_mean + train_std, alpha=0.15, color='r')
        # fill_between()方法表示为测试得分线描边
        ax.fill_between(sizes, test_mean - test_std,
            test_mean + test_std, alpha=0.15, color='g')
        # 在绘图的窗口上添加标题
        ax.set_title('max_depth = {}'.format(depth))
        # 设置 x 轴的标题
        ax.set_xlabel('Number of Training Points')
        # 设置 y 轴的标题
        ax.set_ylabel('Score')
        # 设置 x 轴方向的最小值和最大值
        ax.set_xlim([0, X.shape[0]*0.8])
        # 设置 y 轴方向的最小值和最大值
        ax.set_ylim([-0.05, 1.05])
    # 添加图例
    ax.legend(bbox_to_anchor=(1.05, 2.05), loc='lower left', borderaxespad=0.)
    # 添加图形总标题
    fig.suptitle('Decision Tree Regressor Learning Performances',
            fontsize=16, y=1.03)
    # 自动调整 subplot 符合图的区域的参数的布局。生产环境中不要使用该函数,因为这是一个实验特性函数
    fig.tight_layout()
    # 显示绘图
    fig.show()
ModelLearningGraphMetrics(features, prices)
```

输出的图形如图 1.6 所示。

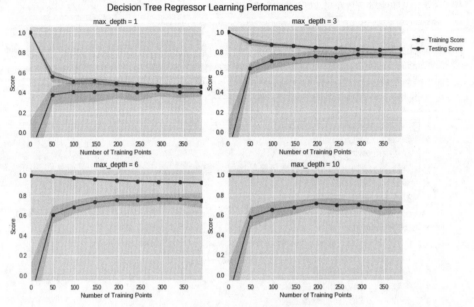

图 1.6　决策树回归器的模型性能表现

通过上面 4 个模型的图形对比,我们发现 max_depth=3 的模型表现不错;max_depth=1 的模型有些欠拟合;max_depth=10 的模型有些过拟合;而 max_depth=6 的模型训练集得分高,但是测试集得分有些低,略微过拟合。

我们还可以通过观察 max_depth 等于 1 到 10 的值来进行模型的性能对比,最后通过输出图形来看它们的表现。比如以后在其他场景中,读者也可以使用下述策略,会更好地找到最佳值模型。

```
# 导入sklearn库的数据清洗与分割函数,验证曲线函数
from sklearn.model_selection import ShuffleSplit, validation_curve
# 定义模型在复杂度高的情况下性能表现的函数
# 随着模型复杂度的增加计算它的性能表现
def ModelComplexityPerformanceMetrics(X, y):
    # 清洗和分割数据对象定义
    # 参数1: n_splits 表示重新清洗和分割数据的迭代次数,默认值为10
    # 参数2: test_size=0.2 表示有 0.2 的数据用于测试,也就是 20%的数据用于测试,80%的数据用于训练
    # 参数3: random_state 表示随机数生成器的种子,如果希望第二次调用ShuffleSplit()方法
    #       的结果和第一次调用的结果一致,那么就可以设置一个值,多少都可以,生产环境不要设值
    cv = ShuffleSplit(n_splits=10, test_size=0.2, random_state=0)
    # 定义从1到10为深度(max_depth)的参数值
    max_depth = np.arange(1,11)
    # 通过不同的 max_depth 的参数值来计算训练集和测试集的分值
    # 参数1:评估器,这里是决策树回归器
    # 参数2:特征样本,房屋特征
    # 参数3:目标标签,房屋价格
```

```
# 参数 4：传入的深度参数名称
# 参数 5：传入的深度参数范围值
# 参数 6：交叉验证生成器，或可迭代对象
# 参数 7：评分器，是一个可调用对象
train_scores, test_scores = \
validation_curve(DecisionTreeRegressor(), X, y, param_name="max_depth",
                 param_range=max_depth, cv=cv, scoring='r2')
# 计算训练集分值和测试集分值的均值
train_mean = np.mean(train_scores, axis=1)
test_mean = np.mean(test_scores, axis=1)

# 计算训练集分值和测试集分值的标准差
train_std = np.std(train_scores, axis=1)
test_std = np.std(test_scores, axis=1)
# 绘制验证分值的曲线图
# figsize 表示要绘制的图形窗口大小，单位是英寸
plt.figure(figsize=(7, 5))
# 在绘制的图形窗口上添加一个标题
plt.title('Decision Tree Regressor Complexity Performance')
# 绘制训练得分线，plot()方法
# 参数 1：x 轴方向的值
# 参数 2：y 轴方向的值
# 参数 3：绘制出来的线的样式风格，比如这里的"o"表示一个圆点标记，而"-"表示实线
# 参数 4：绘制的线的颜色
# 参数 5：图例上的标题
plt.plot(max_depth, train_mean, 'o-', color='r', label='Training Score')
# 绘制测试得分线
plt.plot(max_depth, test_mean, 'o-', color='g', label='Validation Score')
# fill_between()方法表示为两条曲线描边，第一条是训练得分线，第二条是测试得分线
#   参数 1：x 轴方向的值
#   参数 2：y 轴方向的覆盖下限
#   参数 3：y 轴方向的覆盖上限
#   参数 4：设置覆盖区域的透明度
#   参数 5：设置覆盖区域的颜色
plt.fill_between(max_depth, train_mean - train_std, \
    train_mean + train_std, alpha=0.15, color='r')
plt.fill_between(max_depth, test_mean - test_std, \
    test_mean + test_std, alpha=0.15, color='g')
# 图上加标题注解
# 添加图例
plt.legend(loc='lower right')
# 添加 x 轴的标题
plt.xlabel('Maximum Depth')
# 添加 y 轴的标题
plt.ylabel('Score')
# 设置 y 轴方向的最小值和最大值
plt.ylim([-0.05,1.05])
# 显示绘图
plt.show()
ModelComplexityPerformanceMetrics(features, prices)
```

输出的图形如图 1.7 所示。

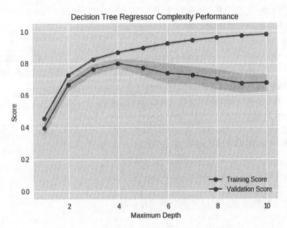

图 1.7 决策树回归器的模型性能表现

从结果图上看，max_depth 的参数值在 1 到 10 中时，max_depth=4 是最佳位置，当 max_depth>4 时模型表现就有些过拟合了。

1.2.3 网格搜索模型

网格搜索最重要的功能就是能够自动调参，只要用户把参数传进去，它就能返回最优的结果模型和参数。它只适用于小数据集，同时还有一个缺点，那就是可能调参到局部最优，而不是全局最优。

```python
# 从 sklearn 库导入网格搜索 VC、数据清洗与分割、决策树和分值计算对象的函数
from sklearn.model_selection import GridSearchCV, ShuffleSplit
from sklearn.tree import DecisionTreeRegressor
from sklearn.metrics import make_scorer
# 定义网格搜索最佳模型函数
def gridSearchVC_fit_model(X, y):
    # 清洗和分割数据对象定义：
    # 参数1：n_splits 表示重新清洗和分割数据的迭代次数，默认值为 10
    # 参数2：test_size=0.2 表示有 0.2 的数据用于测试，也就是 20%的数据用于测试，80%的数据用于训练
    # 参数3：random_state 表示随机数生成器的种子，如果希望第二次调用 ShuffleSplit()方法
    #       的结果和第一次调用的结果一致，那么就可以设置一个值，多少都可以，生产环境不要设值
    cv = ShuffleSplit(n_splits=10, test_size=0.2, random_state=0)
    # 创建决策树回归器对象
    regressor = DecisionTreeRegressor(random_state=0)
    # 创建一个字典，表示 max_depth 的参数值是从 1 到 10
    # 注意：如果代码运行的环境是 Python 2，去掉这个 list()函数调用
    params = { "max_depth" : list(range(1, 10)) }
    # 通过 make_scorer()函数将上面定义的 performance_metric()函数转换成计算分值函数
    scoring_fnc = make_scorer(score_func=performance_metric)
    # 创建网格搜索对象
    # 参数1：评估器，就是回归器，这里表示的是决策树回归器
```

```
    # 参数2：网格搜索参数
    # 参数3：计算分值函数
    # 参数4：cv（Cross-Validation）交叉验证，传入交叉验证生成器，或者可迭代对象
    grid = GridSearchCV(estimator=regressor, param_grid=params,
                        scoring=scoring_fnc, cv=cv)
    # 根据数据计算/训练适合网格搜索对象的最佳模型
    grid = grid.fit(X, y)
    # 返回计算得到的最佳模型
    return grid.best_estimator_
# 网格搜索函数得到最佳模型
reg = gridSearchVC_fit_model(X_train, y_train)
print("参数max_depth={}是最佳模型。".format(reg.get_params()['max_depth']))
```

输出如下。

参数max_depth=4是最佳模型。

1.2.4 波士顿房价预测

在预测波士顿房价前，我们先做一个模拟客户数据来预测，数据如表1.1所示。

表1.1 模拟房屋数据

特 征	客户1	客户2	客户3
房屋内房间总数	3间	7间	9间
社区贫困指数	30%	20%	2%
附近学校的学生和老师比例	10:1	19:1	9:1

```
# 小测验
# 假设有以下3个客户的数据，分别是
client_data = [[3, 30, 10],    # 客户1
               [7, 20, 19],    # 客户2
               [9, 2, 9]]      # 客户3
# 尝试预测
for i, price in enumerate(reg.predict(client_data)):
    print("预测客户{}的销售价格是${:,.2f}".format(i+1, price))
```

输出如下。

预测客户1的销售价格是$320,425.00
预测客户2的销售价格是$410,658.62
预测客户3的销售价格是$938,053.85

定义预测波士顿房价的函数，经过10次迭代，然后看下对比图。

```
# 预测房价函数
def PredictHousingPrice(X, y, fitter):
    # 迭代10次
    epochs = 10
    # 存储预测的价格
    y_predict_test_price = None
```

```python
    # 分割训练集和测试集数据,20%的数据用于测试,80%的数据用于训练
    X_train, X_test, y_train, y_test = train_test_split(X, y,
            test_size=0.2, random_state=0)
    # 迭代训练
    for epoch_i in range(epochs):
        # 根据数据训练模型,并返回最佳模型
        reg = fitter(X_train, y_train)
        # 预测测试集数据
        predicted_price = reg.predict(X_test)
        # 将预测到的结果存起来
        y_predict_test_price = predicted_price
        print("迭代第{}次。".format(epoch_i+1))
    return y_test, y_predict_test_price
y_true_price, y_predict_price = \
PredictHousingPrice(features, prices, gridSearchVC_fit_model)
```

输出如下。

迭代第 1 次。
迭代第 2 次。
迭代第 3 次。
……
迭代第 9 次。
迭代第 10 次。

查看真实房价的前 5 行数据。我们先将 y_true_price 转换成 Series 对象,然后在调用 reset_index()函数时就会把数据创建成一个新的 Series,旧的索引就会被作为数据的一列,再通过 drop()函数删除旧的索引那一列。

```python
pd.Series(y_true_price).reset_index().drop('index', axis=1).head()
```

输出如表 1.2 所示。

表 1.2　查看真实房价的前 5 行数据

	MEDV
0	417900.0
1	632100.0
2	281400.0
3	577500.0
4	474600.0

我们再来看预测房价的前 5 行数据。

```python
pd.Series(y_predict_price).head()
```

输出如表 1.3 所示。

表 1.3 查看预测房价的前 5 行数据

0	436065.000000
1	643455.555556
2	324240.000000
3	248675.000000
4	501352.173913

构建真实房价和预测房价的对比图函数。

```
# 显示真实房价和预测房价对比图
def plotVersusFigure(y_true_price, y_predict_price):
    # 创建一个 10×7 的窗口
    plt.figure(figsize=(10, 7))
    # 绘制的图 1 是真实房价
    X_show = np.rint(np.linspace(1,
                                 np.max(y_true_price),
                                 len(y_true_price))
                     ).astype(int)
    # 绘制图 1 线,plot()方法:
    #   参数 1:x 轴方向的值,真实房价从最低价到最高价
    #   参数 2:y 轴方向的值,真实房价的值
    #   参数 3:绘制出来的线的样式风格,比如这里的"o"表示一个圆点标记,而"-"表示实线
    #   参数 4:绘制的线的颜色,这里是青色
    plt.plot(X_show, y_true_price, 'o-', color='c')
    # 绘制的图 2 是预测房价,叠加在图 1 上
    X_show_predicted = np.rint(np.linspace(1,
                                           np.max(y_predict_price),
                                           len(y_predict_price))
                               ).astype(int)
    # 绘制图 2 线,plot()方法:
    #   参数 1:x 轴方向的值,预测房价从最低价到最高价
    #   参数 2:y 轴方向的值,预测房价的值
    #   参数 3:绘制出来的线的样式风格,比如这里的"o"表示一个圆点标记,而"-"表示实线
    #   参数 4:绘制的线的颜色,这里是洋红色
    plt.plot(X_show_predicted, y_predict_price, 'o-', color='m')
    # 添加标题
    plt.title('Housing Prices Prediction')
    # 添加图例
    plt.legend(loc='lower right', labels=["True Prices", "Predicted Prices"])
    # 添加 x 轴的标题
    plt.xlabel("House's Price Tendency By Array")
    # 添加 y 轴的标题
    plt.ylabel("House's Price")
```

```
# 显示绘制
plt.show()
```

输出打印波士顿真实房价和预测房价的对比图。

```
# 波士顿房屋价格对比图
plotVersusFigure(y_true_price, y_predict_price)
```

输出如图 1.8 所示。

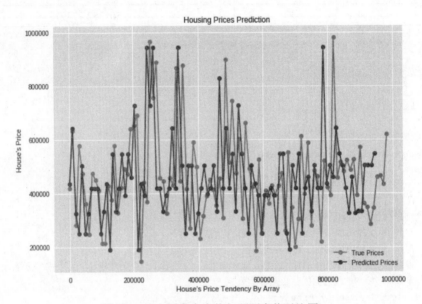

图 1.8 波士顿真实房价与预测房价对比图

通过结果可知，预测房价与真实房价之间是有一定误差的。这需要我们不断地调整参数并重新训练、预测等。

1.2.5 北京房价预测

读取 bj_housing.csv 文件进行北京房价预测，其中一共有 9999 条数据。训练和预测使用的代码都在波士顿房价预测小节中详细介绍过，这里不再重复。所以本小节的代码就是读取数据，训练模型，预测和绘图显示。这里的北京房屋数据是真实的，但它的预测结果肯定受宏观环境等多种因素的影响，可能会造成预测结果不精准。读取数据代码如下。

```
df = pd.read_csv('bj_housing.csv')
df.describe()
```

输出如图 1.9 所示。

	Area	Value	Room	Living	School	Year	Floor
count	9999.000000	9999.000000	9999.000000	9999.000000	9999.000000	9999.000000	9999.000000
mean	92.003900	342.076208	2.156216	1.277628	0.583958	1998.235524	13.326433
std	46.263242	259.406028	0.791407	0.524963	0.492925	13.126885	7.953371
min	14.000000	66.000000	1.000000	0.000000	0.000000	1014.000000	0.000000
25%	61.000000	205.000000	2.000000	1.000000	0.000000	1993.500000	6.000000
50%	83.000000	280.000000	2.000000	1.000000	1.000000	2000.000000	12.000000
75%	110.000000	395.000000	3.000000	2.000000	1.000000	2004.000000	19.000000
max	1124.000000	7450.000000	9.000000	4.000000	1.000000	2015.000000	91.000000

图 1.9 北京房价基本信息统计

数据的各列名称解释如下。
- Area：房屋面积，单位是平方米。
- Value：房屋售价，单位是万元。
- Room：房间数，单位是间。
- Living：厅数，单位是间。
- School：是否为学区房，0（表"否"）或 1（表"是"）。
- Year：房屋建造年份。
- Floor：房屋所处楼层。

然后，将房屋特征数据集和真实价格数据集分开，代码如下。

```
bj_prices = df['Value']
bj_features = df.drop('Value', axis=1)
bj_features.head()
```

房屋特征数据集前 5 行数据输出如图 1.10 所示。

	Area	Room	Living	School	Year	Floor
0	128	3	1	1	2004	21
1	68	1	2	1	2000	6
2	125	3	2	0	2003	5
3	129	2	2	0	2005	16
4	118	3	2	0	2003	6

图 1.10 北京房屋特征数据集前 5 行数据

根据数据训练和预测模型。

```
y_true_bj_price, y_predict_bj_price = \
PredictHousingPrice(bj_features, bj_prices, gridSearchVC_fit_model)
```

输出如下。

迭代第 1 次。
迭代第 2 次。
迭代第 3 次。
......
迭代第 9 次。
迭代第 10 次。

查看北京真实房价的前 5 行数据,先将 y_true_bj_price 转换成 Series 对象,然后在调用 reset_index()时函数就会把数据创建成一个新的 Series,旧的索引就会被作为该数据对象的一列,再通过 drop()函数删除旧的索引那一列。

```
y_true_bj_price.reset_index().drop('index', axis=1).head()
```

输出如表 1.4 所示。

表 1.4 查看真实房价的前 5 行数据

	Value
0	860
1	155
2	140
3	345
4	162

我们再来看预测房价的前 5 行数据。

```
pd.Series(y_predict_bj_price).head()
```

输出如表 1.5 所示。

表 1.5 查看预测房价的前 5 行数据

0	845
1	234
2	122
3	261
4	250

构建真实房价和预测房价的对比图。

```
plotVersusFigure(y_true_bj_price, y_predict_bj_price)
```

绘制的结果如图 1.11 所示。

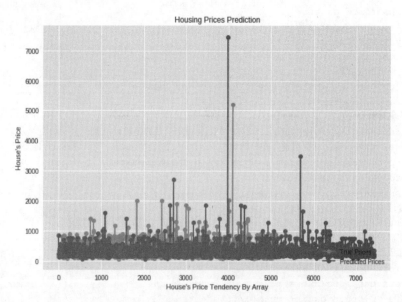

图 1.11　北京真实房价与预测房价对比图

通过对比图我们可以看到，预测房价与真实房价之间还是有一定误差的，这需要我们不断地去调整参数重新训练模型和预测等。

1.3　基于 Keras 实现房价预测

前面我们使用 scikit-learn 机器学习库实现了房价预测。本节将使用神经网络模型来实现房价预测，使用的是 TensorFlow 下的 Keras 的 API 代码。我们将使用 Keras 创建模型、训练模型、预测和对比图预览。

Keras 封装了一套高级的 API 用于构建深度学习模型，常用于快速实现原型设计和高级搜索。Keras 对开发者的友好性、模块化、可组合性和易于扩展的关键特征，使其成为目前使用较为广泛的深度学习开源框架，它的后端之一就是 TensorFlow。Keras 在 2015 年 12 月份就将其以 TensorFlow 为后端的部分 API 融合到了 TensorFlow 框架中，所以目前我们可以通过 tf.keras 或者独立的 Keras 库来使用它。本书后面的章节会大量使用 TensorFlow 和 Keras 来构建深度学习模型。

1.3.1　数据准备

我们通过 TensorFlow 提供的 Keras 接口下的 datasets 模块来加载数据集。数据集由卡内基梅隆大学维护，是波士顿近郊房价数据。读者也可以通过前言中的链接去查看后续更新的数据集。在 tf.keras.datasets 下加载的数据集和我们在链接中看到的数据集的格式是一样的，只是年份可能不同。加载数据集的代码如下。

```python
import tensorflow as tf
# 从 TensorFlow 导入 Keras 模块
from tensorflow import keras
import numpy as np
# 加载波士顿房价数据集
(train_data, train_labels), (test_data, test_labels) = \
    keras.datasets.boston_housing.load_data()
# 清洗训练集数据
# np.random.random()表示返回在0.0到1.0之间指定个数的随机浮点数
# np.argsort()表示返回对数组进行排序的索引
order = np.argsort(np.random.random(train_labels.shape))
train_data = train_data[order]
train_labels = train_labels[order]
# 归一化处理数据
# 对不同的范围和比例进行归一化处理,并且每个元素都要减去均值后再除以标准差
# 虽然模型在没有特征归一化时也可以得到收敛,但是这会让训练更加困难,
# 而且会导致结果模型依赖于训练数据
mean = train_data.mean(axis=0)
std = train_data.std(axis=0)
train_data = (train_data - mean) / std
test_data = (test_data - mean) / std
print("train_data.shape: {}, train_labels.shape: {}."
      .format(train_data.shape, train_labels.shape))
print("test_data.shape: {}, test_labels.shape: {}."
      .format(test_data.shape, test_labels.shape))
```

输出如下。

```
train_data.shape: (404, 13), train_labels.shape: (404,).
test_data.shape: (102, 13), test_labels.shape: (102,).
```

1.3.2 创建神经网络模型

在 Keras 中,我们创建神经网络模型,就是使用 Sequential 类来创建 Keras 模型。本次创建的模型比较简单,通过 Dense 类来创建神经网络层。对于输入层,层的深度是 64 个 units,输入层必须传入 input_shape 参数,表示输入数据的特征维度的大小。

激活函数(activation),我们指定的是修正线性单元(ReLU),它是神经网络中最常用的激活函数。当输入的值为正数时,导数不为 0,返回它本身,这就允许在训练模型时进行基于梯度的学习,也会使计算变得更快;当输入的值为负数时,学习速度可能会变得很慢,甚至会使神经元直接失效,这是因为输入的值是小于 0 的值,计算它的梯度也为 0,从而使其权重无法得到更新,因此在传播到下一个神经网络层的时候,返回的值为 0 就没有什么意义了。

然后我们再添加一个隐藏层,这一层不需要 input_shape 参数,因为在输入层时已经指定了。我们仍然设置层的深度是 64 个 units,激活函数是 ReLu。

输出层只有一个 unit，代码如下。

```python
# 定义创建模型函数
def build_model():
    model = keras.Sequential([
      keras.layers.Dense(64, activation=tf.nn.relu,
                        input_shape=(train_data.shape[1],)),
      keras.layers.Dense(64, activation=tf.nn.relu),
      keras.layers.Dense(1)
    ])
    # 使用 RMSProp（均方根传播）优化器，它可以加速梯度下降，其中学习速度适用于每个参数
    optimizer = tf.train.RMSPropOptimizer(0.001)
    # mse（均方差）一般用于回归问题的损失函数
    # mae（平均绝对误差）一般用于回归问题的测量/评估
    model.compile(loss='mse', optimizer=optimizer, metrics=['mae'])
    return model
model = build_model()
# 查看模型的架构
model.summary()
```

输出的模型架构如图 1.12 所示。全部参数有 5121 个，模型中的每一层参数都可以在表格的 Param 列看到。每一层的输出大小在 Output Shape 列中显示，其中 None 表示是可变的 batch size，它会在训练模型或者模型预测时被自动填充上具体的值。

```
Layer (type)                 Output Shape              Param #
=================================================================
dense (Dense)                (None, 64)                896
dense_1 (Dense)              (None, 64)                4160
dense_2 (Dense)              (None, 1)                 65
=================================================================
Total params: 5,121
Trainable params: 5,121
Non-trainable params: 0
```

图 1.12　一个简单的 Keras 模型架构图

1.3.3　训练网络模型

训练这个模型 500 次，并且将训练精确度和验证精确度记录在 history 对象中，以便绘图预览。我们自定义一个回调对象类，重写 on_epoch_end() 函数，在每次 epoch 结束时会调用该函数。

```python
# 自定义一个回调对象类，在每次 epoch（代）结束时都会调用该函数
class PrintDot(keras.callbacks.Callback):
    def on_epoch_end(self, epoch, logs):
        if epoch % 100 == 0: print('')
        print('.', end='')
EPOCHS = 500
# 训练模型
```

```
# 参数 1：房屋特征数据
# 参数 2：房屋价格数据
# 参数 3：迭代次数
# 参数 4：验证集分割比例，0.2 表示 20%的数据用于验证，80%的数据用于训练
# 参数 5：输出打印日志信息，0 表示不输出打印日志信息
# 参数 6：回调对象，这里我们使用自定义的回调类 PrintDot
history = model.fit(train_data, train_labels, epochs=EPOCHS,
                    validation_split=0.2, verbose=0,
                    callbacks=[PrintDot()])
```

1.3.4 可视化模型的结果

通过 history 对象，我们可以读取该模型训练时的误差数值，以便于观察何时是最佳模型，何时应该停止训练。

```
import matplotlib.pyplot as plt
# 绘制图来显示训练时的 accuracy 和 loss
def plot_history(history):
    plt.figure()
    plt.xlabel('Epoch')
    plt.ylabel('Mean Abs Error [1000$]')
    plt.plot(history.epoch, np.array(history.history['mean_absolute_error']),
             label='Train Loss')
    plt.plot(history.epoch, np.array(history.history['val_mean_absolute_error']),
             label='Val loss')
    plt.legend()
    plt.ylim([0, 5])
    plt.show()
plot_history(history)
```

输出如图 1.13 所示。

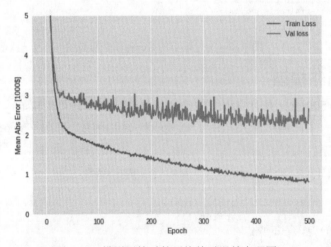

图 1.13　模型训练时的平均绝对误差表现图

可以发现，大约在 150 到 200 次迭代时，训练损失值就没怎么降低了。所以，这里我们要用到一个降低过拟合技术：早期停止（Early Stopping）。它是指在指定的迭代次数内，如果依旧没有损失降低、模型性能提升的话，就自动终止训练。

我们重新构建和训练该模型，在重新构建模型前，请先清除 Keras 的内存状态。最简单的办法就是重新运行 Jupyter Notebook；如果读者使用的是终端，就重新启动该脚本程序。

```
# 重新构建模型
model = build_model()
# 设置早期停止，如果 20 次的迭代依旧没有降低验证损失，则自动停止训练
early_stop = keras.callbacks.EarlyStopping(monitor='val_loss', patience=20)
# 重新训练模型，此时的 callbacks 有两个回调函数，所以我们使用数组的形式传入它们
history = model.fit(train_data, train_labels, epochs=EPOCHS,
                    validation_split=0.2, verbose=0,
                    callbacks=[early_stop, PrintDot()])
# 打印输出历史记录的曲线图
plot_history(history)
```

输出如图 1.14 所示。

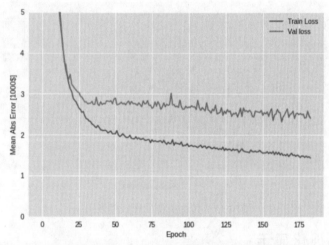

图 1.14　训练和验证模型时的平均绝对误差图（使用早期停止技术）

1.3.5　评估和预测模型

接下来，我们来测试模型在测试集下的表现，这就需要对模型进行评估了。通过 evaluate() 函数，传入测试房屋特征数据集，测试房屋价格数据，计算测试集的平均绝对误差。

```
[loss, mae] = model.evaluate(test_data, test_labels, verbose=0)
print("Testing set Mean Abs Error: ${:7.2f}".format(mae * 1000))
```

输出日志如下。

```
Testing set Mean Abs Error: $2930.86
```

然后预测模型，通过 predict()函数，传入测试房屋特征数据集，返回预测房价；最后将返回的数据通过 flatten()函数进行扁平化处理，以便于绘制散点图。

```
# 使用测试数据集预测模型
test_predictions = model.predict(test_data).flatten()
plt.scatter(test_labels, test_predictions)
plt.xlabel('True Values [1000$]')
plt.ylabel('Predictions [1000$]')
plt.axis('equal')
plt.xlim(plt.xlim())
plt.ylim(plt.ylim())
plt.plot([-100, 100], [-100, 100])
plt.show()
```

输出结果如图 1.15 所示。

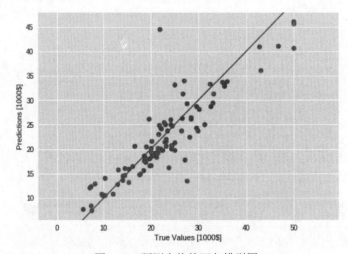

图 1.15　预测房价的回归模型图

1.3.6　预测可视化显示

接下来，我们来看下将真实房价和预测房价进行对比，从而形成的价格差直方图。我们先计算预测房价和真实房价的差价，代码如下。

```
error = test_predictions - test_labels
plt.hist(error, bins=50)
plt.xlabel("Prediction Error [1000$]")
plt.ylabel("Count")
plt.show()
```

输出如图 1.16 所示。

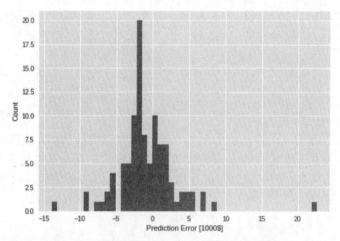

图 1.16　预测房价和真实房价的价格差直方图

最后，我们通过真实房价和预测房价生成一张更直观的图。函数 plotVersusFigure()在上面的代码中已经定义过，这里就不再介绍。

```
plotVersusFigure(test_labels, test_predictions)
```

输出如图 1.17 所示。

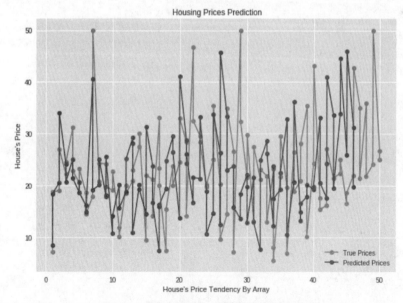

图 1.17　真实房价与预测房价的对比图

通过以上分析可知，不管是用 scikit-learn 的机器学习库来预测房价，还是使用 Keras 的神经网络模型来预测房价，真实房价和预测房价总是有些误差。所以我们能控制的就是在训练神经网络模型时，调整训练的超参数、迭代次数、网络层数和优化器等参数，以得到更好的、适用于该房屋数据的预测模型。这里的数据量比较小，如果数据量更大一些，模型效果会更好。

1.4 小结

我们使用部分开源的波士顿房价数据和北京房价数据进行预测。首先对数据的各个字段进行了分析和绘图查看，然后用 sklearn 开源库的网格搜索交叉验证去搜索最佳的决策树回归器模型，并据此进行预测，最后通过 Keras 构建神经网络模型来预测测试数据集。

第 2 章
泰坦尼克号生还预测

泰坦尼克号生还预测（Titanic Survival Exploration）是一个非常知名的练习项目。当时世界上体积最大、内部设施最豪华的泰坦尼克号游轮，有着"永不沉没"的美誉。但 1912 年 4 月，在它首次从英国南安普顿开往美国纽约的航行中，与一座冰山相撞，造成船体断裂。次日凌晨 2 点多，泰坦尼克号沉入大西洋底部 3700 多米处。在 2224 名船员和乘客中，逾 1500 名不幸遇难。这次沉船事件是和平时期十分惨重的一次海难。

本章将提供 1309 行泰坦尼克号乘客数据，其中 891 行是训练数据，418 行是测试数据，一共有 12 列，其中有一列表示这位乘客是否生还。我们将使用决策树、逻辑回归和神经网络 3 种方法来预测泰坦尼克号乘客的生还情况。

2.1 数据准备

数据集是公开的，读者可以在前言中提供的下载链接找到。数据集文件包括 train.csv、test.csv、gender_submission.csv，还有一个是 titanic_dataset.csv（前 3 个数据集文件的合并）。为了方便数据清洗、缺失值处理和分割，我们就使用 titanic_dataset.csv。本章主要使用 pandas 和 sklearn 作为数据预处理和训练的库，在数据可视化分析查看时使用的是 seaborn 和 matplotlib 库。

2.1.1 环境准备

- numpy = 1.14.5。
- pandas = 0.22.0。
- sklearn = 0.19.2。
- seaborn = 0.7.1。
- matplotlib = 2.1.2。
- keras = 2.1.6。

2.1.2 预处理数据

加载 titanic_dataset.csv 数据集,这是泰坦尼克号公开数据集的完整数据,我们通过 pandas 的 read_csv() 方法获取到了完整的 csv 数据,并且返回 Pandas 的 DataFrame 对象。因为我们的目标是预测该乘客的生还情况,所以 Survived 字段(列)就是 y,即预测目标,而 X 就是特征值,根据这些特征值来预测 y。drop() 方法表示丢弃指定的字段,保留剩下的字段。

```
import pandas as pd
features = pd.read_csv('titanic_dataset.csv')
y_train = features['Survived']
X_train = features.drop('Survived', axis=1)
```

查看前 5 行特征数据,输出如图 2.1 所示。输出显示的列表中,不包含 Survived 字段,因为我们通过调用 drop() 函数将它移除了。

```
X_train.head()
```

	Age	Cabin	Embarked	Fare	Name	Parch	PassengerId	Pclass	Sex	SibSp	Ticket
0	22.0	NaN	S	7.2500	Braund, Mr. Owen Harris	0	1	3	male	1	A/5 21171
1	38.0	C85	C	71.2833	Cumings, Mrs. John Bradley (Florence Briggs Th...	0	2	1	female	1	PC 17599
2	26.0	NaN	S	7.9250	Heikkinen, Miss. Laina	0	3	3	female	0	STON/O2. 3101282
3	35.0	C123	S	53.1000	Futrelle, Mrs. Jacques Heath (Lily May Peel)	0	4	1	female	1	113803
4	35.0	NaN	S	8.0500	Allen, Mr. William Henry	0	5	3	male	0	373450

图 2.1 前 5 行特征数据

查看 X_train 和 y_train 数据的大小。

```
print("X_train.shape={}, y_train.shape={}".format(X_train.shape, y_train.shape))
```

输出如下,变量 features 是有 12 列的,但是 Survived 字段分配给了 y_train,所以 X_train 就剩下 11 列了。

```
X_train.shape=(1309, 11), y_train.shape=(1309,)
```

通过调用 info() 函数查看 X_train 的基本信息,输出如图 2.2 所示。

```
X_train.info()
```

```
<class 'pandas.core.frame.DataFrame'>
RangeIndex: 1309 entries, 0 to 1308
Data columns (total 11 columns):
Age             1046 non-null float64
Cabin            295 non-null object
Embarked        1307 non-null object
Fare            1308 non-null float64
Name            1309 non-null object
Parch           1309 non-null int64
PassengerId     1309 non-null int64
Pclass          1309 non-null int64
Sex             1309 non-null object
SibSp           1309 non-null int64
Ticket          1309 non-null object
dtypes: float64(2), int64(4), object(5)
```

图 2.2 X_train 的基本信息

我们从该输出信息可以查看到，Age、Cabin、Embarked 和 Fare 字段都有缺失值。接下来，将对这些缺失值进行处理，如果不处理将导致无法训练。字段说明如下。

- Age：乘客年龄。
- Cabin：乘客的客舱号。
- Embarked：乘客的登船港（S 表示 Southampton（英国南安普敦），C 表示 Cherbourg-Octeville（法国瑟堡-奥克特维尔），Q 表示 Queenstown（爱尔兰昆士敦））。
- Fare：乘客支付的票价。
- Name：乘客姓名。
- Parch：一起上船的父母和子女人数。
- PassengerId：乘客登记的 ID。
- Pclass：社会阶层（1 表示上层，2 表示中层，3 表示底层）。
- Sex：乘客性别。
- SibSp：一起上船的兄弟姐妹和配偶人数。
- Ticket：乘客的船票号。
- Survived：生还情况（1 表示生还，0 表示未生还）。

2.1.3 缺失值处理

我们先查看 X_train 的缺失值合计数量，代码如下，输出如图 2.3 所示。

```
X_train.isnull().sum()
```

```
Age            263
Cabin         1014
Embarked         2
Fare             1
Name             0
Parch            0
PassengerId      0
Pclass           0
Sex              0
SibSp            0
Ticket           0
dtype: int64
```

图 2.3　查看 X_train 的缺失值数量合计

Age 字段有 263 个有缺失值。我们导入 Seaborn 数据可视化模块，通过以下代码来查看 Age 字段的分布图，输出如图 2.4 所示。

```
import seaborn as sns
sns.distplot(X_train['Age'].dropna(), hist=True, kde=True)
```

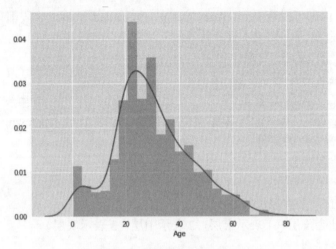

图 2.4 Age 字段的分布图

使用中值来替换缺失值,这样能保证分布图的基本呈现是呈正态分布(Normal Distribution)的。调用 replace() 方法的第一个参数表示要替换的值,np.nan 表示 NaN 数据,np.nanmedian() 方法表示将 NaN 的值都替换成中值,inplace=True 表示在原有的对象上修改且不会产生副本(原有对象的意思是 X_train 是一个类型为 Pandas 下的 Series 的变量,通过指定 inplace 参数的值为 True 后,直接在 X_train 对象上修改要替换的值)。代码如下。

```
X_train['Age'].replace(np.nan, np.nanmedian(X_train['Age']), inplace=True)
```

我们再来看 Age 字段的分布图,代码如下,输出结果如图 2.5 所示。

```
sns.distplot(X_train['Age'], hist=True, kde=True)
```

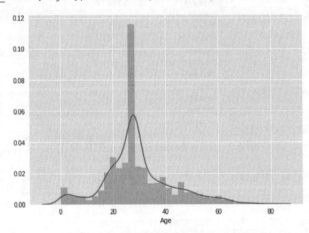

图 2.5 Age 字段里无 NaN 数据的分布图

Cabin 字段有 1014 个缺失值,因为缺失值太多,而且该字段不会影响预测结果,所以删掉就好。

```
X_train.drop("Cabin", axis=1, inplace=True)
```

Embarked 字段有 2 个缺失值。我们先来看 Embarked 字段的直方图情况，代码如下，输出结果如图 2.6 所示。

```
sns.countplot(x='Embarked', data=X_train)
```

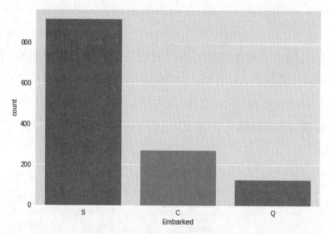

图 2.6　乘客的登船港直方图

大部分乘客的登船港都是英国南安普敦，因此假设缺失值所代表的两位乘客也是从英国南安普敦登船较为合适。我们可以通过调用 replace()函数来替换缺失值，代码如下。

```
X_train['Embarked'].replace(np.nan, 'S', inplace=True)
```

Fare 字段有 1 个缺失值。我们先查询是缺了哪位乘客的票价数据。np.isnan()方法是从指定的数组中查询有 NaN 值的行，然后返回有带 NaN 值的 True 或 False 的数组，再通过 X_train 的切片索引获取值，代码如下。在输出中我们可以看到只有一行数据，如图 2.7 所示。

```
X_train[np.isnan(X_train["Fare"])]
```

Age	Cabin	Embarked	Fare	Name	Parch	PassengerId	Pclass	Sex	SibSp	Ticket
60.5	NaN	S	NaN	Storey, Mr. Thomas	0	1044	3	male	0	3701

图 2.7　Fare 字段有缺失值的乘客数据

我们可以看到该乘客是从英国南安普敦登船，社会阶层是 3。那我们可以查询 Embarked 等于 S、Pclass 等于 3 的所有乘客支付的票价，并且以类型为 DataFrame 的对象返回。使用 query()方法来查询，代码如下。

```
pclass3_fares = X_train.query('Pclass == 3 & Embarked == "S"')['Fare']
```

然后将变量 pclass3_fares 中的 NaN 值修改为 0，代码如下。

```
pclass3_fares = pclass3_fares.replace(np.nan, 0)
```

再对 pclass3_fares 取中值，代码如下。

```
median_fare = np.median(pclass3_fares)
```

最后将该中值变量 median_fare 更新到 X_train 中缺失值的位置。通过 loc 来指定条件以进行赋值，该缺失值对应的乘客的 PassengerId 等于 1044，所以他在 Fare 字段处的值就更新为刚刚计算出的中值了，代码如下。

```
X_train.loc[X_train['PassengerId'] == 1044, 'Fare'] = median_fare
```

对于 Sex 字段，我们也需进行处理，虽然它没有缺失值，但是处理后可以更好地拟合数据。我们将男性（male）赋值为 1，女性（female）赋值为 0，代码如下。

```
X_train['Sex'].replace(['male', 'female'], [1,0], inplace=True)
```

最后，我们来看 X_train 的缺失值情况，代码如下，输出如图 2.8 所示。

```
X_train.isnull().sum()
```

```
Age            0
Embarked       0
Fare           0
Name           0
Parch          0
PassengerId    0
Pclass         0
Sex            0
SibSp          0
Ticket         0
dtype: int64
```

图 2.8　处理后的 X_train 缺失值数量合计

2.1.4　数据清洗与分割

数据清洗与分割需用到 sklearn 的 train_test_split()方法。在分割数据前，我们先对 X_train 的特征数据进行独热编码（one-hot encoding）。而 pandas 下的 get_dummies()函数就是将类别变量转换成虚拟变量（也称为哑变量）；简单地说，就是将类别的变量值转换成 0 和 1，并且最终形成一个矩形表。在该矩形表的同一个类别下，自身的类别用 1 表示，其他的类别用 0 表示。

```
X_train = pd.get_dummies(X_train)
```

然后再次输出 X_train 和 y_train 的 shape，代码如下。

```
print("X_train.shape={}, y_train.shape={}".format(X_train.shape, y_train.shape))
```

输出如下。

```
X_train.shape=(1309, 2246), y_train.shape=(1309,)
```

然后清洗与分割数据。test_size=0.2 表示给测试数据集分配 0.2 的数据，也就是 20%的数据用于测试，80%的数据用于训练，random_state 表示给随机数生成器使用的种子数。

Shuffle=True 表示清洗数据。

```
from sklearn.model_selection import train_test_split
train_X, test_X, train_y, test_y = train_test_split(X_train, y_train, test_size=0.2, random_state=42, shuffle=True)
```

通过 train_test_split() 方法，我们就得到 train_X 和 train_y 的训练集，test_X 和 test_y 的测试集。我们再看训练集和测试集各自分配了多少数据，代码如下。

```
print("train_X.shape={}, train_y.shape={}".format(train_X.shape, train_y.shape))
print("test_X.shape={}, test_y.shape={}".format(test_X.shape, test_y.shape))
```

输出如下。

```
train_X.shape=(1047, 2246), train_y.shape=(1047,)
test_X.shape=(262, 2246), test_y.shape=(262,)
```

2.2 基于决策树模型预测

决策树是用于分类和回归的非参数监督学习方法，目标是创建一个模型，通过学习从数据特征推断出简单的决策规则从而预测目标变量的值。决策树是一种决策支持工具，使用树状图或者决策模型及其可能的结果，包括机会事件结果；它也是仅包含条件控制语句算法的一种方法。决策树通常用于运作研究、决策分析，以帮助我们确定最有可能达到目标的策略。在这些树结构中，叶子表示类标签，分支表示这些类标签的特征的连接。它也是预测建模的方法之一。

使用泰坦尼克号的数据集对乘客的生还情况做出预测，该决策树在每一个条件控制语句运行时都会选择一个与生还情况最相关的特征，这正是决策树的原理。

2.2.1 训练

通过 sklearn 提供的一个便于使用的 DecisionTreeClassifier 类来进行决策树的模型预测，然后只需使用一个 fit() 方法就可以训练模型。使用 fit() 方法只需向之传入特征值 X 和预测值 y 即可。

```
from sklearn.tree import DecisionTreeClassifier
from sklearn.metrics import accuracy_score
# 创建决策树模型
model = DecisionTreeClassifier()
# 训练模型
model.fit(train_X, train_y)
```

2.2.2 预测

预测分为训练集和测试集，以此来对比，然后计算预测后的精确度。我们通过 sklearn 提供的度量函数 accuracy_score() 来计算分值。

```
# 预测训练集数据
train_pred = model.predict(train_X)
# 预测测试集数据
test_pred = model.predict(test_X)
# 计算训练集精确度
train_accuracy = accuracy_score(train_y, train_pred)
# 计算测试集精确度
test_accuracy = accuracy_score(test_y, test_pred)
```

输出预测的分值如下。

```
print('The training accuracy is {}.'.format(train_accuracy))
print('The test accuracy is {}'.format(test_accuracy))
```

输出训练和测试的分值如下。

```
The training accuracy is 1.0.
The test accuracy is 0.854961832061
```

我们再来计算 ROC 曲线的分值,稍后可通过下面的代码输出的可视化窗口进行查看。

```
y_score_dt = model.predict_proba(test_X)
fpr_dt, tpr_dt, thresholds_dt = metrics.roc_curve(test_y, y_score_dt[:,1])
print('Decision Tree Classifier AUC is: {:.3f}'.format(metrics.roc_auc_score(test_y, y_score_dt[:,1])))
```

输出如下。

```
Decision Tree Classifier AUC is: 0.838
```

2.3 基于逻辑回归模型预测

逻辑回归是一种线性回归分析模型,简单地说,就是通过一条线能分类数据的模型。通常来讲,被分类的数据是二元因变量,标记为 1 或者 0。所以当我们想要预测的目标值是二元的时,就可以使用逻辑回归模型,比如学生录取情况,要么被录取,要么被拒。

2.3.1 训练

我们通过 sklearn 提供的一个很便于使用的 LogisticRegression 类来进行逻辑回归的模型预测,同样使用 fit()方法来训练。

```
from sklearn.linear_model import LogisticRegression
from sklearn import metrics
# 创建逻辑回归预测模型
model = LogisticRegression()
# 训练模型
model.fit(train_X, train_y)
```

2.3.2 预测

预测分为训练集和测试集，以此来对比，然后计算预测后的精确度。我们仍通过 sklearn 提供的分值函数 score() 来计算分值。

```
print('Logistic Regression Accuracy for training data is: {:.3f}'.format(model.score(train_X, train_y)))
print('Logistic Regression Accuracy for testing data is: {:.3f}'.format(model.score(test_X, test_y)))
```

输出如下。

```
Logistic Regression Accuracy for training data is: 0.898
Logistic Regression Accuracy for testing data is: 0.863
```

我们可以查看它的曲线下面积的分值，代码如下。

```
y_score_lr = model.decision_function(test_X)
print('Logistic Regression AUC is: {:.3f}'.format(metrics.roc_auc_score(test_y, y_score_lr)))
```

输出如下。

```
Logistic Regression AUC is: 0.897
```

我们来计算逻辑回归的 ROC 曲线的值，稍后可通过下面代码输出的可视化窗口进行查看。

```
fpr_lr, tpr_lr, thresholds_lr = metrics.roc_curve(test_y, y_score_lr)
```

2.4 基于梯度提升分类器模型预测

梯度提升分类器通常用于回归和分类问题，和其他一些提升或增强的方法一样，梯度提升分类器以阶段方式构建模型，并且允许使用任意可微分损失函数来提高。

2.4.1 训练

通过 sklearn 提供的一个很便于使用的 GradientBoostingClassifier 类来进行梯度提升分类器的模型预测，同样使用 fit() 方法。n_estimators 表示执行时提升阶段的数量，对于过拟合，梯度提升相当稳健，因此会产生更好的性能，代码如下。

```
from sklearn.ensemble import GradientBoostingClassifier
model = GradientBoostingClassifier(n_estimators=500)
model.fit(train_X, train_y)
```

2.4.2 预测

预测分为训练集和测试集，以此对比，然后计算预测后的精确度。我们通过 sklearn 提供的精确度分值函数 accuracy_score() 来计算分值，代码如下。

```
train_pred = model.predict(train_X)
test_pred = model.predict(test_X)
print('Gradient Boosting Accuracy for training data is: {:.3f}'.format(accuracy_
score(train_y, train_pred)))
print('Gradient Boosting Accuracy for testing data is: {:.3f}'.format(accuracy_
score(test_y, test_pred)))
```

输出如下。

```
Gradient Boosting Accuracy for training data is: 1.000
Gradient Boosting Accuracy for testing data is: 0.859
```

我们再来计算 ROC 曲线的值，稍后可通过下面代码输出的可视化窗口进行查看，代码如下。

```
y_score_gb = model.predict_proba(test_X)
fpr_gb, tpr_gb, thresholds_gb = metrics.roc_curve(test_y, y_score_gb[:,1])
print('Gradient Boosting Classifier AUC is: {:.3f}'.format(metrics.roc_auc_score
(test_y, y_score_gb[:,1])))
```

输出如下。

```
Gradient Boosting Classifier AUC is: 0.933
```

2.5 基于神经网络模型预测

神经网络模型，也可以称为多层感知器（Multi-Layer Perceptron），在 sklearn 下的神经网络模块中，有已经实现的多层感知器 MLPClassifier 类。这个模型使用 LBFGS 或随机梯度下降来优化对数损失函数。

MLPClassifier 在每个时间步迭代的训练中，计算有关模型参数的损失函数的偏导数以更新参数。它可以在损失函数中添加正则化项，以缩小模型参数，防止过拟合。此实现适用于表示密集的 NumPy 数组或浮点值稀疏的 SciPy 数组的数据。

2.5.1 训练

通过 MLPClassifier 类来进行神经网络模型预测，同样使用 fit() 方法。构造函数的 hidden_layer_sizes 参数表示有多少个隐藏层，batch_size 表示每批次训练的数据大小，max_iter 表示迭代次数，solver 表示优化器。

```
from sklearn.neural_network import MLPClassifier
model = MLPClassifier(hidden_layer_sizes=128, batch_size=64, max_iter=1000, solver=
"adam")
model.fit(train_X, train_y)
```

2.5.2 预测

预测分为训练集和测试集，以此来对比，然后计算预测后的精确度。我们通过 sklearn 提供的精确度分值函数 accuracy_score() 来计算分值，代码如下。

```
train_pred = model.predict(train_X)
test_pred = model.predict(test_X)
print('Neural Network Classifier Accuracy for training data is: {:.3f}'.format
(accuracy_score(train_y, train_pred)))
print('Neural Network Classifier Accuracy for testing data is: {:.3f}'.format
(accuracy_score(test_y, test_pred)))
```

输出如下。

```
Neural Network Classifier Accuracy for training data is: 0.986
Neural Network Classifier Accuracy for testing data is: 0.748
```

我们再来计算 ROC 曲线的值,稍后可通过下面代码输出的可视化窗口进行查看,代码如下。

```
y_score_nn = model.predict_proba(test_X)
fpr_nn, tpr_nn, thresholds_nn = metrics.roc_curve(test_y, y_score_nn[:,1])
print('Neural Network Classifier AUC is: {:.3f}'.format(metrics.roc_auc_score(test_y,
y_score_nn[:,1])))
```

输出如下。

```
Neural Network Classifier AUC is: 0.833
```

2.5.3 绘制曲线图

我们已经通过决策树、逻辑回归、梯度提升分类器和神经网络模型把数据训练过了,并且每次训练完,又计算了 ROC AUC 的分值。现在通过以下代码将它们绘制成一个曲线图。

```
import matplotlib.pyplot as plt
# 创建一个 20×10 的图形
fig = plt.figure(figsize = (20,10))
ax = fig.add_subplot(111)
# 使用 plot()函数的 5 个参数解释
# 参数 1: x 轴的数据
# 参数 2: y 轴的数据
# 参数 3: 绘制的线的颜色
# 参数 4: 线宽
# 参数 5: 本条线的标题
# 绘制决策树的曲线图
ax1 = ax.plot(fpr_dt, tpr_dt, c='c', lw=2, label="Decision Tree")
# 绘制逻辑回归的曲线图
ax2 = ax.plot(fpr_lr, tpr_lr, c='y', lw=2, label="Logistic Regression")
# 绘制梯度提升分类器的曲线图
ax3 = ax.plot(fpr_gb, tpr_gb, c='r', lw=2, label="Gradient Boosting")
# 绘制神经网络的曲线图
ax4 = ax.plot(fpr_nn, tpr_nn, c='b', lw=2, label="Neural Network")
ax.grid()
lns = ax1 + ax2 + ax3 + ax4
# 在左上角添加图例
ax.legend(lns, loc=0)
# 显示绘图
plt.show()
```

通过输出的 ROC 曲线图可清晰地查看它们的走势，如图 2.9 所示。

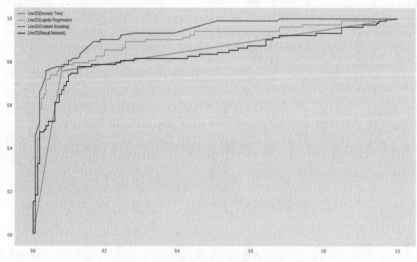

图 2.9　模型训练后的分值曲线图对比

2.6　基于 Keras 的神经网络模型预测

上一节使用基于 sklearn 的神经网络模型来预测，不需要设计自己的网络模型架构。但是，如果使用基于 Keras 的神经网络模型，我们需要自己设计架构，会用到全连接层、激活层、Dropout 和优化器。

2.6.1　训练

首先创建 Keras 的神经网络模型，该模型的第一层的 initializer 使用截断正态分布，输出层的激活函数使用 sigmoid 以二元分类，它的输出值是 0 或者 1。

```
from keras.models import Sequential
from keras.layers import Dense, Dropout, Activation
from keras import utils as np_utils
def createKerasModel(X, y):
    # 创建模型
    model = Sequential()
    # initializer 使用截断正态分布
    initializers = keras.initializers.TruncatedNormal(mean=0.0, stddev=0.05, seed=None)
    # 输入层维度 X.shape[1]，有 128 个单元
    model.add(Dense(input_dim=X.shape[1], units=128, kernel_initializer=initializers,
bias_initializer='zeros'))
    # 添加 ReLU 激活层
    model.add(Activation("relu"))
    # 添加 Dropout 层
    model.add(Dropout(0.2))
```

```python
# 添加全连接层
model.add(Dense(32))
model.add(Activation("relu"))
model.add(Dense(2))
# 输出的结果要么是 1，要么是 0，所以使用 sigmoid 激活函数
model.add(Activation("sigmoid"))
# 编译使用二进制交叉熵，adam 优化器自行调整
model.compile(loss='binary_crossentropy', optimizer='adam', metrics=['accuracy'])
# 将训练数据的 y 进行独热编码
y_train_categorical = np_utils.to_categorical(y)
# 训练模型，epochs 表示要训练 150 次，verbose 表示训练每批次时输出日志信息
model.fit(X.values, y_train_categorical, epochs=150, verbose=1)
return model
keras_model = createKerasModel(train_X, train_y)
```

训练的日志信息如下。一开始，损失值为 2.7388，精确度为 0.5535。当训练到 150 个批次时，我们得到了更小的损失值和几乎接近 1 的精确度。

```
Epoch 1/150
1047/1047 [==========================] - 1s 858us/step - loss: 2.7388 - acc: 0.5535
Epoch 2/150
1047/1047 [==========================] - 0s 282us/step - loss: 1.8277 - acc: 0.5965
Epoch 3/150
1047/1047 [==========================] - 0s 281us/step - loss: 1.3957 - acc: 0.5883
……
Epoch 148/150
1047/1047 [==========================] - 0s 209us/step - loss: 0.0280 - acc: 0.9895
Epoch 149/150
1047/1047 [==========================] - 0s 220us/step - loss: 0.0196 - acc: 0.9933
Epoch 150/150
1047/1047 [==========================] - 0s 220us/step - loss: 0.0236 - acc: 0.9928
```

2.6.2 预测

我们通过 Keras 模型的 evaluate() 函数来评估测试集数据，测试集的目标 y 数据需要做独热编码处理，这是因为我们的训练数据集的目标 y 数据做了独热编码处理，那么测试数据集的目标 y 也需要做同样的处理，计算后得到损失值和精确度。

```python
y_test_categorical = np_utils.to_categorical(test_y)
loss_and_accuracy = keras_model.evaluate(test_X.values, y_test_categorical)
print("Loss={}, Accuracy={}.".format(loss_and_accuracy[0], loss_and_accuracy[1]))
```

输出如下。

```
Loss=0.617491665687, Accuracy=0.839694656489.
```

最后，我们用预测的结果对照 PassengerId，以对比乘客的生还预测情况，代码如下。

```python
# 使用测试集的数据来预测乘客的生还情况
predictions_classes = keras_model.predict_classes(test_X.values)
# 将预测的结果和乘客的 ID 对齐
```

```
submission = pd.DataFrame({
    "PassengerID": test_X["PassengerId"],
    "Survived": predictions_classes})
# 查看前 15 条信息
print(submission[0:15])
```

输出如图 2.10 所示。

```
     PassengerId  Survived
            1149         0
            1050         0
             983         0
             809         0
            1196         1
             241         1
            1119         1
             597         1
             925         1
              66         0
            1177         0
            1187         0
            1058         0
             880         1
             997         0
```

图 2.10 Keras 模型预测乘客生还情况的前 15 条结果显示

2.7 小结

我们使用开源的泰坦尼克号的 2000 多名船员和乘客的数据，进行数据分析，对 NaN 数据进行处理和替换，通过 sklearn 开源库的决策树、逻辑回归、梯度提升分类器、多层感知器和基于 Keras 构建的神经网络模型来预测乘客的生还情况。

第 3 章

共享单车使用情况预测

共享单车在这几年非常火热,其目的是解决"最后一千米"问题。无论国内还是国外,提供共享单车服务的公司都非常多。那么读者平时在上班或上学的路途中是否骑过共享单车呢?有没有想过为什么共享单车总是大量集中在某些地方,如某些地铁站出口、商业街道旁或某些写字楼下,而有些地铁站出口或者写字楼下却没有共享单车可骑呢?还有就是为什么天气好的时候路边的共享单车非常抢手,而天气不好的时候,单车多得没人骑呢?

3.1 数据准备

我们将使用 Capital Bikeshare 公司提供的开源数据,为读者可视化地分析和讲解该公司提供的共享单车在所服务的部分地区的使用情况。数据集的下载地址可以在前言中找到。该数据集是文件 hour.csv,共有 17000 多条数据。

3.1.1 环境准备

- numpy = 1.14.6。
- pandas = 0.22.0。
- matplotlib = 2.1.2。
- seaborn = 0.7.1。
- tensorflow = 1.11.0。

3.1.2 数据可视化

我们先来看文件 hour.csv 里有哪些值。最终我们要根据骑行数据的一些条件来预测骑行人数,通过 pandas 的 read_csv()函数加载数据集,共有 17 列。

```
import numpy as np
import pandas as pd
rides = pd.read_csv('hour.csv')
rides.head()
```

输出如图 3.1 所示。

	instant	dteday	season	yr	mnth	hr	holiday	weekday	workingday	weathersit	temp	atemp	hum	windspeed	casual	registered	cnt
0	1	2011-01-01	1	0	1	0	0	6	0	1	0.24	0.2879	0.81	0.0	3	13	16
1	2	2011-01-01	1	0	1	1	0	6	0	1	0.22	0.2727	0.80	0.0	8	32	40
2	3	2011-01-01	1	0	1	2	0	6	0	1	0.22	0.2727	0.80	0.0	5	27	32
3	4	2011-01-01	1	0	1	3	0	6	0	1	0.24	0.2879	0.75	0.0	3	10	13
4	5	2011-01-01	1	0	1	4	0	6	0	1	0.24	0.2879	0.75	0.0	0	1	1

图 3.1 共享单车骑行数据预览

查看该数据的维度的代码如下。

```
rides.shape
```

输出如下。

```
(17379, 17)
```

我们首先解释各个字段（列）所代表的意思如下。

- instant：索引。
- dteday：日期。
- season：季节（1 表示春；2 表示夏；3 表示秋；4 表示冬）。
- yr：年份（0 表示 2011 年；1 表示 2012）。
- mnth：月份（1～12）。
- hr：时（0～23）。
- holiday：当天是否是节假日（0 表示否；1 表示是）。
- weekday：当天是星期几（0（星期日）～6（星期六））。
- workingday：当天是否是工作日（0 表示否；1 表示是）。
- weathersit：天气情况。
  - ——1：清澈、少量云、部分多云。
  - ——2：雾+多云、雾+碎云、雾+少量云、雾。
  - ——3：小雪、小雨+雷暴+散云、小雨+散云。
  - ——4：大雨+冰雹+雷暴+雾、雪+雾。
- temp：摄氏温度，最大值为 41。
- atemp：体感摄氏温度，最大值为 50。
- hum：湿度，最大值为 100。
- windspeed：风速，最大值为 67。
- casual：临时用户数量。
- registered：已注册用户数量。

- cnt：骑行用户总共的租车数量，包含临时用户和已注册用户。

然后，我们查看前 10 天的数据，假设每天的骑行人数是 24（这并不精确，只是为了大致了解）。从 DataFrame 变量 rides 里取出"24×10"的数据，然后通过 plot 函数绘图，它的参数 *x* 和 *y* 分别是变量 rides 里的其中一个字段，代码如下。

```
rides[:24*10].plot(x='dteday', y='cnt')
```

输出如图 3.2 所示。

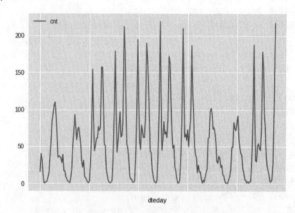

图 3.2　假设是前 10 天的数据

前两章在绘图时，图的 *X* 轴和 *y* 轴上的标题文字都是英文，如果我们希望这些标题都显示中文，那么该怎么做呢？在 matplotlib 库中，有一个字体管理器（Font Manager），通过 FontProperties 类可以加载字体，设置字体库、字体大小、权重和风格等属性。本章我们使用宋体字体库，即 Songti.ttc，将该字体库拷贝到本代码文件同级目录。

```
# 如果读者希望绘制的图的标注采用中文，可以使用如下方式
from matplotlib.font_manager import FontProperties
# 变量 font_path 是字体地址，读者使用自己的计算机或者服务器上的中文字体即可
font_path = 'Songti.ttc'
font = FontProperties(fname=font_path, size="large", weight="medium")
```

在后面的绘图中我们就可以使用变量 font 作为标题的字体对象了。接下来，我们来可视化分析温度、感觉温度、风速和湿度的散点图，并且 *X* 轴和 *y* 轴的标题和图题采用中文。注意中文的字符串前面要加上一个小写"u"表示 unicode 字符，代码如下。

```
import matplotlib.pyplot as plt
# 设置 matplotlib 在绘图时的默认样式
plt.style.use('default')
import seaborn as sns
# 创建两个绘图
fig,(ax1,ax2) = plt.subplots(ncols=2)
# 设置两个绘图的总容器大小
fig.set_size_inches(10, 5)
# regplot()函数的参数说明如下
# 参数 1：X 轴的值
```

```
# 参数 2：y 轴的值
# 参数 3：原始数据
# 参数 4：绘图的对象
# 参数 5：绘图时每个点使用的标记符号
# 显示温度的骑行数据
# 这里是：温度值 = 温度值（概率）* 温度最大值 41
sns.regplot(x=rides['temp'] * 41, y='cnt', data=rides, ax=ax1, marker='+')
ax1.set_xlabel(u"温度", fontproperties=font)
ax1.set_ylabel(u"骑行人数", fontproperties=font)
# 显示体感温度的骑行数据
# 这里是：实际体感温度值 = 体感温度值（概率）* 体感温度最大值 50
sns.regplot(x=rides['atemp'] * 50, y='cnt', data=rides, ax=ax2, marker='^')
ax2.set_xlabel(u"体感温度值", fontproperties=font)
ax2.set_ylabel(u"骑行人数", fontproperties=font)
# 创建两个绘图
fig,(ax3,ax4) = plt.subplots(ncols=2)
# 设置两个绘图的总容器大小
fig.set_size_inches(10, 5)
# 显示风速的骑行数据
# 这里是：实际风速值 = 风速值（概率）* 风速最大值 67
sns.regplot(x=rides['windspeed'] * 67, y='cnt', data=rides, ax=ax3, marker='*')
ax3.set_xlabel(u"实际风速值", fontproperties=font)
ax3.set_ylabel(u"骑行人数", fontproperties=font)
# 显示湿度的骑行数据
# 这里是：实际湿度值 = 湿度值（概率）* 湿度最大值 100
sns.regplot(x=rides['hum'] * 100, y='cnt', data=rides, ax=ax4, marker='.')
ax4.set_xlabel(u"实际湿度值", fontproperties=font)
ax4.set_ylabel(u"骑行人数", fontproperties=font)
```

输出如图 3.3 所示。

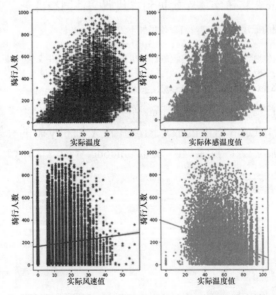

图 3.3　温度、体感温度、风速和湿度的散点图

我们可以通过热图（heat map）来分析一下相关性。热图一般通过区域和颜色这种更容易理解的方式直观地呈现一些原本不易理解或表达的数据，比如密度、频率、温度等。这里我们来看因变量受这些特征数值影响的共同点与因变量之间存在的相关性。通过 corr()函数就可以计算值的相关性，corr 的完整单词是 correlation。

```
# 我们来看温度、体感温度、临时骑行用户、已注册用户、湿度、风速和总骑行人数的相关性热图
corrMatt = rides[["temp","atemp","casual","registered","hum","windspeed","cnt"]].corr()
mask = np.array(corrMatt)
# np.tril_indices_from()表示返回数组的下三角形的索引
mask[np.tril_indices_from(mask)] = False
# 创建一个绘图
fig, ax= plt.subplots()
# 设置绘图的大小，20×10
fig.set_size_inches(20, 10)
# 用一个个彩色的小矩形组成一个大矩形
# heatmap()函数的参数说明
# 参数1：二维的矩形数据集
# 参数2：如果数据中有缺失值的cell就自动被屏蔽
# 参数3：如果是True，表示cell的宽和高相等
# 参数4：在每个cell上标出实际的数值
sns.heatmap(corrMatt, mask=mask, square=True, annot=True)
```

输出如图 3.4 所示。

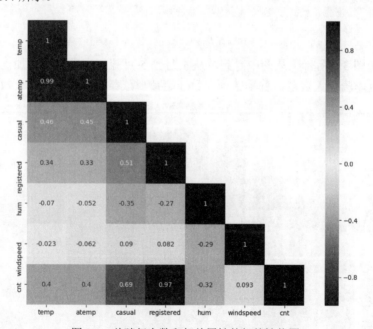

图 3.4　总骑行人数和相关属性的相关性热图

接下来，我们来可视化地分析月份和骑行人数的关系，用柱形图表示每月的骑行人数。这里绘制的图的 X 轴和 y 轴标题也使用中文显示，就是把变量 font 传递给参数 fontproperties。

对于字段 mnth（月份），它的原始值是从 1 到 12，这里正好只有 12 行数据，所以直接把 1 到 12 的数值转换成具体的月份字符串。

```
# 创建一个绘图
fig, ax_mnth = plt.subplots(1)
# 设置绘图的大小
fig.set_size_inches(13, 7)
# 对不同月份的骑行人数分组，然后计算均值
# 通过 reset_index()函数重置索引并创建一个新的 DataFrame 或者 Series 对象
mnth_cnt = pd.DataFrame(rides.groupby("mnth")["cnt"].mean()).reset_index()
# 对 DataFrame 根据月份从小到大排序，使用 ascending 参数
mnth_cnt = mnth_cnt.sort_values(by="mnth", ascending=True)
# 将月份数字转换成具体的月份字符串
mnth_cnt['mnth'] = ["January","February","March","April","May","June",
                    "July","August","September","October","November","December"]
# 绘制柱状图
# 参数 1：所有要绘制的数据
# 参数 2：x 表示 x 轴的数据，数据字段名是 mnth，数据在 data 里
# 参数 3：y 表示 y 轴的数据，数据字段名是 cnt，数据在 data 里
# 参数 4：被绘图的对象
sns.barplot(data=mnth_cnt, x='mnth', y='cnt', ax=ax_mnth)
# 设置图的 x 轴标题
ax_mnth.set_xlabel(u"月份", fontproperties=font)
# 设置图的 y 轴标题
ax_mnth.set_ylabel(u"骑行人数", fontproperties=font)
# 设置图题
ax_mnth.set_title(u"一年里每月平均骑行人数", fontproperties=font)
```

输出结果如图 3.5 所示。从图中看到，从 1 月到 5 月骑行人数持续增加，10 月后骑行人数开始减少；夏季骑行人数多，秋季骑行人数开始减少，春季和冬季骑行人数最少。

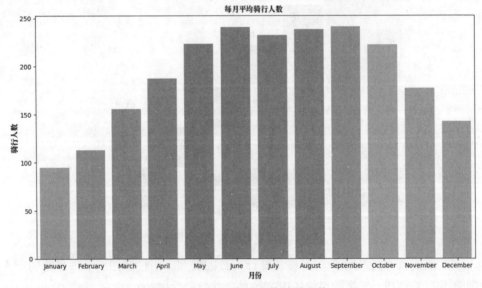

图 3.5　一年里每月平均骑行人数

然后，可视化地分析季节和时间（单位：时）与骑行人数的关系。这里用散点图来绘制，对比字段是 season、hr 和 cnt。我们把季节的原始数据 1、2、3 和 4 转换成具体的名称。

```python
# 创建一个绘图
fig, ax_hr = plt.subplots(1)
# 设置绘图的大小
fig.set_size_inches(13, 7)
# 将季节数字转换成具体的季节名称字符串
ride_feature_copied = rides.copy()
ride_feature_copied["season"] = \
ride_feature_copied.season.map({ 1 : "Spring", 2 : "Summer",
                                 3 : "Autumn", 4 : "Winter" })
# 对各个季节一天里的每小时的骑行人数分组和排序，然后计算均值
# 通过 reset_index() 函数重置索引并创建一个新的 DataFrame 或者 Series 对象
hr_cnt = pd.DataFrame(ride_feature_copied.groupby(["hr", "season"],
                      sort=True)["cnt"].mean()).reset_index()
# 绘制散点图
# 参数 1：所有要绘制的数据
# 参数 2：x 表示 X 轴的数据，数据字段名是 hr，数据在 data 里
# 参数 3：y 表示 y 轴的数据，数据字段名是 cnt，数据在 data 里
# 参数 4：根据指定字段的数据来绘制彩色的点
# 参数 5：点与点之间是否绘制线来连接
# 参数 6：被绘图的对象
# 参数 7：用不同的标记符号绘制不同的类别
sns.pointplot(data=hr_cnt,
              x=hr_cnt["hr"],
              y=hr_cnt["cnt"],
              hue=hr_cnt["season"],
              join=True,
              ax=ax_hr,
              markers=['+','o','*','^'])
# 设置图的 x 轴标题
ax_hr.set_xlabel(u"一天里的每时", fontproperties=font)
# 设置图的 y 轴标题
ax_hr.set_ylabel(u"骑行人数", fontproperties=font)
# 设置图标题
ax_hr.set_title(u"在四季里，以时为计数单位的每小时平均骑行人数", fontproperties=font)
```

输出结果如图 3.6 所示。从时间上观察，早高峰时间是 7:00、8:00 和 9:00，晚高峰时间是 17:00、18:00 和 19:00。从 20:00 到 1:00，有少量人用车；到了深夜，几乎没人用车。而且春季骑行人数明显很少。

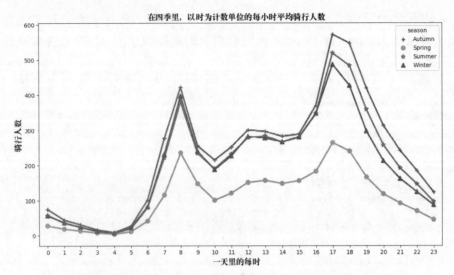

图 3.6 在四季里，以时为计数单位的每小时平均的骑行人数图

我们继续可视化地分析工作日和时间（单位：时）与骑行人数的关系。对于字段 weekday，我们希望使用具体的星期几来表示，所以将数值 0 到 6 对应转换成星期日到星期六，代码如下。

```
# 创建一个绘图
fig,(ax_hr_weekday) = plt.subplots(1, 1)
# 设置绘图的大小
fig.set_size_inches(13, 7)
# 将工作日数字转换成对应星期几字符串
ride_feature_copied["weekday"] = ride_feature_copied.weekday.map(
                                { 0 : "Sunday", 1 : "Monday",
                                  2 : "Tuesday", 3 : "Wednesday",
                                  4 : "Thursday", 5 : "Friday",
                                  6 : "Saturday" })
# 对工作日和休息日各骑行人数分组和排序，然后计算均值
# 通过 reset_index() 函数重置索引并创建一个新的 DataFrame 或者 Series 对象
hr_weekday_cnt = pd.DataFrame(ride_feature_copied.groupby(["hr", "weekday"],
                              sort=True)["cnt"].mean()).reset_index()
# 绘制散点图
# 参数 1：所有要绘制的数据
# 参数 2：x 表示 X 轴的数据，数据字段名是 hr，数据在 data 里
# 参数 3：y 表示 y 轴的数据，数据字段名是 cnt，数据在 data 里
# 参数 4：根据指定字段的数据来绘制彩色的点
# 参数 5：点与点之间是否绘制线来连接
# 参数 6：被绘图的对象
# 参数 7：用不同的标记符号绘制不同的类别
sns.pointplot(data=hr_weekday_cnt,
              x=hr_weekday_cnt["hr"],
              y=hr_weekday_cnt["cnt"],
              hue=hr_weekday_cnt["weekday"],
```

```
                    join=True,
                    ax=ax_hr_weekday,
                    markers=['+','o','*','^','x','h','s'])
# 设置图的 x 轴文字
ax_hr_weekday.set_xlabel(u"一天里的每时", fontproperties=font)
# 设置图的 y 轴文字
ax_hr_weekday.set_ylabel(u"骑行人数", fontproperties=font)
# 设置图的标题文字
ax_hr_weekday.set_title(u"在工作日和休息日里,以时为计数单位的每小时平均骑行人数",
                        fontproperties=font)
```

输出结果如图 3.7 所示,很明显的是休息日(星期六和星期日)用车时间不在早高峰或晚高峰,而是集中在中午,在工作日(星期一到星期五)用车时间很符合上下班的时间。

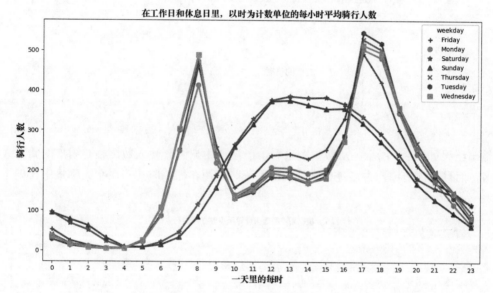

图 3.7 在工作日和休息日里,以时为计数单位的每小时平均骑行人数图

最后,我们可视化地分析已注册/临时用户和时间(单位:时)与骑行人数的关系。

```
# 创建一个绘图
fig, ax_hr_cas_reg = plt.subplots(1)
# 设置绘图的大小
fig.set_size_inches(13, 7)
# pd.melt()方法表示从宽格式到长格式的转换数据,可选择设置标识符变量
# 这里的标识符是字段 hr,变量字段是 casual 或 registered 的新字段名 variable,而值是 value
hr_cas_reg_data = pd.melt(rides[["hr","casual","registered"]],
                          id_vars=['hr'],
                          value_vars=['casual', 'registered'])
# 然后通过 reset_index()函数重置索引并创建一个新的 DataFrame 或者 Series 对象
# variable 表示已注册/临时用户,而 value 表示骑行人数
hr_cas_reg_data = pd.DataFrame(hr_cas_reg_data.groupby(
        ["hr","variable"], sort=True)["value"].mean()).reset_index()
```

```
# 绘制散点图
# 参数 1：所有要绘制的数据
# 参数 2：x 表示 X 轴的数据，数据字段名是 hr，数据在 data 里
# 参数 3：y 表示 y 轴的数据，数据新字段名是 value，其实就是骑行人数，数据在 data 里
# 参数 4：会根据指定字段的数据来绘制彩色的点
# 参数 5：绘制的色彩的顺序
# 参数 6：点与点之间是否绘制线来连接
# 参数 7：被绘图的对象
# 参数 8：用不同的标记符号绘制不同的类别
sns.pointplot(data=hr_cas_reg_data,
              x=hr_cas_reg_data["hr"],
              y=hr_cas_reg_data["value"],
              hue=hr_cas_reg_data["variable"],
              hue_order=["casual", "registered"],
              join=True,
              ax=ax_hr_cas_reg,
              markers=['p','s'])
# 设置图的 x 轴标题
ax_hr_cas_reg.set_xlabel(u"一天里的每时", fontproperties=font)
# 设置图的 y 轴标题
ax_hr_cas_reg.set_ylabel(u"骑行人数", fontproperties=font)
# 设置图题
ax_hr_cas_reg.set_title(u"根据用户类型来计算每小时平均骑行人数", fontproperties=font)
```

输出结果如图 3.8 所示。从数据图来看，已注册用户的骑行人数比临时用户的骑行人数要多得多，而且已注册用户大多是在常规的上下班时间用车，临时用户用车大都集中在偏下午时间段。

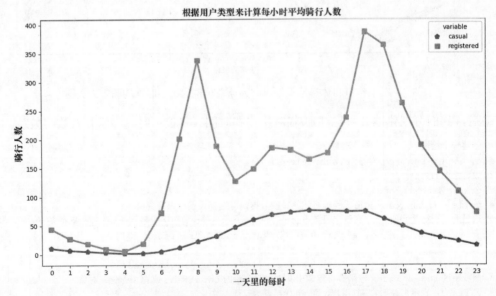

图 3.8　根据用户类型来计算每小时平均骑行人数图

3.1.3 预处理数据

我们先对部分字段的数据进行哑变量处理,pandas 提供的 get_dummies()函数直接可以对字段 season、weathersit、mnth、hr 和 weekday 进行处理。将处理完的数据矩阵再合并到主数据里(也就是变量 rides 里)。这里,我们需要丢弃进行哑变量处理前的几个字段的数据;除此之外,字段 instant、atemp 和 workingday 也不需要参与训练,所以我们通过 drop()函数将其删掉。

```
dummy_fields = ['season', 'weathersit', 'mnth', 'hr', 'weekday']
for each in dummy_fields:
    dummies = pd.get_dummies(rides[each], prefix=each, drop_first=False)
    rides = pd.concat([rides, dummies], axis=1)
fields_to_drop = ['instant', 'season', 'weathersit', 'weekday',
                  'atemp', 'mnth', 'workingday', 'hr']
rides = rides.drop(fields_to_drop, axis=1)
```

我们的目标是预测未来使用单车的骑行人数,所以在训练模型时,骑行人数是我们要预测的目标,剩下的都是特征。我们可以单独对未来的骑行总人数进行预测,也可以对未来会注册的用户和未来的临时用户进行预测。为了训练模型时更加容易些,需要对一些字段的数据进行标准化处理,使其符合标准正态分布(Normal Distribution),代码如下。

```
scaled_features = {}
quant_features = ['casual', 'registered', 'cnt', 'temp', 'hum', 'windspeed']
for each in quant_features:
    # 计算这几个字段的均值和标准差
    mean, std = rides[each].mean(), rides[each].std()
    scaled_features[each] = [mean, std]
    # 数据减去均值后再除以标准差等于标准分值(Standard Score),这样处理是为了使数据符合标准正态分布
    rides.loc[:, each] = (rides[each] - mean) / std

# 分开 features 和 target
target_cols = ['cnt', 'casual', 'registered']
y_labels = rides[target_cols]
# features 不保留骑行人数字段
X_features = rides.drop(target_cols, axis=1)
X_features.head()
```

输出结果如图 3.9 所示。经过哑变量处理和删除一些字段,最终我们得到的数据从原先的 17 列变为现在的 58 列。

	yr	holiday	temp	hum	windspeed	season_1	season_2	season_3	season_4	weathersit_1	...	hr_23	weekday_0	weekday_1	weekda
0	0	0	-1.334609	0.947345	-1.553844	1	0	0	0	1	...	0	0	0	
1	0	0	-1.438475	0.895513	-1.553844	1	0	0	0	1	...	0	0	0	
2	0	0	-1.438475	0.895513	-1.553844	1	0	0	0	1	...	0	0	0	
3	0	0	-1.334609	0.636351	-1.553844	1	0	0	0	1	...	0	0	0	
4	0	0	-1.334609	0.636351	-1.553844	1	0	0	0	1	...	0	0	0	

5 rows × 58 columns

图 3.9 数据处理后的前 5 条数据

3.1.4 数据清洗与分割

训练模型前,我们先对数据进行分割,将之分为训练集、验证集和测试集。通过 train_test_split()函数只能把数据分割成训练集和测试集,所以我们把测试集划分为两部分,一部分(10%)数据作为验证数据集,另一部分(10%)数据作为测试数据集。

```
from sklearn.model_selection import train_test_split
X_train, X_test, y_train, y_test = train_test_split(X_features,
                                                    y_labels,
                                                    test_size=0.2,
                                                    random_state=42)
test_size = X_test.shape[0] / 2
X_valid = X_test[:test_size]
y_valid = y_test[:test_size]
X_test = X_test[test_size:]
y_test = y_test[test_size:]
print("X_train.shape={}, y_train.shape={}.".format(X_train.shape, y_train.shape))
print("X_valid.shape={}, y_valid.shape={}.".format(X_valid.shape, y_valid.shape))
print("X_test.shape={}, y_test.shape={}.".format(X_test.shape, y_test.shape))
```

输出如下。

```
X_train.shape=(13903, 58), y_train.shape=(13903, 3).
X_valid.shape=(1738, 58), y_valid.shape=(1738, 3).
X_test.shape=(1738, 58), y_test.shape=(1738, 3).
```

3.2 基于 TensorFlow 的长短期记忆网络模型预测

长短期记忆网络(LSTM)是一种用于深度学习领域的人工递归神经网络架构,与标准的前馈神经网络不同,LSTM 具有反馈连接,使其成为"通用目的计算机"(也就是图灵机所能做的任何事情)。LSTM 是有状态的,很适合处理基于序列数据的预测。

本节我们将用 LSTM 来预测单车的骑行数据是比较合适的。在上面的代码中,我们已经准备好训练集、验证集和测试集。在训练之前,我们会对 X_train、X_valid 和 X_test 进行数据序列拆分,这里使用 100 的长度。

3.2.1 处理序列

我们首先定义一个函数,该函数可对数据按序列长度进行拆分,最后通过 yield 关键字返回一个 Python 生成器(可迭代的对象),在训练、验证和测试时都会用到。

```
def get_batches(X, y, num_seqs, num_steps):
    # 使用每个序列大小乘以序列步骤数得出每批次
    per_batch = num_seqs * num_steps
    # 用总个数除以每批次得出需要多少个批次
    num_batches = len(X)//per_batch
```

```python
# 将 X 和 y 的值转换成 numpy, 然后去掉索引列
X = X.reset_index().values[:,1:]
y = y.reset_index().values[:,1:]
# 取出最终的数据
X, y = X[:num_batches*per_batch], y[:num_batches*per_batch]
# 将 X 和 y 数据分别分割成 num_steps 个
dataX = []
dataY = []
for i in range(0, num_batches*per_batch, num_steps):
    dataX.append(X[i:i+num_steps])
    dataY.append(y[i:i+num_steps])

# 将 X 和 y 数据转换成 ndarray
X = np.asarray(dataX)
y = np.asarray(dataY)
# 将 X 数据分割成[samples, time steps, features]的元素,并返回生成器对象
for i in range(0,(num_batches*per_batch)//num_steps, num_seqs):
    # 使用 yield 关键字表示生成可迭代对象并返回
    yield X[i:i+num_seqs,:,:], y[i:i+num_seqs,:,:]
```

3.2.2 参数准备

本次预测的目标列数是 3 列,设置 LSTM 的单元大小是 256,有两层 LSTM 网络,序列的步长是 100,代码如下。

```python
# 获取训练数据集列数
num_features = X_train.shape[1]
# 预测目标的列数是 3 列,分别有 cnt、casual、registered
num_targets = 3
# 小批次训练
batch_size = 10
# 每个批次时希望序列能记住的步长是 100
num_steps = 100
# 设置 LSTM Cell 单元的大小为 256
lstm_size = 256
# 设置两层 LSTM
num_layers = 2
# 学习率
learning_rate = 0.0005
# 保留率
keep_prob_val = 0.75
```

3.2.3 创建 LSTM 模型

我们根据超参数来创建 LSTM 模型,输入层有 num_features 列,输出层有 num_targets 列;创建输入值(inputs)、目标值(targets)、保留率和学习率的占位符;创建 LSTM 的 cell;通过 tf.name_scope()将这些 LSTM 的层加上命名空间,也就是加一个前缀;然后使用均方差

来计算损失值；设置 Adam 优化器。

```python
import tensorflow as tf
# 创建输入值和目标值的占位符，之后会动态地传递数据到 TensorFlow 的计算图中
inputs = tf.placeholder(tf.float32, [batch_size, None, num_features], name='inputs')
targets = tf.placeholder(tf.float32, [batch_size, None, num_targets], name='targets')
# 创建保留率的占位符
keep_prob = tf.placeholder(tf.float32, name='keep_prob')
# 创建学习率的占位符
learningRate = tf.placeholder(tf.float32, name='learningRate')
# 定义创建 LSTM 的单元函数
def lstm_cell():
    # 创建基础 LSTM cell
    lstm = tf.contrib.rnn.BasicLSTMCell(lstm_size, reuse=tf.get_variable_scope().reuse)
    # 添加 dropout 层到 cell 上
    return tf.contrib.rnn.DropoutWrapper(lstm, output_keep_prob=keep_prob)
# 添加命名范围，相当于给计算图上的 tensors 的 RNN 层添加一个前缀 RNN_layers
with tf.name_scope("RNN_layers"):
    # 参数 state_is_tuple 等于 True 表示接收并返回状态，状态是 N 维元组
    cell = tf.contrib.rnn.MultiRNNCell([lstm_cell() for _ in range(num_layers)],
    state_is_tuple=True)
    # 初始化 Cell 的状态
    initial_state = cell.zero_state(batch_size, tf.float32)
# 创建由 RNNCell 指定的循环神经网络，执行完全动态的输入展开
outputs, final_state = tf.nn.dynamic_rnn(cell, inputs, dtype=tf.float32)
# 添加全连接输出层，输出层数为 3, activation_fn 设置为 None 表示使用线性激活，默认激活函数是 ReLU
predictions = tf.contrib.layers.fully_connected(outputs, 3, activation_fn=None)
# 使用均方差计算损失函数值
cost = tf.losses.mean_squared_error(targets, predictions)
# 设置优化器为 Adam
optimizer = tf.train.AdamOptimizer(learning_rate).minimize(cost)
# 计算验证精确度
correct_pred = tf.equal(tf.cast(tf.round(predictions), tf.int32), tf.cast(tf.round(targets), tf.int32))
accuracy = tf.reduce_mean(tf.cast(correct_pred, tf.float32))
```

3.2.4 训练模型

我们对所有的数据进行训练 150 次的运算，在训练过程中会保存训练损失值和验证精确度，以便于训练后绘图输出。训练时每 5 次打印一次输出记录，每 25 次验证一次并输出，最后保存训练时的最佳检查点，代码如下。

```python
epochs = 150
# 创建保存训练时的检查点对象
saver = tf.train.Saver()
# 先创建两个数组，在训练时保存训练记录，训练后绘图输出
val_accuracy=[]
training_loss=[]
# 创建 Session 会话
```

```python
with tf.Session() as sess:
    # 初始化全局变量
    sess.run(tf.global_variables_initializer())
    
    iteration = 1
    # 开始根据 epochs 值进行训练
    for e in range(epochs):
        # 先计算初始化 RNN Cell 的状态
        state = sess.run(initial_state)
        
        # 通过 get_batches() 返回一个可迭代对象，
        # 对象里的每个元素的形状是[samples, time steps, features]
        for ii, (x, y) in enumerate(get_batches(X_train, y_train, batch_size, num_steps),1):
            # 训练参数，现在将之前创建的占位符赋上值
            feed = {inputs: x,
                    targets: y,
                    keep_prob: keep_prob_val,
                    initial_state: state}
            # 训练模型
            # 根据损失函数计算得出损失值
            # 根据 RNN Cell 的状态对象计算得出状态值
            loss, state, _ = sess.run([cost, final_state, optimizer], feed_dict=feed)
            # 每训练（循环）5 次，就打印一次记录，并记录到变量 training_loss 中
            if iteration%5==0:
                print("Epoch: {}/{}".format(e, epochs),
                      "Iteration: {}".format(iteration),
                      "Train loss: {:.3f}".format(loss))
                training_loss.append(loss)
            
            # 每训练（循环）25 次，就做一次验证
            if iteration%25==0:
                val_acc = []
                # 在验证前，先恢复 RNN Cell 的状态
                val_state = sess.run(cell.zero_state(batch_size, tf.float32))
                # 验证整个验证集的验证得分，最后计算均值
                for x, y in get_batches(X_valid, y_valid, batch_size, num_steps):
                    # 验证参数，现在将之前创建的占位符赋上值
                    feed = {inputs: x,
                            targets: y,
                            keep_prob: 1,
                            initial_state: val_state}
                    # 开始计算验证
                    batch_acc, val_state=sess.run([accuracy, final_state], feed_dict=feed)
                    # 将每次计算所得分添加到数组，以便于输出打印最后的均值
                    val_acc.append(batch_acc)
                # 并将均值存储到验证得分数组，稍后通过绘图输出
                val_accuracy.append(np.mean(val_acc))
                print("Val acc: {:.3f}".format(np.mean(val_acc)))
```

```
            iteration +=1
    # 将训练的检查点保存到文件 bike-sharing.ckpt 里
    saver.save(sess, "checkpoints/bike-sharing.ckpt")
```

输出如下。

```
('Epoch: 0/150', 'Iteration: 5', 'Train loss: 0.953')
('Epoch: 0/150', 'Iteration: 10', 'Train loss: 1.008')
('Epoch: 1/150', 'Iteration: 15', 'Train loss: 0.933')
('Epoch: 1/150', 'Iteration: 20', 'Train loss: 0.886')
('Epoch: 1/150', 'Iteration: 25', 'Train loss: 1.006')
Val acc: 0.370
……
('Epoch: 148/150', 'Iteration: 1930', 'Train loss: 0.045')
('Epoch: 148/150', 'Iteration: 1935', 'Train loss: 0.045')
('Epoch: 149/150', 'Iteration: 1940', 'Train loss: 0.046')
('Epoch: 149/150', 'Iteration: 1945', 'Train loss: 0.042')
('Epoch: 149/150', 'Iteration: 1950', 'Train loss: 0.042')
Val acc: 0.801
```

可以看到验证精确度从一开始的 0.370 变为最终的 0.801。

3.2.5　模型预览与测试

因为在训练时，我们分别保存了训练损失值和验证精确度到不同的数组变量中，所以可以对它们进行可视化显示。首先输出精确度曲线，代码如下。

```
plt.plot(val_accuracy, label='Accuracy')
plt.legend()
_ = plt.ylim()
```

输出如图 3.10 所示。

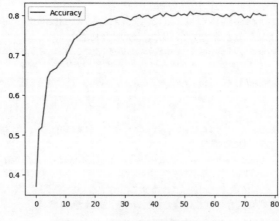

图 3.10　训练时的精确度曲线图

输出损失值曲线的代码如下。

```
plt.plot(training_loss, label='Loss')
plt.legend()
_ = plt.ylim()
```

输出如图 3.11 所示。

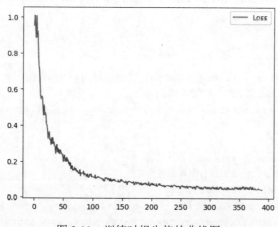

图 3.11　训练时损失值的曲线图

最后，我们来评估模型在测试集上的表现。先从文件中恢复检查点，然后对测试集数据进行批量训练，最后将 TensorFlow 计算的每个批次的分值添加到数组并取均值就是测试得分。

```
test_acc = []
# 获取 TensorFlow 的会话
with tf.Session() as sess:
    # 从文件中恢复检查点
    saver.restore(sess, tf.train.latest_checkpoint('checkpoints'))
    # 评估模型前重新初始化 RNN Cell 的状态
    test_state = sess.run(cell.zero_state(batch_size, tf.float32))
    # 再次通过 get_batches() 函数获得一个测试数据集的可迭代对象
    for ii, (x, y) in enumerate(get_batches(X_test, y_test, batch_size, num_steps), 1):
        feed = {inputs: x,
                targets: y,
                keep_prob: 1,
                initial_state: test_state}
        # 开始评估模型
        # 计算得出测试集表现的分值
        batch_acc, test_state = sess.run([accuracy, final_state], feed_dict=feed)
        # 将分值添加到数组中，最后取均值就是测试得分
        test_acc.append(batch_acc)
    print("Test accuracy: {:.3f}".format(np.mean(test_acc)))
```

得到的最终测试集的分值如下。

```
Test accuracy: 0.794
```

3.2.6 对比预测值模型预览

最后,我们来比较测试的值和原始的值——对比前 600 条数据。请注意,预测时我们用的是变量 predictions 在 TensorFlow 计算图中运算,变量 predictions 是 LSTM 网络层中的最后一个输出层,代码如下。

```
with tf.Session() as sess:
    # 从文件中恢复检查点
    saver.restore(sess, tf.train.latest_checkpoint('checkpoints'))
    # 测试模型前重新初始化 RNN Cell 的状态
    test_state = sess.run(cell.zero_state(batch_size, tf.float32))

    # 设置绘图的大小
    fig, ax = plt.subplots(figsize=(12, 6))
    # 变量 scaled_features 是在分割数据时创建的
    mean, std = scaled_features['cnt']
    # 获取测试数据集的一个批次的训练数据
    batch = get_batches(X_test, y_test, batch_size, num_steps)
    x, y = next(batch)
    feed = {inputs: x,
            targets: y,
            keep_prob: 1,
            initial_state: test_state}
    # 计算预测值
    pred = sess.run([predictions], feed_dict=feed)
    # 取出前 600 条数据进行预测
    pred = pred[0].reshape(600, -1)
    # 因为之前我们将骑行人数减去均值再除以标准差
    # 那么现在需要还原显示,就需要乘以标准差再加上均值
    pred[:,0] *= std
    pred[:,0] += mean
    predicted_val = pred[:,0]
    # 绘制预测的值的图
    ax.plot(predicted_val, label='Predicted Value')
    # 绘制原始的值的图
    ax.plot((y_test['cnt'][:600]*std + mean).values, label='Original Value')
    # 设置右侧限制
    ax.set_xlim(right=len(predicted_val))
    # 显示图例
    ax.legend()
```

输出结果如图 3.12 所示。

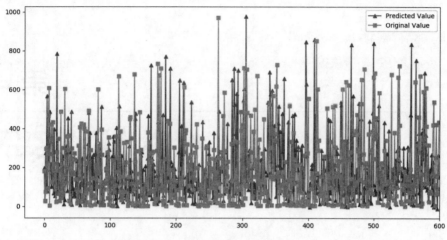

图 3.12　预测的值和原始的值对比图

3.3　小结

本章使用开源的美国共享单车公司 Capital Bikeshare 的用户骑行数据，首先对用户的骑行数据进行月份、季节、工作日和用户类型的绘图分析和查看，然后通过基于 TensorFlow 构建 LSTM 长短期记忆网络模型来训练，验证和预测数据。

第 4 章

福彩 3D 中奖预测

福彩 3D 是很多"彩民"日常"玩"的彩票之一，它是一种以 3 位自然数为投注号码的彩票，即从 000～999 的数字中挑选一个 3 位数来进行投注。它有统一的中奖规则，全国统一中奖号码，并且以随机的形式中奖，基本每天开一次奖（除法定节假日外）。由于中奖具有随机性，所以我们使用决策树、多层感知器、长短期记忆网络和时间序列来预测未来中奖号码是有一定难度的。

4.1 数据准备

数据集中包含从 2004 年 10 月 18 日到 2018 年 7 月 24 日的中奖记录，共有 4927 条。该数据集是从某彩票网站上获取的，如果读者有兴趣，也可以自己写个简单的程序，获取数据。由于本书是关于深度学习的，所以就不介绍程序是如何获取数据的。我们将会通过走势图和词云图来分析和预测中奖走势和常见的中奖号码。

4.1.1 环境准备

- numpy = 1.14.6。
- pandas = 0.22.0。
- sklearn = 0.19.2。
- matplotlib = 2.1.2。
- wordcloud = 1.5.0。
- keras = 2.2.4。

4.1.2 数据准备

数据集已经准备好，读者可通过前言中提供的 Github 地址下载。数据集是一个 csv 文件，可以通过 pandas 库的 read_csv()函数来加载，加载后返回一个 DataFrame 对象，代码如下。

```
import pandas as pd
df = pd.read_csv('3d_lottery.csv')
df.head()  # 默认输出前 5 条记录
```

输出结果如图 4.1 所示。

	期号	中奖号码	总和	总销售额	直选注数	直选奖金	组选3注数	组选3奖金	组选6注数	组选6奖金	开奖日期
0	2018198	3 7 9	19	44,710,690	19178	1,040	NaN	NaN	29671.0	173.0	2018-07-24
1	2018197	1 5 0	6	43,907,342	14712	1,040	NaN	NaN	22994.0	173.0	2018-07-23
2	2018196	0 0 5	5	46,401,818	7886	1,040	11089.0	346.0	NaN	NaN	2018-07-22
3	2018195	7 3 6	16	46,324,112	20491	1,040	NaN	NaN	29380.0	173.0	2018-07-21
4	2018194	3 7 5	15	46,948,712	22647	1,040	NaN	NaN	47504.0	173.0	2018-07-20

图 4.1　前 5 条中奖记录

查看福彩 3D 从 2004 年第一期到 2018 年 7 月的中奖个数，从 2004 年 10 月到 2018 年 7 月的中奖记录。

```
min_lottery_date = min(df["中奖日期"])
max_lottery_date = max(df["中奖日期"])
lottery_len = len(df)
print("从{}到{}的福彩 3D 中奖记录，共计{}条。".format(min_lottery_date, max_lottery_date, lottery_len))
```

输出如下。

从 2004-10-18 到 2018-07-24 的福彩 3D 中奖记录，共计 4927 条。

4.1.3　数据预处理

由于要预测中奖号码，所以我们要更加关注对变量 df 中"中奖号码"这一列的处理。从图 4.1 中可以看出，中奖号码的 3 个数字之间有空格，那它肯定是字符串类型，现在我们将空格去掉，然后将之转换成整数来处理。

```
win_number_list = [[n if len(n) > 0 else '' for n in num_str.split(' ')] for num_str in df["中奖号码"]]
print(win_number_list[:10])  # 输出前 10 条进行查看
```

输出如下。

[['3', '7', '9'], ['1', '5', '0'], ['0', '0', '5'], ['7', '3', '6'], ……, ['8', '9', '3'], ['5', '0', '3'], ['5', '5', '6'], ['7', '6', '0']]

将中奖号码转换成整数来处理。

```
won_number_list = [int(''.join(num)) for num in win_number_list]
print(won_number_list[:10])  # 输出前 10 条进行查看
```

输出如下。

[379, 150, 5, 736, 375, 61, 893, 503, 556, 760]

将转换后的整数数值更新到变量 df 中的"中奖号码"这一列。

```
df["中奖号码"] = won_number_list
df.head()
```

输出结果如图 4.2 所示。

	期号	中奖号码	总和	总销售额	直选注数	直选奖金	组选3注数	组选3奖金	组选6注数	组选6奖金	开奖日期
0	2018198	379	19	44,710,690	19178	1,040	NaN	NaN	29671.0	173.0	2018-07-24
1	2018197	150	6	43,907,342	14712	1,040	NaN	NaN	22994.0	173.0	2018-07-23
2	2018196	5	5	46,401,818	7886	1,040	11089.0	346.0	NaN	NaN	2018-07-22
3	2018195	736	16	46,324,112	20491	1,040	NaN	NaN	29380.0	173.0	2018-07-21
4	2018194	375	15	46,948,712	22647	1,040	NaN	NaN	47504.0	173.0	2018-07-20

图 4.2 中奖号码转换为整数数值后的表格

可能有读者会产生一些疑问，比如当中奖号码为 "005" 时，为什么转换后在图 4.2 上显示为 5？这是因为 "005" 是一个字符串，将其转换为整数后，"5" 前面的 "00" 就不能再显示了。

4.1.4 数据可视化

我们定义根据日期查询数据的函数，以便在走势图中查看。这里将展示 2018 年和 2017 年的走势图。通过走势图和词云图来展示中奖号码走势和最常见的中奖号码，代码如下。

```
import matplotlib.pyplot as plt
# 设置 matplotlib 在绘图时的默认样式
plt.style.use('default')
def query_by_date(df, begin_date_str, end_date_str):
    """
    定义根据日期查询中奖号码的函数
    """
    # 设置查询时中奖日期的条件
    mask = (df['中奖日期'] >= begin_date_str) & (df['中奖日期'] <= end_date_str)
    # 通过 loc 切片获取查询数据
    return df.loc[mask]
def draw_line_chart(X, y, title):
    """
    绘图，X 表示中奖期号，y 表示中奖号码，
    """
    # 设置绘图对象大小
    plt.figure(figsize = (14, 5))
    # 根据 X 和 y 来绘图
    plt.plot(X, y)
    # 设置标题
    plt.title(title)
    # 设置 X 轴的标签
    plt.xlabel("Number of Periods")
    # 设置 y 轴的标签
    plt.ylabel("Won Numbers")
    # 显示网格
```

```
    plt.grid()
    # 显示图形
    plt.show()

# 起始日期
start_date = '2018-01-01'
# 结束日期
end_date = '2018-07-31'
# 查询 2018 年的中奖号码记录
year_of_2018_df = query_by_date(df, start_date, end_date)
# 显示前 5 条中奖号码记录
year_of_2018_df.head()
```

输出结果如图 4.3 所示。

	期号	中奖号码	总和	总销售额	直选注数	直选奖金	组选3注数	组选3奖金	组选6注数	组选6奖金	开奖日期
0	2018198	379	19	44,710,690	19578	1,040	NaN	NaN	29671.0	173.0	2018-07-24
1	2018197	150	6	43,907,342	14712	1,040	NaN	NaN	22994.0	173.0	2018-07-23
2	2018196	5	5	46,401,818	7886	1,040	11089.0	346.0	NaN	NaN	2018-07-22
3	2018195	736	16	46,324,112	20491	1,040	NaN	NaN	29380.0	173.0	2018-07-21
4	2018194	375	15	46,948,712	22647	1,040	NaN	NaN	47504.0	173.0	2018-07-20

图 4.3 2018 年的中奖记录

查看共有多少条中奖记录，代码如下。

```
print("{}到{}共有{}条中奖记录。".format(start_date, end_date, len(year_of_2018_df)))
```

输出如下。

2018-01-01 到 2018-07-31 共有 198 条中奖记录。

绘制从 2018 年 1 月到 7 月的中奖走势图，代码如下。

```
title_2018 = "Lottery 3D Won Numbers, Date from {} to {}.".format(start_date, end_date)
draw_line_chart(year_of_2018_df["期号"], year_of_2018_df["中奖号码"], title_2018)
```

输出结果如图 4.4 所示。

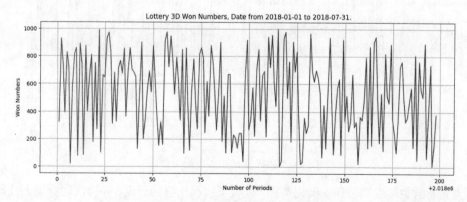

图 4.4 2018 年的中奖号码走势图

查询 2017 年从 1 月到 12 月的中奖号码记录，代码如下。

```
start_date_2017 = '2017-01-01'
end_date_2017 = '2017-12-31'
year_of_2017_df = query_by_date(df, start_date_2017, end_date_2017)
year_of_2017_df.head()
```

输出结果如图 4.5 所示。

	期号	中奖号码	总和	总销售额	直选注数	直选奖金	组选3注数	组选3奖金	组选6注数	组选6奖金	开奖日期
198	2017358	612	9	45,404,058	14879	1,040	NaN	NaN	28116.0	173.0	2017-12-31
199	2017357	665	17	46,328,368	14010	1,040	18408.0	346.0	NaN	NaN	2017-12-30
200	2017356	647	17	47,229,832	24328	1,040	NaN	NaN	43443.0	173.0	2017-12-29
201	2017355	31	4	46,945,714	12371	1,040	NaN	NaN	29519.0	173.0	2017-12-28
202	2017354	294	15	48,270,332	7927	1,040	NaN	NaN	26860.0	173.0	2017-12-27

图 4.5　2017 年的中奖记录

查看共有多少条中奖记录，代码如下。

```
print("{}到{}共有{}条中奖记录。".format(start_date_2017, end_date_2017, len(year_of_2017_df)))
```

输出如下。

2017-01-01 到 2017-12-31 共有 358 条中奖记录。

绘制从 2017 年 1 月到 12 月的中奖走势图，代码如下。

```
title_2017 = "Lottery 3D Won Numbers, Date from {} to {}.".format(start_date_2017, end_date_2017)
draw_line_chart(year_of_2017_df["期号"], year_of_2017_df["中奖号码"], title_2017)
```

输出结果如图 4.6 所示。

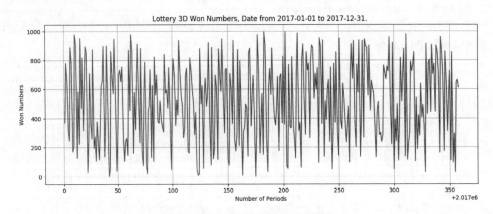

图 4.6　2017 年的中奖号码走势图

若要查看其他年度的整体中奖走势图，读者可修改相应的日期，并再次运行以上代码即可。

通过 Counter 对象来查看在历年中最常见的中奖号码是哪些,它提供了便利、快速的计数器功能,属于 Python 内置的 collections 模块。变量 won_number_list 在 4.1.3 节中出现过,代码如下。

```
from collections import Counter
counter = Counter(won_number_list)
# 查看最常见的前 15 组数字
counter.most_common(15)
```

输出如下。

```
[(369, 14),
 (573, 13),
 (209, 12),
 (393, 12),
 (827, 12),
 (139, 12),
 (669, 11),
 (960, 11),
 (731, 11),
 (899, 11),
 (784, 11),
 (144, 11),
 (667, 11),
 (363, 10),
 (20, 10)]
```

可以看出,其中"369""573""209""393""827""139"等号码中奖最频繁。那么从历史的中奖记录来看,一共开了多少种组合的号码呢?我们通过 most_common()函数可以做到,该函数会统计变量 counter 里所有的 key 和 key 的个数,代码如下。

```
most_common_list = counter.most_common()
print("开过奖的数字组合有{}种。".format(len(most_common_list)))
```

输出如下。

开过奖的数字组合有 994 种。

可以断定的是,从"000"到"999"的数字中,一共有 1000 种组合。但是从历年中奖记录来看,只有 994 种组合中过奖,那么剩下的 6 种组合是什么?

```
other_numbers = []
# 先对 1000 以内的数字进行遍历
for i in range(0, 1000):
    found = False
    # 再将之与中过奖的数字进行对比
    for num in most_common_list:
        if i == num[0]:
            found = True
            break
```

```python
        if not found:
            other_numbers.append(i)

# 在整型的中奖号码前面填充字符串"00"或"0"
def padding_zero(other_numbers):
    results = []
    for num in other_numbers:
        if len(str(num)) == 1:
            results.append("00" + str(num))
        elif len(str(num)) == 2:
            results.append("0" + str(num))
        else:
            results.append(str(num))
    return results
print("这6种数字组合从未中奖,它们分别是:{}。".format(padding_zero(other_numbers)))
```

输出如下。

这6种数字组合从未中奖,它们分别是:['025', '089', '121', '445', '629', '953']。

查看中奖号码的词云图,越常出现的号码,其数字字号越大,反之越小。我们使用的是 WordCloud 开源库来生成词云图,但是中奖号码都是由数字组成,无法生成词云图,所以作者在每个中奖号码前面加上了一个字母 i,以便生成词云图。

```python
from wordcloud import WordCloud
import matplotlib.pyplot as plt
def draw_wordcloud_image(data_text):
    # 生成词云图 WordCloud 对象
    # 参数 min_font_size:生成的文字的最小字体号
    # 参数 max_font_size:生成的文字的最大字体号
    # 参数 width:生成的图的宽度
    # 参数 height:生成的图的高度
    wordcloud = WordCloud(min_font_size=5,
                          max_font_size=200,
                          width=1200,
                          height=1000,
                          ).generate(data_text)
    # 绘图
    plt.figure()
    plt.imshow(wordcloud, interpolation='bilinear')
    # 绘图时不显示 x 和 y 轴的标尺
    # plt.axis("off")
    # 显示绘图
    plt.show()
draw_wordcloud_image(" ".join(["i" + str(n) for n in won_number_list]))
```

输出如图 4.7 所示。

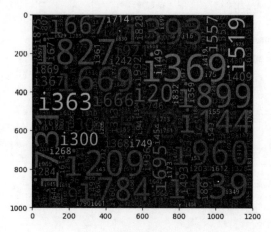

图 4.7 中奖号码词云图

当我们预测未来的中奖号码时，其中有几列的值作为特征值其实是噪点，会影响我们预测，这些列是期号、总销售额、直选注数、直选奖金、组选 3 注数、组选 3 奖金、组选 6 注数、组选 6 奖金、中奖日期。其中"中奖号码"既作为特征值，也作为预测目标。为了方便训练，我们定义了一个 split_sequence() 方法用来将"中奖号码"分割成 X 和 y，代码如下。

```
features = df["中奖号码"]
def split_sequence(sequence, n_steps):
    """
    分割单变量序列（univariate sequence）
    """
    # 将数据反转，因为我们要从 2004 年的中奖号码开始，向上预测
    sequence = sequence[::-1]
    X, y = list(), list()
    for i in range(len(sequence)):
        # 找到指定 n_steps 长度的模式，X 加上 y 的长度
        end_ix = i + n_steps
        # 如果最后一个模式的长度超出了总长度，就忽略最后一个模式
        if end_ix > len(sequence)-1:
            break
        # 获取本次的 X 和 y 的值，
        # X 的获取方式：i 表示起始值，end_ix 表示该 n_steps 模式的长度的值
        # y 的获取方式：就是 end_ix 本身
        seq_x, seq_y = sequence[i:end_ix], sequence[end_ix]
        # 添加 X 和 y 到数组中
        X.append(seq_x)
        y.append(seq_y)
    return np.array(X), np.array(y)

def preview_sequence(X, y, top=10):
    """
    查看前"top"个的特征和目标序列值
```

```
"""
    _X = X[len(X) - top:]
    _y = y[len(X) - top:]
    for i, v in enumerate(_X):
        print(v, _y[i])
```

分割 X 和 y 的数据，每 3 个一组，预测第 4 个组合的值，代码如下。

```
X, y = split_sequence(features.values, 3)
preview_sequence(X, y)
```

输出如下。

```
[141 807  48] 760
[807  48 760] 556
[ 48 760 556] 503
[760 556 503] 893
[556 503 893] 61
[503 893  61] 375
[893  61 375] 736
[ 61 375 736] 5
[375 736   5] 150
[736   5 150] 379
```

手动分割训练集和测试集的数据，代码如下。

```
# 测试数据集的比例是 15%
test_ratio = 0.15
# 整个数据集的长度
feature_len = len(X)
# 测试集的长度
feature_test_length = int(feature_len * test_ratio)
# 训练集的数据使用后面的一段
X_train, y_train = X[feature_test_length:], y[feature_test_length:]
# 测试集的数据使用前面的一段
X_test, y_test = X[:feature_test_length], y[:feature_test_length]
# 输出
print("X_train.shape={}, y_train.shape={}".format(X_train.shape, y_train.shape))
print("X_test.shape={}, y_test.shape={}".format(X_test.shape, y_test.shape))
```

输出如下。

```
X_train.shape=(4186, 3), y_train.shape=(4186,)
X_test.shape=(738, 3), y_test.shape=(738,)
```

4.2 基于神经网络模型预测

我们使用决策树、多层感知器和时间序列来预测。在使用时间序列预测前，我们会先以一种基础的方式来演示构建和训练的代码，然后再训练福彩 3D 中奖号码的模型以进行预测。

4.2.1 决策树

首先使用决策树（Decision Tree Classifier）来预测中奖号码，训练时的表现精确度很高，但是泛化能力不好，在测试集上的表现则非常差。

```
from sklearn.tree import DecisionTreeClassifier
from sklearn.metrics import accuracy_score
def predict_by_DecisionTreeClassifier(train_X, train_y, test_X, test_y):
    # 创建决策树模型对象
    model = DecisionTreeClassifier()
    # 训练模型
    model.fit(train_X, train_y)
    # 预测训练集
    train_pred = model.predict(train_X)
    # 预测测试集
    test_pred = model.predict(test_X)
    # 计算训练集精确度
    train_accuracy = accuracy_score(train_y, train_pred)
    # 计算测试集精确度
    test_accuracy = accuracy_score(test_y, test_pred)
    # 输出
    print('训练精确度是：{}.'.format(train_accuracy))
    print('测试精确度是：{}'.format(test_accuracy))
predict_by_DecisionTreeClassifier(X_train, y_train, X_test, y_test)
```

输出如下。

```
训练精确度是：1.0.
测试精确度是：0.0040650406504065045
```

4.2.2 多层感知器

然后使用多层感知器（Multi-Layer Perceptron，MLP）来预测中奖号码。设置超参数有128个隐藏层，批次大小是64，迭代1000次训练，并使用adam优化器。

```
from sklearn.neural_network import MLPClassifier
def predict_by_MLPClassifier(train_X, train_y, test_X, test_y):
    # 创建多层感知器模型对象
    model = MLPClassifier(hidden_layer_sizes=128, batch_size=64, max_iter=1000, solver="adam")
    # 训练模型
    model.fit(train_X, train_y)
    # 预测训练集
    train_pred = model.predict(train_X)
    # 预测测试集
    test_pred = model.predict(test_X)
    # 输出
    print('训练集的神经网络分类器精确度是：{}。'.format(accuracy_score(train_y, train_pred)))
```

```
print('测试集的神经网络分类器精确度是：{}。'.format(accuracy_score(test_y, test_pred)))

predict_by_MLPClassifier(X_train, y_train, X_test, y_test)
```

输出如下。

训练集的神经网络分类器精确度是：0.08767319636884854。
测试集的神经网络分类器精确度是：0.0013550135501355014。

不管是训练集还是测试集表现都很差。

4.2.3 时间序列基础

在使用时间序列（Time Series）预测前，我们先写基础的构建和训练代码。由于我们在训练时，可能会反复构建和训练模型，所以需要经常清空 Keras 的会话，以便让 Keras 每次都重新构建模型，不受上一次构建模型的影响。

```
from keras import backend as K
def clear_session():
    K.clear_session()
```

自定义训练数据，比如训练数据是[10, 20, 30]，那训练目标是 40。我们构建一个在此序列上更好的拟合模型，让模型能预测序列中的下一个值。

```
import numpy as np
X = np.array([[10, 20, 30], [20, 30, 40], [30, 40, 50], [40, 50, 60]])
y = np.array([40, 50, 60, 70])
```

基于单变量（Univariate）的多层感知器模型预测，一个输入层，一个输出层。

```
import numpy as np
from keras.models import Sequential
from keras.layers import Dense
def train_mlp_predictor(X, y, need_clear_session=False):
    # 是否需要清空 Keras 会话
    if need_clear_session:
        clear_session()
    # 定义模型
    model = Sequential()
    # 输入序列的长度是 3
    model.add(Dense(100, activation='relu', input_dim=3))
    # 输出序列的长度是 1
    model.add(Dense(1))
    # 编译模型
    model.compile(optimizer='adam', loss='mse')
    model.summary()
    # 训练模型
    model.fit(X, y, epochs=2000, verbose=0)
    return model
mlp_predictor_model = train_mlp_predictor(X, y, True)
# 测试
```

```
x_input = np.array([50, 60, 70])
x_input = x_input.reshape((1, 3))
yhat = mlp_predictor_model.predict(x_input, verbose=0)
print("预测的值: {}。".format(yhat))
```

输出如下。

预测的值：[[80.05862]]。

输出结果如图 4.8 所示。

```
Layer (type)                 Output Shape              Param #
=================================================================
dense_1 (Dense)              (None, 100)               400
_____
dense_2 (Dense)              (None, 1)                 101
=================================================================
Total params: 501
Trainable params: 501
Non-trainable params: 0
```

图 4.8　单变量 MLP 网络模型概要

基于单变量的卷积神经网络（CNN）模型预测。该模型的输入层是 1D 卷积，并对其添加了最大池化层和扁平化层，然后添加一个有 50 个深度的全连接层，最后添加的输出层只有一个深度，因为预测的结果 y 仅有一个计算值。

```
import numpy as np
from keras.models import Sequential
from keras.layers import Dense
from keras.layers import Flatten
from keras.layers.convolutional import Conv1D
from keras.layers.convolutional import MaxPooling1D
# 将训练数据从[samples, timesteps]大小转换到[samples, timesteps, features]格式
X = X.reshape((X.shape[0], X.shape[1], 1))
def train_cnn_predictor(X, y, need_clear_session=False):
    # 是否需要清空 Keras 会话
    if need_clear_session:
        clear_session()
    # 定义模型
    model = Sequential()
    # 添加 1D 卷积层，64 的深度，内核大小为 2，使用 relu 来调节权重和误差，输入大小是(3,1)
    model.add(Conv1D(filters=64, kernel_size=2, activation='relu', input_shape=(3, 1)))
    model.add(MaxPooling1D(pool_size=2))
    model.add(Flatten())
    model.add(Dense(50, activation='relu'))
    # 输出层数是 1
    model.add(Dense(1))
    model.compile(optimizer='adam', loss='mse')
    model.summary()
    # 训练模型
    model.fit(X, y, epochs=1000, verbose=0)
    return model
```

```
cnn_predictor_model = train_cnn_predictor(X, y, True)
# 测试
x_input = np.array([50, 60, 70])
x_input = x_input.reshape((1, 3, 1))
yhat = cnn_predictor_model.predict(x_input, verbose=0)
print("预测的值: {}。".format(yhat))
```

输出如下。

预测的值: [[80.908264]]。

输出结果如图 4.9 所示。

```
Layer (type)                 Output Shape              Param #
=================================================================
conv1d_1 (Conv1D)            (None, 2, 64)             192
_____
max_pooling1d_1 (MaxPooling1 (None, 1, 64)             0
_____
flatten_1 (Flatten)          (None, 64)                0
_____
dense_1 (Dense)              (None, 50)                3250
_____
dense_2 (Dense)              (None, 1)                 51
=================================================================
Total params: 3,493
Trainable params: 3,493
Non-trainable params: 0
```

图 4.9　单变量 CNN 网络模型概要

基于单变量的长短期记忆网络（LSTM）模型预测，一个输入层，一个输出层。

```
import numpy as np
from keras.models import Sequential
from keras.layers import LSTM
from keras.layers import Dense
def train_lstm_predictor(X, y, need_clear_session=False):
    # 是否需要清空 Keras 会话
    if need_clear_session:
        clear_session()
    # 定义模型
    model = Sequential()
    model.add(LSTM(50, activation='relu', input_shape=(3, 1)))
    model.add(Dense(1))
    model.summary()
    model.compile(optimizer='adam', loss='mse')
    # 训练模型
    model.fit(X, y, epochs=1000, verbose=0)
    return model
X = X.reshape((X.shape[0], X.shape[1], 1))
lstm_predictor_model = train_lstm_predictor(X, y, True)
# 预测
x_input = np.array([50, 60, 70])
x_input = x_input.reshape((1, 3, 1))
yhat = lstm_predictor_model.predict(x_input, verbose=0)
```

```
print("预测的值：{}。".format(yhat))
```

输出如下。

预测的值：[[83.401054]]。

输出结果如图 4.10 所示。

```
Layer (type)                 Output Shape              Param #
=================================================================
lstm_1 (LSTM)                (None, 50)                10400
_____
dense_1 (Dense)              (None, 1)                 51
=================================================================
Total params: 10,451
Trainable params: 10,451
Non-trainable params: 0
```

图 4.10　单变量 LSTM 网络模型概要

不管是使用上述的 3 个方法 MLP、CNN 和 LSTM 中的哪一个来训练模型，它们最终的结果都差不多，但是不会每次都一模一样。

4.2.4　时间序列预测

现在使用 MLP 和 LSTM 来进行福彩 3D 中奖号码的预测，分别使用的方式是 train_mlp_predictor()和 train_lstm_predictor()。

```
import math
# 训练模型
mlp_predictor_model = train_mlp_predictor(X_train, y_train, True)
# 预测
y_hat = mlp_predictor_model.predict(X_test, verbose=0)
# 输出前 10 个号码
for i in range(10):
    print("真实中奖号码={} : {}=预测中奖号码.".format(y_test[i], math.ceil(y_hat[i])))
```

输出如下。

```
真实中奖号码=791 : 434=预测中奖号码.
真实中奖号码=941 : 513=预测中奖号码.
真实中奖号码=611 : 535=预测中奖号码.
真实中奖号码=144 : 492=预测中奖号码.
真实中奖号码=471 : 532=预测中奖号码.
真实中奖号码=505 : 527=预测中奖号码.
真实中奖号码=828 : 516=预测中奖号码.
真实中奖号码=607 : 461=预测中奖号码.
真实中奖号码=878 : 510=预测中奖号码.
真实中奖号码=477 : 522=预测中奖号码.
```

使用 LSTM 模型来训练中奖号码的预测，代码如下。

```
X_train = X_train.reshape((X_train.shape[0], X_train.shape[1], 1))
# 训练模型
lstm_predictor_model = train_lstm_predictor(X_train, y_train, True)
```

```
# 预测
X_test = X_test.reshape((X_test.shape[0], X_test.shape[1], 1))
y_hat = lstm_predictor_model.predict(X_test, verbose=0)
# 输出前 10 个号码
for i in range(10):
    print("真实中奖号码={} : {}=预测中奖号码.".format(y_test[i], math.ceil(y_hat[i])))
```

输出如下。

```
真实中奖号码=791 : 409=预测中奖号码.
真实中奖号码=941 : 528=预测中奖号码.
真实中奖号码=611 : 620=预测中奖号码.
真实中奖号码=144 : 536=预测中奖号码.
真实中奖号码=471 : 581=预测中奖号码.
真实中奖号码=505 : 644=预测中奖号码.
真实中奖号码=828 : 535=预测中奖号码.
真实中奖号码=607 : 473=预测中奖号码.
真实中奖号码=878 : 466=预测中奖号码.
真实中奖号码=477 : 519=预测中奖号码.
```

4.2.5 根据中奖号码单变量单个位数预测

在上面的代码中，我们是通过完整的 3 位数的中奖号码来预测的，这里将使用单变量单个位数来预测。比如中奖号码[369, 567, 123]，那么预测分组就是[351, 662, 973]。其中变量 win_number_list 是在 4.1.3 节中创建的，代码如下。

```
# 将每个数字都转换成整型
win_number_list_2 = [[int(n) for n in t] for t in win_number_list]
# 取出第一列的值
first_column = np.array(win_number_list_2)[:,0]
# 取出第二列的值
second_column = np.array(win_number_list_2)[:,1]
# 取出第三列的值
third_column = np.array(win_number_list_2)[:,2]
```

进行数据分割，将数据中的 3 列都分割开来形成独立的序列数据，再对各独立的序列数据进行分割，分割的长度依然是 3。

```
def split_data_features(X, n_steps=3):
    X, y = split_sequence(X, n_steps)
    return X, y

# 数据分割与准备
# 分割第一列数据的 X 和 y
first_X, first_y = split_data_features(first_column)
# 分割第二列数据的 X 和 y
second_X, second_y = split_data_features(second_column)
# 分割第三列数据的 X 和 y
third_X, third_y = split_data_features(third_column)
```

训练模型,我们采用 LSTM 的训练方法来训练福彩 3D 中奖号码。

```python
def train_my_model(X, y, modelpath):
    X = X.reshape(X.shape[0], X.shape[1], 1)
    # 使用 LSTM 方法来训练
    lstm_predictor_model = train_lstm_predictor(X, y, True)
    lstm_predictor_model.save(modelpath)

# 定义存储训练模型的文件名
first_model_path = "first_model.h5"
second_model_path = "second_model.h5"
third_model_path = "third_model.h5"
# 训练模型
train_my_model(first_X, first_y, first_model_path)
train_my_model(second_X, second_y, second_model_path)
train_my_model(third_X, third_y, third_model_path)
```

首先从当前目录加载模型,然后分割 10 条中奖数据用来测试,最后把真实数据和预测数据进行对比。

```python
from keras.models import load_model
# 加载第一列序列的训练模型
first_data_model = load_model(first_model_path)
# 加载第二列序列的训练模型
second_data_model = load_model(second_model_path)
# 加载第三列序列的训练模型
third_data_model = load_model(third_model_path)
# 分割测试数据,在训练过的数据中选择 10 条中奖记录,然后每 3 个一组,最终会产生 7 组
first_test_X, first_test_y = split_data_features(first_column[:10])
second_test_X, second_test_y = split_data_features(second_column[:10])
third_test_X, third_test_y = split_data_features(third_column[:10])
# 使用 LSTM 网络训练的模型,在数据传入前,需要 reshape 把数据
# 从 [samples, timesteps] 修改成 [samples, timesteps, features]
first_test_X = first_test_X.reshape(first_test_X.shape[0], first_test_X.shape[1], 1)
second_test_X = second_test_X.reshape(second_test_X.shape[0], second_test_X.shape[1], 1)
third_test_X = third_test_X.reshape(third_test_X.shape[0], third_test_X.shape[1], 1)
# 预测数据
first_predicted_val = first_data_model.predict(first_test_X, verbose=0)
second_predicted_val = second_data_model.predict(second_test_X, verbose=0)
third_predicted_val = third_data_model.predict(third_test_X, verbose=0)
import math
# 我们来对比下预测数据和真实数据,会发现预测数据错得离谱
# 将这 3 列的序列数据水平堆砌起来
final_predicted_X = np.vstack([first_predicted_val[:,0], second_predicted_val[:,0], third_predicted_val[:,0]])
final_target_X = np.vstack([first_test_y[:10].tolist(), second_test_y[:10].tolist(), third_test_y[:10].tolist()])
# 通过 transpose() 转置数据数组,便于打印输出
final_predicted_X = final_predicted_X.transpose()
final_target_X = final_target_X.transpose()
for i, v in enumerate(final_predicted_X):
```

```
        print("P: {} vs T: {}.".format([math.ceil(p) for p in final_predicted_X[i]],
final_target_X[i]))
```

输出如下。

```
P: [5, 5, 4] vs T: [8 9 3].
P: [4, 5, 5] vs T: [0 6 1].
P: [5, 5, 5] vs T: [3 7 5].
P: [4, 4, 5] vs T: [7 3 6].
P: [3, 5, 5] vs T: [0 0 5].
P: [5, 5, 5] vs T: [1 5 0].
P: [5, 6, 6] vs T: [3 7 9].
```

P 表示预测数据，T 表示真实数据。

4.3 小结

本章开始介绍了对福彩 3D 中奖数据的获取，然后对福彩 3D 中奖数据进行走势图分析和查看。我们使用基于 sklearn 库的决策树和多层感知器进行预测；介绍了用 Keras 构建基于时间序列的基础用法，通过基于单变量的 MLP 模型、卷积神经网络（CNN）模型、长短期记忆网络（LSTM）模型来对 3 个号码进行整体预测和基于 LSTM 模型对分开的数字进行独立预测。通过本章的介绍，我们学到了如何使用部分的时间序列技术来预测。

第 5 章

股票走势预测

股市有风险,投资需谨慎。我们经常看到这句话,事实也确实如此。经常看到大盘上涨,但是个股总在波动,投还是不投总是让人拿捏不定。那么有没有一种技术可以帮助我们计算、评估、推测未来的股票走势呢?这当然是有的。本章将会讲解在纳斯达克上市的两家世界知名企业——百度和微软的历年股票走势数据,并且进行分析和预测。

5.1 数据准备

截至 2018 年,百度股票的历年数据有 3000 多条,微软股票的历年数据有 8000 多条。我们知道,股市会在休息日(周六和周日)、法定节假日休市,所以一年只有约 250 个工作日。这里的股票走势数据也是开市时一天一条,包含开盘价、收盘价、最高价和最低价等。

5.1.1 环境准备

- numpy = 1.14.6。
- pandas = 0.22.0。
- matplotlib = 2.1.2。
- quandl = 3.4.5。
- fbprophet = 0.3。

5.1.2 数据集说明

数据集已经准备好,读者可以通过本书前言中提供的 Github 地址下载。数据集是两个 csv 文件,分别是 BaiduStock.csv 和 MicrosoftStock.csv,读者可以通过 pandas 库的 read_csv()函数来加载,也可以通过 quandl 库来请求获取数据,但是需要申请注册账号。

5.2 百度股票预测

百度在纳斯达克上市的日期是 2005 年 8 月 5 日，所以数据就是从 2005 年开始的。我们将会分析数据、训练模型和计算假如自己购买的话收益如何。假设我们在百度刚上市时就持有 100 股，到了 2018 年，收益会有 22000 多美元，长期投资该股会有数百倍收益。

5.2.1 数据准备

我们先使用 quandl 库来获取百度股票的历年数据。在开始之前，需要先去 quandl 官网注册一个账号，拿到 API Key，然后方可获取数据。我们在编写正式应用时，一般会隐藏 API Key 这样的关键敏感数据，然后通过一个函数加载，这样做的目的是避免敏感数据直接暴露在代码中。这里的 save_key()函数就是先将 API Key 保存到路径为$HOME/.quandl_apikey 文件中。

```
def init_api_key():
    quandl.save_key("Your API Key")
    print(quandl.ApiConfig.api_key)
init_api_key()
```

输出 API Key，我们可以通过 Linux 的 cat 命令查看该文件内容。

```
cat $HOME/.quandl_apikey
```

如果是在 Jupyter Notebook 的 Cell 中执行该命令，需要在 cat 命令前面加上一个英文感叹号，也可以通过 Python 的接口函数来读取 API Key。

```
quandl.read_key()
print(quandl.ApiConfig.api_key)
```

quandl 库封装好了请求数据接口的函数，我们来获取百度的股票数据，百度在纳斯达克的股票代码为 BIDU。

```
def init_stock(stock_name):
    # 获取百度股票数据
    stock = quandl.get("WIKI/{}".format(stock_name))
    # 设置列 Date 为第一列
    stock = stock.reset_index(level=0)
    return stock

# 初始化百度股票所有数据
stock_name = "BIDU"
baiduStock = init_stock(stock_name)
baiduStock.head()
```

输出如图 5.1 所示。

	Date	Open	High	Low	Close	Volume	Ex-Dividend	Split Ratio	Adj. Open	Adj. High	Adj. Low	Adj. Close	Adj. Volume
0	2005-08-05	66.00	151.21	60.00	122.54	22681100.0	0.0	1.0	6.600	15.121	6.000	12.254	226811000.0
1	2005-08-08	137.75	153.98	115.24	115.50	15488900.0	0.0	1.0	13.775	15.398	11.524	11.550	154889000.0
2	2005-08-09	120.50	125.30	95.69	96.10	8667700.0	0.0	1.0	12.050	12.530	9.569	9.610	86677000.0
3	2005-08-10	101.00	103.50	88.30	91.75	4963800.0	0.0	1.0	10.100	10.350	8.830	9.175	49638000.0
4	2005-08-11	91.20	100.50	90.60	97.90	7324800.0	0.0	1.0	9.120	10.050	9.060	9.790	73248000.0

图 5.1　百度股票从上市开始的前 5 条数据

数据各列名称的解释如下。

- Date：股市开市日期。
- Open：开盘价。
- High：最高价。
- Low：最低价。
- Close：收盘价。
- Volume：成交量。
- Ex-Dividend：无股息。
- Split：股票分拆。
- Ratio：比率。
- Adj. Open：调整开盘价。
- Adj. High：调整最高价。
- Adj. Low：调整最低价。
- Adj. Close：调整收盘价。
- Adj. Volume：调整成交量。

查看百度股票一共有多少条记录，代码如下。

```
print("baiduStock 共计{}条。".format(len(baiduStock)))
```

输出如下。

baiduStock 共计 3180 条。

查看股票数据的开始日期和终止日期，代码如下。

```
min_date = min(baiduStock['Date'])
max_date = max(baiduStock['Date'])
print("百度的股票数据从{}到{}。".format(min_date, max_date))
```

输出如下。

百度的股票数据从 2005-08-05 00:00:00 到 2018-03-27 00:00:00。

在这里，我们想把这个数据保存到文件中，以便下次使用时不需要再次请求获取。正确的做法是先检查 baiduStock 对象是什么类型的，然后对它调用保存到文件的函数。先检查对象类型，代码如下。

```
print(type(baiduStock))
```

输出如下。

```
<class 'pandas.core.frame.DataFrame'>
```

它是一个 DataFrame 类，那么我们可以使用 pandas 中的 to_csv()函数保存到文件。

```
baiduStock.to_csv("baiduStock.csv", index=False)
```

index=False 表示保存时不保存 DataFrame 数据中的索引列。此时我们去代码文件的同级目录查看，就可以看到 baiduStock.csv 文件了，然后直接通过文本编辑器就可以打开并查看内容。将保存的数据文件从本地加载到内存，代码如下。

```
import pandas as pd
baidu_df = pd.read_csv("baiduStock.csv")
baidu_df.head()
```

输出结果和图 5.1 所示是一样的。

5.2.2　数据可视化

绘制从 2005 年到 2018 年的百度个股走势图，取 Adj.Close 字段（列）为走势图的每个点的值，代码如下。

```
import numpy as np
import matplotlib.pyplot as plt
def plot_basic_stock_history(df, start_date, end_date, stock_name):
    # 定义已调整收盘价作为统计字段
    stats_Ajd_Close = 'Adj. Close'
    # 最低已调整收盘价
    stat_min = min(df[stats_Ajd_Close])
    # 最高已调整收盘价
    stat_max = max(df[stats_Ajd_Close])
    # 已调整收盘价均价
    stat_mean = np.mean(df[stats_Ajd_Close])
    # 找出最小已调整收盘价的日期
    date_stat_min = df[df[stats_Ajd_Close] == stat_min]['Date']
    # 取出该日期，并转换 Timestamp 类型
    date_stat_min = date_stat_min[date_stat_min.index[0]].date()
    # 找出最大已调整收盘价的日期
    date_stat_max = df[df[stats_Ajd_Close] == stat_max]['Date']
    # 取出该日期，并转换 Timestamp 类型
    date_stat_max = date_stat_max[date_stat_max.index[0]].date()
    # 输出打印基本信息
    print("{}在{}最小，价格是：{}美元。".format(stats_Ajd_Close, date_stat_min, stat_min))
    print("{}在{}最高，价格是：{}美元。".format(stats_Ajd_Close, date_stat_max, stat_max))
    # df.index[-1]表示最后一个索引值
    print("{}在{}当前价格是：{}美元。".format(stats_Ajd_Close, end_date.date(), df.loc[df.index[-1], 'Adj. Close']))
```

```python
# 设置绘图背景颜色风格
plt.style.use("default")
# 绘图
# 参数 1：x 轴是日期
# 参数 2：y 轴是已调整收盘价
# 参数 3：线条颜色是红色
# 参数 4：线宽是 3
# 参数 5：标签
plt.plot(df["Date"],
         df[stats_Ajd_Close],
         color='r',
         linewidth=3,
         label=stats_Ajd_Close)
# 设置 x 轴的标签是 Date
plt.xlabel("Date")
# 设置 y 轴的标签是 US $
plt.ylabel("US $")
# 设置标题是 BIDU Stock History
plt.title("{} Stock History".format(stock_name))
# 显示网格
plt.grid()
# 显示绘图
plt.show()
# 基本走势图显示
start_date = min_date
end_date = max_date
plot_basic_stock_history(baiduStock, start_date, end_date, stock_name)
```

输出如下。

Adj. Close 在 2006-02-07 最小，价格是：4.515 美元。
Adj. Close 在 2017-10-16 最高，价格是：272.82 美元。
Adj. Close 在 2018-03-27 当前价格是：230.96 美元。

输出的百度个股历年走势图如图 5.2 所示。

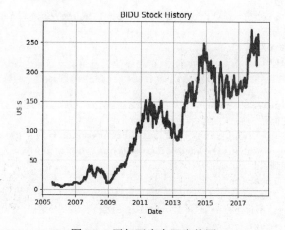

图 5.2　历年百度个股走势图

5.2.3 计算购买的股票收益

假设我们在 2005 年百度刚上市时购买了 100 股百度股票,开盘价为 66 美元,共花费 6600 美元。那么到了 2018 年 3 月 27 日,我们看看这 100 股涨了多少?

```python
import pandas as pd
def plot_potential_profit(df,
                          start_date,
                          end_date,
                          stock_name,
                          line_color,
                          text_color,
                          myshares=1):
    # 获取刚上市时的开盘价
    start_price = float(df[df["Date"] == start_date]["Adj. Open"])
    # 获取最后一次获取该数据时的已调整收盘价
    end_price = float(df[df["Date"] == end_date]["Adj. Close"])
    # 计算每一次开市的收益 =(当天已调整的收盘价 - 刚上市时的开盘价)* 购买的股票数量
    df["profits"] = (df["Adj. Close"] - start_price) * myshares
    # 计算总收益
    total_hold_profit = (end_price - start_price) * myshares
    # 输出基本信息打印
    print("从{}到{},购买{}股,总收益是:{}。".format(start_date.date(),
                                            end_date.date(),
                                            myshares,
                                            total_hold_profit))
    # 设置绘图背景颜色为黑色
    plt.style.use("default")
    # 绘图
    # 参数 1:X 轴是日期
    # 参数 2:y 轴是收益
    # 参数 3:线的颜色
    # 参数 4:线的宽度
    plt.plot(df["Date"], df["profits"], color=line_color, linewidth=3)
    # 设置 X 轴的日期标签
    plt.xlabel("Date")
    # 设置 y 轴的收益标签
    plt.ylabel("Profit $")
    # 设置图题
    plt.title("My Shares From {} to {}.".format(start_date.date(), end_date.date()))
    # 计算显示数字的位置
    text_location_x = (end_date - pd.DateOffset(months=1)).date()
    text_location_y = total_hold_profit + (total_hold_profit / 40)
    # 将收益数字绘制到图上
    plt.text(text_location_x,
             text_location_y,
             "${}".format(int(total_hold_profit)),
             color=text_color,
             size=15)
```

```
    # 显示网格
    plt.grid()
    # 显示绘图
    plt.show()
# 绘制收益图
start_date = min_date
end_date = max_date
# m 表示线的颜色是洋红色
# g 表示收益文本的颜色是绿色
# 这里假设当初购买了 100 股
plot_potential_profit(baiduStock, start_date, end_date, stock_name, 'm', 'g', 100)
```

输出的文字信息如下。

从 2005-08-05 到 2018-03-27，购买 100 股，总收益是：22436.0 美元。

输出的图如图 5.3 所示。

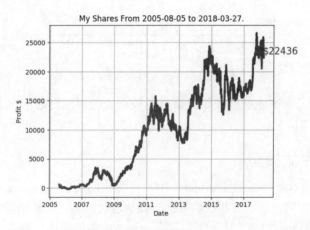

图 5.3　百度股票收益走势图

通过以上代码和收益走势图的分析，我们可知若当初购买 100 股并持股到 2018 年，确实可挣很多，会翻几倍。但是我们同时也看到在中间的几年时间里该只股票下跌得有些严重，并且持续了两三年。我们再来看从 2012 年 8 月 7 日到 2013 年 3 月 5 日持有该股的收益状态。

```
# 设置计算起始日期
start_date = pd.to_datetime("2012-08-07")
# 设置计算截止日期
end_date = pd.to_datetime("2013-03-05")
# 获取这段日期的股票数据
baiduStockLowerPricePhase = baiduStock[
                    (baiduStock['Date'] >= start_date.date()) &
                    (baiduStock['Date'] <= end_date.date())
                    ]
# 绘图
plot_potential_profit(baiduStockLowerPricePhase, start_date, end_date, stock_name, 'c', 'r', 100)
```

输出如下。

从 2012-08-07 到 2013-03-05，购买 100 股，总收益是：-3853.999999999999 美元。

输出结果如图 5.4 所示。

图 5.4　百度股票部分时段的收益走势图

很多股民会在看见一家企业的往年股票市场数据表现得很好的情况下，贸然持股，就可能会面对图 5.4 所示的情况。

5.2.4　训练和评估模型

我们将通过 Facebook 开源的 fbprophet 库来训练我们的模型。通过它可实现基于可加性模型的时间序列数据预测，适于对非线性数据的趋势进行预测，也能够很好地处理异常值。此模型仅接受一个 DataFrame 对象，由一列日期和一列数据数值组成，列名分别为 ds 和 y。

```
def train_model(stock_history, days=0, weekly_seasonality=False, monthly_seasonality=False):
    # 创建一个 Prophet 模型对象
    model = fbprophet.Prophet(daily_seasonality=False,
                              weekly_seasonality=False,
                              yearly_seasonality=True,
                              changepoint_prior_scale=0.05)
    # 添加月份季节性
    if monthly_seasonality:
        model.add_seasonality(name='monthly', period=30.5, fourier_order=5)
    # 训练模型
    model.fit(stock_history)
    # 设置根据指定的 days（天数）来预测未来多少天的数据的 DataFrame
    future = model.make_future_dataframe(periods=days)
    # 预测
    future = model.predict(future)
    # 返回模型和预测的数据
```

```python
        return model, future

    def create_prophet_model(df,
                             stock_name,
                             days=0,
                             weekly_seasonality=False,
                             monthly_seasonality=False):
        # 取出最近 3 年的记录
        stock_history = df[df["Date"] > (max_date - pd.DateOffset(years=3)).date()]
        # 训练模型
        model, future = train_model(stock_history, days, weekly_seasonality, monthly_seasonality)
        # 设置绘图背景是默认的
        plt.style.use("default")
        # 初始化一个绘图对象
        fig, ax = plt.subplots(1, 1)
        # 设置绘图的画布大小
        fig.set_size_inches(10, 5)
        # 绘制真实的值
        # 参数 1：x 轴的日期显示
        # 参数 2：y 轴的价格数值
        # 参数 3：线条的样式，线中间有实心圆圈
        # 参数 4：线宽
        # 参数 5：透明度
        # 参数 6：线条样式的大小（marker size）
        # 参数 7：图例标签的文本
        ax.plot(stock_history['ds'],
                stock_history['y'],
                'v-',
                linewidth=1.0,
                alpha=0.8,
                ms=1.8,
                label='Observations')
        # 绘制预测的值
        ax.plot(future['ds'],
                future['yhat'],
                'o-',
                linewidth=1.,
                label='Modeled')
        # 使用带状绘制一个不确定的区间值
        ax.fill_between(future['ds'].dt.to_pydatetime(),
                        future['yhat_upper'],
                        future['yhat_lower'],
                        alpha=0.3,
                        facecolor='g',
                        edgecolor='k',
                        linewidth=1.0,
                        label='Confidence Interval')
        # 设置图例，loc 等于 2 表示 "upper left"（右上），字体大小是 10
        plt.legend(loc=2, prop={'size': 10})
```

```
    # 设置标题
    plt.title("{} Historical and Modeled Stock Price".format(stock_name))
    # 设置 X 轴的标签日期
    plt.xlabel('Date')
    # 设置 y 轴的标签价格
    plt.ylabel('Price $')
    # 添加网格,线宽是 0.6,透明度是 0.6
    plt.grid(linewidth=0.6, alpha=0.6)
    # 绘图显示
    plt.show()
    # 返回模型和预测的值
    return model, future
# 将开市日期作为 ds 列,真实数据作为 y 列
baiduStock["ds"] = baiduStock['Date']
baiduStock["y"] = baiduStock['Adj. Close']
model, future_data = create_prophet_model(baiduStock, stock_name, monthly_seasonality=True)
```

输出结果如图 5.5 所示。

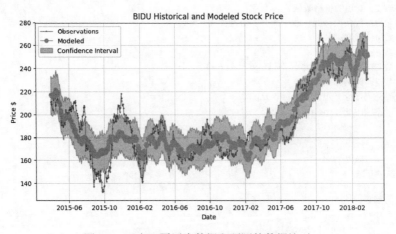

图 5.5　百度股票历史数据和预测的数据比对

第一条线 Observations 表示真实的已调整的收盘价;第二条线 Modeled 表示预测的收盘价;第三条线 Confidence Inteval 表示置信区间,意思是以第二条线为基准预测的值的范围。

基于时间序列的数据,刚刚我们预测出的值用变量 future_data 表示,现在我们再来看趋势、月份季节性和年度季节性的显示图。代码如下。

```
model.plot_components(future_data)
plt.show()
```

输出结果分别如图 5.6、图 5.7、图 5.8 所示。

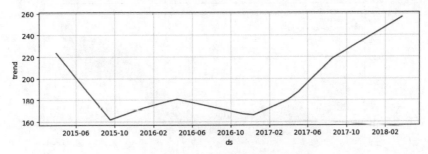

图 5.6　时间序列百度股票数据的趋势

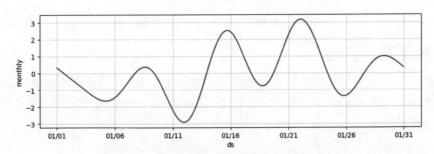

图 5.7　时间序列百度股票数据的月份季节性

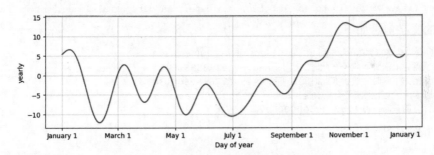

图 5.8　时间序列股票数据的年度季节性

5.2.5　股票预测

通过上面的模型训练和分析,我们将基于时间序列预测未来 180 天的百度股票价格。将参数 days 设为 180,代码如下。

```
model, future = create_prophet_model(baiduStock, stock_name, days=180)
```

输出结果如图 5.9 所示。

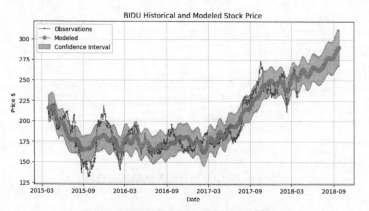

图 5.9 预测百度未来 180 天的股票价格

5.2.6 股票买入策略

我们使用 3 年的股票走势数据来训练模型，然后使用最后一年的数据做测试。代码文件保存在 prophet_evaluator.py 文件中，在执行下面的代码前，先导入该文件，文件中只有一个函数，就是 evaluator()。

```
import prophet_evaluator
baiduStock["ds"] = baiduStock['Date']
baiduStock["y"] = baiduStock['Adj. Close']
prophet_evaluator.evaluator(baiduStock, min_date, max_date, train_model, stock_name, 1000)
```

输出的文字信息如下。

```
BIDU 的股票从 2017-03-27 到 2018-03-27，购买 1000 股。
当模型预测到上涨时，股价在此时间点上涨 56.69291338582677%。
当模型预测到下跌时，股价在此时间点下跌 49.586776859504134%。
使用 Prophet 模型预测整个收益是：$38780.00000000006。
买入并持有的策略：$61150.00000000001。
```

输出结果如图 5.10 所示。

图 5.10 预测和买入百度股票持仓收益

5.3 微软股票预测

微软在纳斯达克上市的日期是 1986 年 3 月 13 日，所以数据的起始日期就是这一天。与 5.2 节类似，我们将分析数据、训练模型并计算假如自己购买的话收益如何。累计这 32 年的 8000 多条微软股票市场开市记录。本节用到的一些函数均已在 5.2 节中执行过。

5.3.1 数据准备

微软在纳斯达克的股票代码是 MSFT，通过在 5.2 节中定义的 init_stock() 函数来加载数据，代码如下。

```
stock_name = "MSFT"
microsoftStock = init_stock(stock_name)
microsoftStock.head()
```

输出结果如图 5.11 所示。

	Date	Open	High	Low	Close	Volume	Ex-Dividend	Split Ratio	Adj. Open	Adj. High	Adj. Low	Adj. Close	Adj. Volume
0	1986-03-13	25.50	29.25	25.5	28.00	3582600.0	0.0	1.0	0.058941	0.067609	0.058941	0.064720	1.031789e+09
1	1986-03-14	28.00	29.50	28.0	29.00	1070000.0	0.0	1.0	0.064720	0.068187	0.064720	0.067031	3.081600e+08
2	1986-03-17	29.00	29.75	29.0	29.50	462400.0	0.0	1.0	0.067031	0.068765	0.067031	0.068187	1.331712e+08
3	1986-03-18	29.50	29.75	28.5	28.75	235300.0	0.0	1.0	0.068187	0.068765	0.065876	0.066454	6.776640e+07
4	1986-03-19	28.75	29.00	28.0	28.25	166300.0	0.0	1.0	0.066454	0.067031	0.064720	0.065298	4.789440e+07

图 5.11 微软股票走势数据预览

查看数据共有多少条，代码如下。

```
print("Microsoft Stock 共计{}条。".format(len(microsoftStock)))
```

输出如下。

```
Microsoft Stock 共计 8076 条。
```

查看开始日期和终止日期，代码如下。

```
min_date = min(microsoftStock['Date'])
max_date = max(microsoftStock['Date'])
print("微软的股票数据从{}到{}。".format(min_date, max_date))
```

输出如下。

微软的股票数据从 1986-03-13 00:00:00 到 2018-03-27 00:00:00。

5.3.2 数据可视化

绘制从 1986 年到 2018 年的历年微软个股走势图，取 Adj.Close 列为走势图的每个点的值，代码如下。

```
microsoftStock["Date"] = pd.to_datetime(microsoftStock["Date"])
start_date = min_date
end_date = max_date
plot_basic_stock_history(microsoftStock, start_date, end_date, stock_name)
```

输出的文字信息如下。

Adj. Close 在 1986-03-24 最小，价格是：0.060097123934714 美元。
Adj. Close 在 2018-03-12 最高，价格是：96.77 美元。
Adj. Close 在 2018-03-27 当前价格是：89.47 美元。

输出结果如图 5.12 所示。

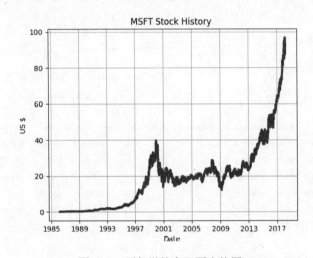

图 5.12　历年微软个股票走势图

5.3.3　计算购买的股票收益

假设我们在 1986 年微软刚上市时购买了 100 股微软股票，开盘价为 25.5 美元，共花费 2550 美元。那么到了 2018 年 3 月 27 日，我们看看这 100 股涨了多少呢？

```
start_date = min_date
end_date = max_date
plot_potential_profit(microsoftStock, start_date, end_date, stock_name, 'm', 'g', 100)
```

输出的文字信息如下。

从 1986-03-13 到 2018-03-27，购买 100 股，总收益是：8941.105858998711 美元。

输出结果如图 5.13 所示。

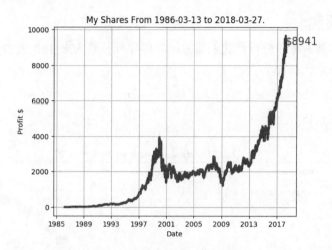

图 5.13 微软股票收益走势图

5.3.4 训练和评估模型

训练和评估函数已经在 5.2.4 节中定义和执行过了，这里直接调用。

```
microsoftStock["ds"] = microsoftStock['Date']
microsoftStock["y"] = microsoftStock['Adj. Close']
# 将create_prophet_model()函数里的第二个plot()方法里的绘图线条样式"o-"修改成"-"，使绘出的
图更美观
model, future_data = create_prophet_model(microsoftStock, stock_name, monthly_seaso
nality=True)
```

输出结果如图 5.14 所示。

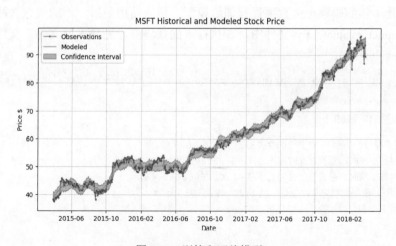

图 5.14 训练和评估模型

5.3.5 股票预测

通过上面的模型训练和分析,我们将基于时间序列预测未来 180 天的微软股票价格。将参数 days 设为 180,代码如下。

```
model, future = create_prophet_model(microsoftStock, stock_name, days=180)
```

输出结果如图 5.15 所示。

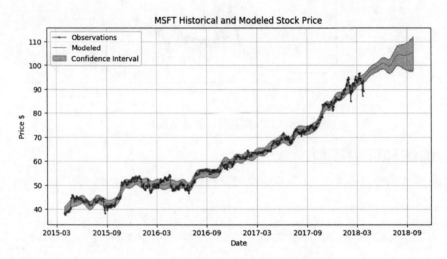

图 5.15 预测微软未来 180 天的股票价格

5.3.6 股票买入策略

我们使用 3 年的股票走势数据进行训练模型,然后使用最后一年的数据做测试。代码文件保存在 prophet_evaluator.py 文件中,在执行下面代码前,先导入该文件,文件中只有一个函数,就是 evaluator()。

```
prophet_evaluator.evaluator(microsoftStock, min_date, max_date, train_model, stock_name, 1000)
```

输出的文字信息如下。

```
MSFT 的股票从 2017-03-27 到 2018-03-27,购买 1000 股。
当模型预测到上涨时,股价在此时间点上涨 55.932203389830505%。
当模型预测到下跌时,股价在此时间点下跌 41.66666666666667%。
使用 Prophet 模型预测整个收益是: $12827.423107539924。
买入并持有的利润: $23111.007152734004。
```

输出结果如图 5.16 所示。

图 5.16　预测和买入微软股票的持仓收益

5.4　小结

我们通过 quandl 开源库获取两家企业——百度和微软在纳斯达克的股票价格走势数据集，当然也可以通过作者已经下载好的数据集获取数据。然后使用了 Facebook 开源的 fbprophet 库来进行股票价格预测，并且分析了两种股票的买入策略。

第 6 章

垃圾邮件预测

我们在生活和工作中经常收到垃圾邮件,打开邮箱常常看到一堆无用的推广邮件,甚至还有一些欺诈邮件,而一些计算机新手用户就可能就会陷入垃圾邮件的欺诈中。为了防止此类欺诈案件的发生,各大邮箱平台服务公司都研发了自己的邮件反欺诈系统并植入用户所使用的邮件系统中。为了一探究竟,本章将讲解如何训练我们自己的垃圾邮件过滤器模型。

将垃圾邮件分类使用的是一种文本分类(Text Classification)算法。使用该算法可以处理大规模文本数据,比如对论坛中用户的评论是否合法合规进行标记、邮件是否需要被过滤到广告邮件或者欺诈邮件的分类箱里。

6.1 数据准备

本章将用 NLTK 库来对邮件正文的内容进行分析和处理。NLTK 库是一个处理文本数据的开源库,可用于文本分类、标记、解析、词义还原和语义推理等。读者需要先通过 pip install nltk 命令来下载安装该库。

6.1.1 环境准备

- numpy = 1.14.6。
- pandas = 0.22.0。
- matplotlib = 2.1.2。
- sklearn = 0.20.1。
- nltk = 3.2.5。
- wordcloud = 1.5.0。
- tensorflow = 1.12.0。

6.1.2 数据准备

数据集已经准备好，读者可通过本书前言中提供的 Github 地址下载。这是一个 spam_emails.csv 文件，包含 5728 封邮件，其中垃圾邮件有 1368 封，正常邮件有 4360 封。读者可以通过 Pandas 库的 read_csv()函数来加载，代码如下。

```
import pandas as pd
df = pd.read_csv('spam_emails.csv')
df.head()
```

输出如图 6.1 所示。

	text	spam
0	Subject: naturally irresistible your corporate...	1
1	Subject: the stock trading gunslinger fanny i...	1
2	Subject: unbelievable new homes made easy im ...	1
3	Subject: 4 color printing special request add...	1
4	Subject: do not have money , get software cds ...	1

图 6.1　前 5 封邮件预览

数据中的两个列的解释如下。

- text：邮件正文内容。
- spam：是否是垃圾邮件，1 表示是，0 则不是。

统计邮件的数量，代码如下。

```
spam_email_df = df[df["spam"] == 1]
notspam_email_df = df[df["spam"] == 0]
print("共计{}封邮件，正常邮件有{}封，垃圾邮件有{}封。".format(df.shape[0],
                                                    len(notspam_email_df),
                                                    len(spam_email_df)))
```

输出如下。

共计 5728 封邮件，正常邮件有 4360 封，垃圾邮件有 1368 封。

6.1.3 数据预处理

在训练模型前，我们需要对数据集进行预处理以便于更好地了解数据、分析数据和处理数据。现在我们查看数据分布情况，由于 spam 列的值只能是 0 和 1，所以通过 groupby()函数分组会产生两条记录，一个是 spam 等于 0 的总个数，另一个是 spam 等于 1 的总个数。

```
import matplotlib.pyplot as plt
# 设置matplotlib的默认样式
plt.style.use("default")
groupeddf = df.groupby(["spam"]).count()
groupeddf.head()
```

输出结果如图 6.2 所示。

	text
spam	
0	4360
1	1368

图 6.2 邮件数据分组后的图

通过该分组数据，我们绘制直方图以显示数据分布。变量 *X* 表示将分组后的两个数据的数值对应的标签名称添加到数组中，这是为了方便绘图；变量 *y* 表示获取分组数据的数值（即各自总量），代码如下。

```
X = []
X.append("not spam" if groupeddf.index.values[0] == 0 else "")
X.append("spam" if groupeddf.index.values[1] == 1 else "")
# 或者直接写元素的名称作为标签
# X = ["not spam", "spam"]
y = groupeddf.values[:,0]
# 创建一个 figure 实例对象
fig, ax = plt.subplots()
# 设置绘图窗口的大小
fig.set_size_inches(10, 5)
# 绘制直方图
ax.bar(X, y)
# 设置图题
plt.title('Spam Email Bar Graph')
# 设置 y 轴标题
plt.ylabel('Number of Emails')
# 网格显示
plt.grid()
# 显示绘图
plt.show()
```

输出结果如图 6.3 所示。

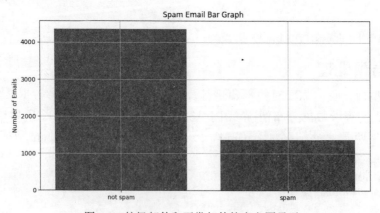

图 6.3 垃圾邮件和正常邮件的直方图显示

我们看到邮件的正文内容都以"Subject: "开始,这个不仅对训练没有用,还会产生噪点,因此将其删除。对于每封邮件的内容长度,我们添加一个新列叫作 length,并对此进行处理。

```
df["text"] = [text[len("Subject: "):] for text in df["text"]]
df['length'] = df['text'].map(lambda text: len(text))
df.head()
```

输出结果如图 6.4 所示。

	text	spam	length
0	naturally irresistible your corporate identity...	1	1475
1	the stock trading gunslinger fanny is merrill...	1	589
2	unbelievable new homes made easy im wanting t...	1	439
3	4 color printing special request additional i...	1	491
4	do not have money , get software cds from here...	1	226

图 6.4 有新列 length 的邮件数据预览

检查数据为 null 的情况。代码如下。

```
df.isnull().sum()
```

输出如下。

```
text      0
spam      0
length    0
dtype: int64
```

然后,我们对垃圾邮件和正常邮件绘制一个基本的数据分布图。

```
import numpy as np
# 取出 DataFrame 里的所有垃圾邮件的正文内容长度的数据
spam_df = df[df["spam"] == 1]["length"].values
# 取出 DataFrame 里的所有正常邮件的正文内容长度的数据
notspam_df = df[df["spam"] == 0]["length"].values
def plot_basic(spam_df, notspam_df):
    fig, ax = plt.subplots()
    fig.set_size_inches(10, 5)
    # 绘制垃圾邮件的正文内容长度的图
    # 参数 1: X 轴,根据真实的 DataFrame 的值设置长度的数组
    # 参数 2: y 轴,经过排序后的真实的数值
    # 参数 3: 线的颜色
    # 参数 4: 透明度
    # 参数 5: 图例的显示文字
    ax.plot(
        np.array(range(spam_df.shape[0])),
        np.sort(spam_df),
        color='r',
        alpha=0.5,
```

```
            label='Spam')
    # 绘制正常邮件的正文内容长度的图
    ax.plot(
            np.array(range(notspam_df.shape[0])),
            np.sort(notspam_df),
            color='g',
            alpha=0.5,
            label='Not Spam')
    plt.title('Email Spam or Not Spam Distribution Diagram')
    # legend 是用来设置图例的
    # loc 等于 1 表示图例右上角显示,size 表示图例的文字大小
    plt.legend(loc=1, prop={'size': 12})
    # 网格显示
    plt.grid()
    plt.show()
plot_basic(spam_df, notspam_df)
```

输出结果如图 6.5 所示。

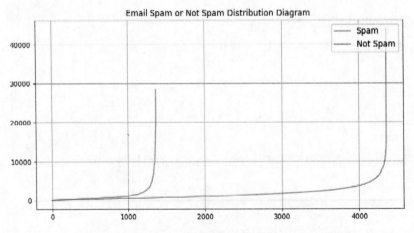

图 6.5　垃圾邮件和正常邮件的正文内容长度分布图

接下来,我们将使用 NLTK 库对 DataFrame 的数据进行处理。使用 NLTK 库可以对数据进行标记化(Tokenization)、停用词(StopWords)和词性还原(Lemmatization)的操作。在标记化前,我们需要先下载 punkt 组件,它是一个标记器模型(Tokenizer Model)。

```
import nltk
nltk.download('punkt')
```

下载完后,我们就将邮件内容中的每个句子转换成独立的单词,这就是 token。将 token 添加到 DataFrame 中,成为一个新列。

```
df['tokens'] = df['text'].map(lambda text: nltk.tokenize.word_tokenize(text))
print(df['tokens'][0])
```

输出如下。

```
['naturally', 'irresistible', 'your', 'corporate', 'identity', 'lt', 'is', 'really',
'hard', 'to', 'recollect', 'a', 'company', ':', ……
```

处理停用词,它在使用前也是需要下载的。

```
nltk.download('stopwords')
```

如果停用词是英语,有哪些常见的停用词。

```
stop_words = nltk.corpus.stopwords.words('english')
print("停用词共有{}个,前 15 个是: {}。".format(len(stop_words), stop_words[:15]))
```

输出如下。

```
停用词共有 179 个,前 15 个是: ['i', 'me', 'my', 'myself', 'we', 'our', 'ours', 'ourselves',
'you', "you're", "you've", ……
```

对数据进行停用词处理,我们新添加一个 text_stopwords 列,这一列表示已经经过了停用词处理的列数据。在 map 函数的遍历过程中,我们将每句话的数组合并成为一个字符串。

```
df['text_stopwords'] = df['tokens'].map(lambda tokens: " ".join([t for t in tokens
if t not in stop_words]))
print(df['text_stopwords'][0])
```

输出如下。

```
naturally irresistible corporate identity lt really hard recollect company : market
full suggestions information ……
```

去掉标点符号,因为标点符号对于我们的训练毫无用处。再将去掉标点符号的值存到新建的列中,列名为 text_punctuation。

```
import string
print("string.punctuation 所包含的标点符号有: {}。".format(string.punctuation))
df['text_punctuation'] = df['text_stopwords'].map(lambda text: "".join([w for w in
text if w not in string.punctuation]))
print(df['text_punctuation'][0])
```

输出如下。

```
string.punctuation 所包含的标点符号有: !"#$%&'()*+,-./:;<=>?@[\]^_`{|}~。
naturally irresistible corporate identity lt really hard recollect company  market
full suggestions information ……
```

我们都知道英语的语法会将单词变体,包括一般时、进行时、完成时、完成进行时等时态,每个时态的动词变体不一样。名词的单复数形式也有很多特殊变体。那么词性还原,就是将这些变体的单词转换成原来的状态,处理过的数据存储到新列 processed_text 中。

```
# 下载 wordnet 以使词性还原
nltk.download('wordnet')
lemmatizer = nltk.WordNetLemmatizer()
```

```
df['processed_text'] = df['text_punctuation'].map(lambda text: lemmatizer.lemmatize
(text))
print(df['processed_text'][0])
```

输出如下。

```
naturally irresistible corporate identity lt really hard recollect company  market
full suqgestions information ……
```

绘制垃圾邮件的词云图，让我们来看看什么单词在垃圾邮件中出现得最频繁。绘制词云图将使用 wordcloud 库。

```
from wordcloud import WordCloud
# 定义绘制词云图的函数
def plot_wordcloud(email_text):
    # 创建一个词云图实例对象，窗口大小为 300×300 英寸
    email_wordcloud = WordCloud(width=300, height=300).generate(email_text)
    # 创建一个 matplotlib 的绘图对象
    plt.figure(figsize=(8, 6), facecolor='k')
    # 将词云图对象在 matplotlib 的对象上显示
    plt.imshow(email_wordcloud)
    plt.show()

spam_words = ''.join(list(df[df['spam']==1]['processed_text']))
plot_wordcloud(spam_words)
```

输出结果如图 6.6 所示。

图 6.6　垃圾邮件数据的词云图

绘制正常邮件的词云图，同样让我们来看看什么单词在正常邮件中出现得最频繁。

```
notspam_words = ''.join(list(df[df['spam']==0]['processed_text']))
plot_wordcloud(notspam_words)
```

输出结果如图 6.7 所示。

图 6.7　正常邮件数据的词云图

接下来,我们将数据分割成两份,分别是训练集(占 85%)和测试集(占 15%)。通过 train_test_split()函数把数据集分割成训练集和测试集,代码如下。

```
# 导入sklearn中的训练集和测试集的分割函数
from sklearn.model_selection import train_test_split
features = df["processed_text"]
labels = df["spam"]
X_train, X_test, y_train, y_test = train_test_split(features, labels, test_size=0.15)
print("X_train.shape={}, y_train.shape={}.".format(X_train.shape, y_train.shape))
print("X_test.shape={}, y_test.shape={}.".format(X_test.shape, y_test.shape))
```

输出如下。

```
X_train.shape=(4868,), y_train.shape=(4868,).
X_test.shape=(860,), y_test.shape=(860,).
```

6.2　基于多项式朴素贝叶斯分类器的邮件分类

多项式朴素贝叶斯分类器(Multinomial Naive Bayes Classifier)一般适用于文本分类任务中的以整数特征计数。比如使用 TF IDF 就可以达到类似的效果。因为数据是文本文档,所以训练前我们需要将其转换为 token 计数的稀疏矩阵来表示。训练完模型后,我们会通过自定义的邮件文本内容数据来测试该模型的表现。

6.2.1　数据处理

通过 CountVectorizer 类,我们对训练集和测试集进行向量化处理,代码如下。

```python
from sklearn.feature_extraction.text import CountVectorizer
# 创建向量化计数对象
count_vectorizer = CountVectorizer()
# 将训练集数据适配和转换，以便于训练模型
X_train_text_counts = count_vectorizer.fit_transform(X_train.values)
# 将测试集数据转换为文档项矩阵，以便于在测试时可以从原始文本文档中提取标记计数
X_test_text_counts = count_vectorizer.transform(X_test.values)
```

6.2.2　创建和训练模型

在使用 sklearn 中的模型时，我们不需要编写该模型算法和架构，只需要创建实例对象，然后调用 fit()方法即可训练模型。

```python
from sklearn.naive_bayes import MultinomialNB
# 创建多项式朴素贝叶斯模型
classifierNB = MultinomialNB()
# 训练模型
classifierNB.fit(X_train_text_counts, y_train)
```

6.2.3　测试模型

通过模型的 predict()方法即可测试模型的性能，代码如下。

```python
from sklearn.metrics import accuracy_score
# 测试结果
predictions = classifierNB.predict(X_test_text_counts)
acc = accuracy_score(y_test, predictions)
print("测试精确度是：{:.7}%。".format(acc * 100))
```

输出如下。

测试精确度是：99.47644%。

现在，我们自己来编写两封邮件，然后测试该模型在真实情况下的表现。先测试一封正常邮件在该模型下的结果。

```python
a_good_mail_test_text = ["Zhang Qiang, it's important that our meeting will open at 3pm tommorrow."]
a_good_mail_test_text_counts = count_vectorizer.transform(a_good_mail_test_text)
prediction1 = classifierNB.predict(a_good_mail_test_text_counts)
print("测试结果等于{}表示这是一封正常邮件。".format(prediction1))
```

输出如下。

测试结果等于[0]表示这是一封正常邮件。

然后测试一封垃圾邮件在该模型下的结果。

```python
a_spam_mail_test_text = ["Due to her husband passed away, she's looking for somebody to inherit the legacy of $100 million."]
a_spam_mail_test_text_counts = count_vectorizer.transform(a_spam_mail_test_text)
```

```
prediction2 = classifierNB.predict(a_spam_mail_test_text_counts)
print("测试结果等于{}表示这是一封垃圾邮件。".format(prediction2))
```

输出如下。

测试结果等于[1]表示这是一封垃圾邮件。

每封邮件内容在测试前,都需要经过 transform()方法的文档项矩阵处理。

6.3 基于 TensorFlow 的神经网络模型的邮件分类

Keras 早期是一个独立的开源项目,后端引擎可以是 TensorFlow,也可以是 Theano。后来 Keras 被集成为 TensorFlow 的一部分,作为高级 API 来实现和使用,即 tf.keras,我们这里实现的神经网络模型就是通过 tf.keras 来实现的。

6.3.1 构建 N-Gram 向量化数据

我们通过 TF-IDF 对训练集和验证集的数据进行向量化处理。

```
import tensorflow as tf
from sklearn.feature_extraction.text import TfidfVectorizer
from sklearn.feature_selection import SelectKBest
from sklearn.feature_selection import f_classif
def ngram_vectorize(X_train, y_train, X_test):
    # 创建 TF-IDF 的向量器的关键参数
    kwargs = {
            'ngram_range': (1, 2),  # n-gram 的范围
            'dtype': 'int32',
            # 可以是 ascii 或 unicode,unicode 表示可以处理任意字符,处理速度比 ascii 稍慢些
            'strip_accents': 'unicode',
            'decode_error': 'replace',  # 字符解码时的容错处理,默认是 strict
            'analyzer': 'word', # N-Gram 可以是 word,也可以是 char,这里我们使用 word(单词)来分割
            'min_df': 2, # 最小文本的频率
    }
    vectorizer = TfidfVectorizer(**kwargs)
    # 对训练集的邮件内容文本进行向量化处理
    x_train = vectorizer.fit_transform(X_train)
    # 对验证集的邮件内容文本进行向量化处理
    x_test = vectorizer.transform(X_test)
    # 根据 k 个最高分选择特征
    # f_classif 表示计算提供的样本的方差分析 f 值
    selector = SelectKBest(f_classif, k=min(20000, x_train.shape[1]))
    selector.fit(x_train, y_train)
    x_train = selector.transform(x_train)
    x_test = selector.transform(x_test)
    x_train = x_train.astype(np.float32)
    x_test = x_test.astype(np.float32)
    return x_train, x_test
```

6.3.2 创建模型

用 TensorFlow 提供的高级 API 创建 Keras 模型,我们通过构建一个函数来创建它。这是一个两层的 MLP(多层感知器)模型,它具有简单、容易理解、计算量少、精确度高等特点。

```python
from tensorflow.python.keras.models import Sequential
from tensorflow.python.keras.layers import Dense
from tensorflow.python.keras.layers import Dropout
def create_mlp_model(layers, units, dropout_rate, input_shape, num_classes):
    # 创建 Sequential 模型
    model = Sequential()
    # 添加输入层
    model.add(Dropout(rate=dropout_rate, input_shape=input_shape))
    # 添加隐藏层
    for _ in range(layers-1):
        model.add(Dense(units=units, activation='relu'))
        model.add(Dropout(rate=dropout_rate))
    # 添加输出层,因为最终预测的结果要么是 1,要么是 0,所以输出层的激活函数就用 sigmoid
    op_units = num_classes - 1
    model.add(Dense(units=op_units, activation='sigmoid'))
    return model
```

6.3.3 训练模型

用 TensorFlow 提供的高级 API 创建 Keras 模型,我们通过构建一个函数来创建它,函数包含处理邮件正文文本内容数据,模型创建和训练。这里用到了 Keras 的 EarlyStopping 的回调,我们设置的迭代次数是 1000,但是实际上模型只训练了 19 次就达到我们需要的模型结果,早期停止回调就自动终止了训练。模型训练结束后,我们将模型保存到了当前目录下,名为 spam_email_classifier_model.h5。当下次使用时,我们直接加载模型即可。

```python
def train_model(X_train, y_train, X_test, y_test):
    # 学习率,此处使用的是科学计数法,即 0.001
    learning_rate = 1e-3
    # 一共需要迭代训练多少次
    epochs = 1000
    # 每个 batch 的大小
    batch_size = 128
    # 隐藏层的层数
    layers = 2
    # 输出空间的维度
    units = 64
    # 每个网络层的丢弃率
    dropout_rate = 0.2
    # 只有两种结果,要么是 0,要么是 1
    num_classes = 2
    # 对邮件内容文本向量化
    x_train, x_test = ngram_vectorize(X_train, y_train, X_test)
    # 创建神经网络模型
```

```python
    model = create_mlp_model(layers=layers,
                             units=units,
                             dropout_rate=dropout_rate,
                             input_shape=x_train.shape[1:],
                             num_classes=num_classes)
    # 编译模型
    optimizer = tf.keras.optimizers.Adam(lr=learning_rate)
    model.compile(optimizer=optimizer, loss='binary_crossentropy', metrics=['accuracy'])
    model.summary()
    # 创建训练模型在验证时回调为早期停止,即在给定的 2 次内,如果损失值没有降低,则停止训练
    callbacks = [tf.keras.callbacks.EarlyStopping(monitor='val_loss', patience=2)]
    # 训练和验证模型
    history = model.fit(
            x_train,
            y_train,
            epochs=epochs,
            callbacks=callbacks,
            validation_data=(x_test, y_test),
            verbose=2,
            batch_size=batch_size)
    # 打印输出
    _history = history.history
    val_acc = _history['val_acc']
    _val_acc = val_acc[np.argmax(val_acc)]
    val_loss = _history['val_loss']
    _val_loss = val_loss[np.argmax(val_loss)]
    print('验证精确度是:{},验证损失是:{}。'.format(_val_acc, _val_loss))
    # 保存模型
    model.save('spam_email_classifier_model.h5')
    return model, _history
# 训练和验证模型
model, history = train_model(X_train, y_train, X_test, y_test)
```

输出如下。

```
Layer (type)                 Output Shape              Param #
=================================================================
dropout (Dropout)            (None, 20000)             0
_____
dense (Dense)                (None, 64)                1280064
_____
dropout_1 (Dropout)          (None, 64)                0
_____
dense_1 (Dense)              (None, 1)                 65
=================================================================
Total params: 1,280,129
Trainable params: 1,280,129
Non-trainable params: 0
_____
Train on 4582 samples, validate on 573 samples
Epoch 1/1000
 - 3s - loss: 0.5612 - acc: 0.8514 - val_loss: 0.4096 - val_acc: 0.9145
```

```
Epoch 2/1000
……
Epoch 19/1000
 - 2s - loss: 0.0095 - acc: 0.9991 - val_loss: 0.0424 - val_acc: 0.9860
验证精确度是：0.9860383944153578，验证损失是：0.4243479267048794。
```

模型训练结束后，返回了一个 history 对象，其中包含了训练损失、训练精确度、验证损失和验证精确度。它是一个字典，所以可以通过 keys()函数查看。

```
print(history.keys())
```

输出如下。

```
dict_keys(['val_loss', 'val_acc', 'loss', 'acc'])
```

6.3.4 可视化训练结果

模型训练结束后，会产生训练验证时的值，通过绘图，我们可以清晰地看到训练和验证时的走势。以下 plot()方法中的 marker 参数表示绘制的线的样式，全部的样式可以在 matplotlib 官网查阅。

```
import matplotlib
# 绘制训练时的验证精确度曲线
plt.plot(history["val_acc"],
         np.array(range(len(history["val_acc"]))),
         marker=matplotlib.markers.CARETDOWNBASE,
         label="Val Acc")
# 绘制训练时的验证损失值曲线
plt.plot(history["val_loss"],
         np.array(range(len(history["val_loss"]))),
         marker=matplotlib.markers.TICKRIGHT,
         label="Val Loss")
# 绘制训练时精确度曲线
plt.plot(history["acc"],
         np.array(range(len(history["acc"]))),
         marker='h',
         label="Train Acc")
# 绘制训练时损失曲线
plt.plot(history["loss"],
         np.array(range(len(history["loss"]))),
         marker='d',
         label="Train Loss")
# 设置图题
plt.title("Accuracy and Loss")
# 设置 x 轴是验证或损失值
plt.xlabel("Accuracy or Loss values")
# 设置 y 轴标题
plt.ylabel("Epochs")
# 设置图例在左上角显示
plt.legend(loc=2)
# 显示绘图
plt.show()
```

输出结果如图 6.8 所示。

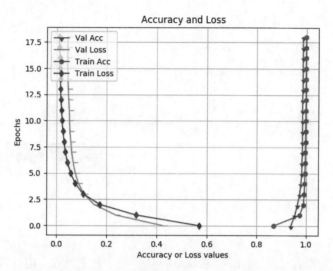

图 6.8　训练和验证时的精确度和损失值图

在训练调参时，如果不能一次达到比较好的效果，读者再次运行创建模型代码时，Keras 不会清空当前的会话状态，这时再次运行创建 Keras 模型，会发现模型的层数名称后面的值的序号是从上一次创建模型时的最后一个序号开始的，会导致模型的不正确。为了解决这个问题，tf.keras 提供了一个清空 Keras 会话状态的函数，由此开发者在重新运行创建模型的代码时就可以调用清空 Keras 会话的代码了。

```
tf.keras.backend.clear_session()
```

6.4　小结

本章通过对垃圾邮件的数据集进行基本分析，可以查看垃圾邮件和正常邮件的表现及其关键字有哪些。然后采用 sklearn 的多项式朴素贝叶斯模型进行邮件分类，其测试的表现还不错。最后我们使用 TensorFlow 下的 Keras 模块来构建多层感知器模型，训练和验证的精确度达到了较高的水平。

第 7 章

影评的情感分析

情感分析在自然语言处理（NLP）领域是很复杂的，有主观的，也有客观的。基于当前环境，针对不同的人或物，我们应该做出什么样的情感反应。本章将讲解如何通过分析情感文本数据，预测出说话者在当时的情况下的情绪状态是积极的，还是消极的。

生活中就有很多例子，比如在京东、淘宝等电商平台购物后，用户都会被请求对收到的货物进行拍照、点赞、评论和评价星级等。平台收集这些数据后去做情感分析，从而通过了解买家对于产品的喜好和满意度来改善产品和服务。这为平台提供了一些潜在的用户会购买哪些产品的数据。

我们将使用循环神经网络（RNN）来编写该神经网络模型的代码，创建此网络模型会使用到长短期记忆网络（LSTM）和嵌入层（Embedding Layers），最后的输出层会使用 sigmoid 激活函数，因为我们预测的结果要么是积极的，要么是消极的。

7.1 数据准备

读者可以通过前言中提到的 Github 地址下载数据集。数据集分为 reviews.txt 和 labels.txt，分别表示评论和标签。每一条评论的文本都对应着一个积极或者消极的标签。我们主要使用 numpy 来做数据预处理，使用 TensorFlow 来训练模型和预测评论。

7.1.1 环境准备

- numpy = 1.14.6。
- tensorflow = 1.11.0。

7.1.2 预处理数据

评论数据是来自于《Bromwell High》系列动画喜剧影片的观后影评，每条评论都对应着一个积极或者消极的标签。我们将对该数据集进行特征向量（Feature Vector）的处理和标签编码。通过以下代码读取影评和标签，读取数据使用 python 内置的 open()函数。

```python
import numpy as np
# 定义加载数据的函数
def loadData():
    # 加载评论（字符串）
    with open('reviews.txt', 'r') as f:
        reviews = f.read()
    # 加载评论（字符串）的对应标签，是积极的还是消极的
    with open('labels.txt', 'r') as f:
        labels = f.read()
    # 返回评论和标签
    return reviews, labels
# 调用函数
reviews, labels = loadData()
```

然后，通过以下代码来看评论字符串的前 150 个字符是什么。

```
reviews[:150]
```

输出如下。

'bromwell highisacartooncomedy .itranatthesametimeassomeotherprogramsaboutschoollifesuchasteachers .myyearsin the teaching '

通过以下代码再来看看前 150 个标签字符串是什么。

```
labels[:150]
```

输出如下。

'positive\nnegative\npositive\nnegative\npositive\nnegative\npositive\nnegative\npositive\nnegative\npositive\nnegative\npositive\nnegative\npositive\nnegative\npositi'

我们可以看到，影评文本中有很多标点符号，其实这些标点符号对于影评预测没有什么用，那我们现在就从影评中去掉这些标点符号。Python 内置的 String 模块下有一个 punctuation 对象，该对象包含了常见的标点符号。

```python
from string import punctuation
# 定义数据预处理函数
def dataPreprocess(reviews_str):
    # 通过列表推导式将 reviews_str 字符串里包含的各种标点符号去掉，并返回一个字符组成的数组
    # 然后通过 join()函数将数组里的元素都连接成一个长长的字符串
    all_text = ''.join(
        [review for review in reviews_str if review not in punctuation])
    # 通过\n 换行符将该字符串分割成数组
    review_list = all_text.split('\n')
    # 通过空格将数组里的元素连接起来，形成一个长长的字符串
    all_text = ' '.join(review_list)
    # 然后使用 split()函数的默认分隔符—— 空格将字符串分割成一个个单词的数组
    words = all_text.split()
    return review_list, all_text, words
```

```python
# 调用函数
reviews, all_text, words = dataPreprocess(reviews)
```

然后，我们来看变量 words 里前 20 个元素的值。

```
['bromwell',
 'high',
 'is',
 'a',
 'cartoon',
 'comedy',
 'it',
 'ran',
 'at',
 'the',
 'same',
 'time',
 'as',
 'some',
 'other',
 'programs',
 'about',
 'school',
 'life',
 'such']
```

words 变量里的值都是一个个单词。再来看变量 all_text 的前 150 个字符的值，代码如下。

```
all_text[:150]
```

输出如下，我们可看见标点符号都没有了。

```
'bromwell highisacartooncomedyitranatthesametimeassomeotherprogramsaboutschoollifes
uchasteachersmy    yearsinthe teaching pr'
```

7.1.3 数据集编码

嵌入层要求传递给神经网络的值是数值，但是影评是一个个英文单词，因此我们需要对影评的单词字符串进行编码处理。最简单的方式就是创建一个字典，映射单词到整型数值之间的关系，这样做的好处就是方便数值在神经网络模型中传递和对之进行计算。

```python
from collections import Counter
# 统计单词的重复个数
word_counter = Counter(words)
# 根据默认顺序将变量 word_counter 进行逆序排序（从大到小）。使用 sorted 方法，逆序设置参数 reverse=True
sorted_vocab = sorted(word_counter, key=word_counter.get, reverse=True)
# 定义显示前 10 个单词以及它的重复个数的函数
def showTop10Item(dict_obj):
    word_index = 0
    for k, v in dict_obj.items():
```

```
        if word_index >= 10:
            break
    print("{}:{}".format(k, v))
    word_index+=1
```

我们使用到 Python 内置的集合模块 collections。在该集合模块下的 Counter 类,可用于计算对象的个数,由于它从类型 dict 继承而来,所以 Counter 里的元素的值是无序的。然后,我们来看变量 word_counter 的前 10 个元素。

```
showTop10Item(word_counter)
```

输出如下。

```
bromwell:8
high:2161
is:107328
a:163009
cartoon:545
comedy:3246
it:96352
ran:238
at:23513
the:336713
```

元素的值的格式是{单词:数量}。我们再来看出现次数最多的前 15 个单词及数量,可以通过 Counter 类里的 most_common 实例方法实现,代码如下。

```
word_counter.most_common(15)
```

输出如下。

```
[('the', 336713),
 ('and', 164107),
 ('a', 163009),
 ('of', 145864),
 ('to', 135720),
 ('is', 107328),
 ('br', 101872),
 ('it', 96352),
 ('in', 93968),
 ('i', 87623),
 ('this', 76000),
 ('that', 73245),
 ('s', 65361),
 ('was', 48208),
 ('as', 46933)]
```

再来看将变量 word_counter 排序后的前 15 个单词,代码如下。

```
sorted_vocab[:15]
```

输出如下。

```
['the',
 'and',
 'a',
 'of',
 'to',
 'is',
 'br',
 'it',
 'in',
 'i',
 'this',
 'that',
 's',
 'was',
 'as']
```

在单词的个数统计和排序后,我们需要给单词添加一个对应的索引关系,代码如下。

```
# 创建单词对应的索引关系字典
vocab_to_int = {word: i for i, word in enumerate(sorted_vocab, 1)}
# 然后显示前 10 个单词以及它的个数
showTop10Item(vocab_to_int)
```

输出如下。

```
the:1
and:2
a:3
of:4
to:5
is:6
br:7
it:8
in:9
i:10
```

从输出的日志可以看到,单词排列是从大到小的顺序,且单词的索引是从 1 开始,而不是 0。创建一个变量 reviews_ints 用来承载每个单词对应的索引。我们对变量 reviews 进行遍历,得到每一个评论,再通过变量 vocab_to_int 就能找到该单词对应的索引。

```
reviews_ints = []
for review in reviews:
    reviews_ints.append([vocab_to_int[word] for word in review.split()])
# 查看第一个评论的索引数值数据是多少
print(reviews_ints[:1])
```

输出如下。

```
[[21025, 308, 6, 3, 1050, …… 2211, 12, 8, 215, 23]]
```

我们再来看一共有多少个单词索引。

```
len(reviews_ints)
```

接下来，我们对数据集的对应标签进行编码。我们的标签数据不是积极（positive），就是消极（negative），为了方便在神经网络中计算标签，我们将 positive 转换成 1，将 negative 转换成 0。

```
# 对positive编码为1, negative为0
labels = labels.split('\n')
labels = np.array([1 if label == 'positive' else 0 for label in labels])
# 查看前10个编码标签值
labels[:10]
```

输出如下。

```
array([1, 0, 1, 0, 1, 0, 1, 0, 1, 0])
```

接着，我们要过滤掉影评里为空的字符串，以及将评论字符串里最长的长度裁切到 200 以内，不够长度的在前面用 0 填充。

```
# 过滤掉评论的字符串长度为0的情况，形成数组并返回长度非零的索引
non_zero_idx = [i for i, review in enumerate(reviews_ints) if len(review) != 0]
# 去掉字符串长度为0的情况后，还有多少个评论
print(len(non_zero_idx))
# 通过变量non_zero_idx索引数组，过滤掉变量reviews_ints里的字符串为0的情况
reviews_ints = [reviews_ints[i] for i in non_zero_idx]
# 由于过滤掉了上面的字符串长度为0的那一行评论，所以它对应的标签也需要被过滤掉
labels = np.array([labels[i] for i in non_zero_idx])
```

输出如下。

```
25000
```

紧接着，我们需要创建一个特征向量（Feature Vector），用变量 features 表示，该特征向量就是我们要传递到神经网络中去的。

```
# 定义一个评论的字符串最大长度是200
seq_len = 200
# 创建一个矩阵，里面的值都默认是0
features = np.zeros((len(reviews_ints), seq_len), dtype=int)
# 将reviews_ints里的值都截断在200的长度，并填充到变量features里。
# 不足200的长度，就是它本身长度
for i, row in enumerate(reviews_ints):
    # 评论长度不足200的，我们在前面使用0来填充
    features[i, -len(row):] = np.array(row)[:seq_len]
# 查看第一个评论的索引值
features[0:1]
```

输出如图结果 7.1 所示。我们可见前面都是 0，因为该文本长度不够 200 个，所以前面都用 0 来填充。

```
array([[    0,     0,     0,     0,     0,     0,     0,     0,     0,
            0,     0,     0,     0,     0,     0,     0,     0,     0,
            0,     0,     0,     0,     0,     0,     0,     0,     0,
            0,     0,     0,     0,     0,     0,     0,     0,     0,
            0,     0,     0,     0,     0, 21025,   308,     6,
            3,  1050,   207,     8,  2138,    32,     1,   171,    57,
           15,    49,    81,  5785,    44,   382,   110,   140,    15,
         5194,    60,   154,     9,     1,  4975,  5852,   475,    71,
            5,   260,    12, 21025,   308,    13,  1978,     6,    74,
         2395,     5,   613,    73,     6,  5194,     1, 24103,     5,
         1983, 10166,     1,  5786,  1499,    36,    51,    66,   204,
          145,    67,  1199,  5194, 19869,     1, 37442,     4,     1,
          221,   883,    31,  2988,    71,     4,     1,  5787,    10,
          686,     2,    67,  1499,    54,    10,   216,     1,   383,
            9,    62,     3,  1406,  3686,   783,     5,  3483,   180,
            1,   382,    10,  1212, 13583,    32,   308,     3,   349,
          341,  2913,    10,   143,   127,     5,  7690,    30,     4,
          129,  5194,  1406,  2326,     5, 21025,   308,    10,   528,
           12,   109,  1448,     4,    60,   543,   102,    12, 21025,
          308,     6,   227,  4146,    48,     3,  2211,    12,     8,
          215,    23]])
```

图 7.1 特征向量里的第一个向量

最后来看下变量 features 的大小,代码如下。

```
print(features.shape)
```

输出如下。

```
(25000, 200)
```

7.1.4 数据集分割

我们得到了影评的特征向量的变量 features 和标签 labels,现在要对该数据集进行训练集、验证集和测试集分割。这次采用手动分割数据集的方式,训练集为 80%,验证集和测试集各自是 10%,代码如下。

```
# 定义80%的数据用于训练
split_train_ratio = 0.8
# 特征向量的长度
features_len = len(features)
# 训练集的个数
train_len = int(features_len * split_train_ratio)
# 分割出训练集和验证集的数据
train_x, val_x = features[:train_len], features[train_len:]
train_y, val_y = labels[:train_len], labels[train_len:]
# 将验证集的数量折半
val_x_half_len = int(len(val_x) / 2)
# 将验证集数据分成两半,一半做验证集,另一半做测试集
val_x, test_x = val_x[:val_x_half_len], val_x[val_x_half_len:]
val_y, test_y = val_y[:val_x_half_len], val_y[val_x_half_len:]
# 输出打印
print("\t\t\tFeature Shapes:")
```

```
print("Train set: \t\t{}".format(train_x.shape),
      "\nValidation set: \t{}".format(val_x.shape),
      "\nTest set: \t\t{}".format(test_x.shape))
```

输出如下。

```
                Feature Shapes:
Train set:              (20000, 200)
Validation set:         (2500, 200)
Test set:               (2500, 200)
```

7.2 基于 TensorFlow 的长短期记忆网络实现影评的情感分析

长短期记忆网络（Long Short Term Memory，LSTM）。在 LSTM 中，用来处理信息的那一块功能称为 Cell。该 Cell 设置了 4 个门，分别是遗忘门（Forget Gate）、记忆门（Remember Gate）、学习门（Learn Gate）和使用门（Use Gate），它们的关系和计算流程如图 7.2 所示。

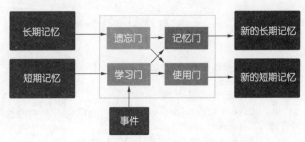

图 7.2　长短期记忆网络的 4 个门的关系图

7.2.1　参数准备

在创建神经网络模型前，要先准备好参数，便于之后再次调整参数，4 个参数的解释分别如下。

- lstm_size 表示在 LSTM Cell 里隐藏层的数量，设置大一些的值通常会使之有比较好的性能表现。常规的值是 128、256 和 512 等。
- lstm_layers 表示在神经网络中 LSTM 的层数。一般设置为 1，如果欠拟合，就往上加。这里我们设置为 2。
- batch_size 表示在一个批次的训练中，将指定大小的数据送入神经网络，一般就根据计算机内存的大小来设置。常规的值是 32、64、128、256、512 和 1024。这里我们设置为 512。
- learning_rate 表示学习率。

```
lstm_size = 256
lstm_layers = 2
batch_size = 512
learning_rate = 0.01
```

7.2.2　创建 LSTM 模型

首先，我们会创建输入特征占位符、输入标签占位符和保留率占位符。保留率（Keep Probability）在训练期间设置为 0.5，验证和测试期间设置为 1.0。保留率就是在模型训练、验证和测试期间用来控制丢弃率（Dropout Rate）的，以防止过拟合。

```
import tensorflow as tf
# 获取单词的总长度
n_words = len(vocab_to_int) + 1
# 创建默认计算图对象
tf.reset_default_graph()
# 给计算图上的张量的输入占位符添加一个前缀 inputs
with tf.name_scope('inputs'):
    # 输入特征占位符
    inputs_ = tf.placeholder(tf.int32, [None, None], name="inputs")
    # 输入标签占位符
    labels_ = tf.placeholder(tf.int32, [None, None], name="labels")
    # 保留率占位符
    keep_prob = tf.placeholder(tf.float32, name="keep_prob")
# 嵌入向量的大小
embed_size = 300
# 给计算图上的张量的嵌入层变量和查找表添加一个前缀 Embeddings
with tf.name_scope("Embeddings"):
    # 均匀分布初始化嵌入层的变量，范围是-1 到 1
    embedding = tf.Variable(tf.random_uniform((n_words, embed_size), -1, 1))
    # 将输入特征占位符传入嵌入查找表
    embed = tf.nn.embedding_lookup(embedding, inputs_)
```

接下来，添加 LSTM Cell、输出层和精准度。

```
def lstm_cell():
    # 创建基础 LSTM Cell
    lstm = tf.contrib.rnn.BasicLSTMCell(lstm_size, reuse=tf.get_variable_scope().reuse)
    # 添加 dropout 层到 Cell 上
    return tf.contrib.rnn.DropoutWrapper(lstm, output_keep_prob=keep_prob)
# 给 graph 上的 tensors 的 RNN 层添加一个前缀 RNN_layers
with tf.name_scope("RNN_layers"):
    # 创建多个 LSTM 层
    cell = tf.contrib.rnn.MultiRNNCell([lstm_cell() for _ in range(lstm_layers)])
    
    # 获取一个初始化状态，默认值都是 0
    initial_state = cell.zero_state(batch_size, tf.float32)
with tf.name_scope("RNN_forward"):
    # 通过 dynamic_rnn 可以返回每一步的输出和隐藏层的最后状态
    outputs, final_state = tf.nn.dynamic_rnn(cell, embed, initial_state=initial_state)

with tf.name_scope('predictions'):
    # 创建输出层，由于我们预测的输出是 1 或者 0，所以 sigmoid 激活函数是最好的选择
    predictions = tf.contrib.layers.fully_connected(outputs[:, -1], 1, activation_fn=tf.sigmoid)
```

```python
with tf.name_scope('cost'):
    # 定义均方差训练损失函数
    cost = tf.losses.mean_squared_error(labels_, predictions)
with tf.name_scope('train'):
    # 定义训练优化器
    optimizer = tf.train.AdamOptimizer(learning_rate).minimize(cost)

with tf.name_scope('validation'):
    # 计算验证精确度
    correct_pred = tf.equal(tf.cast(tf.round(predictions), tf.int32), labels_)
    accuracy = tf.reduce_mean(tf.cast(correct_pred, tf.float32))
```

先对数据集 X 和 y 分批次，以便于在训练时更容易获取训练和测试数据集，我们使用 yield 关键字来生成可迭代对象，然后在使用时，就可以直接通过 for 循环来遍历取出数据了。batch_size 默认是 100，我们传入的值是 512。

```python
# 定义获取数据批次的生成器函数
def get_batches(x, y, batch_size=100):
    # 计算得出有多少个批次，这里是整除，所以假如 x 的总数不能被 batch_size 整除，
    # 那么会剩下很小的一部分数据暂时会被丢弃
    n_batches = len(x)//batch_size
    # 然后再次确定 x 和 y 的数据集的数据
    x, y = x[:n_batches*batch_size], y[:n_batches*batch_size]
    # 通过 for 循环，使用 yield 关键字构建生成器函数
    for ii in range(0, len(x), batch_size):
        yield x[ii:ii+batch_size], y[ii:ii+batch_size]
```

7.2.3 训练模型

在循环神经网络的模型的架构设计好后，我们就要进行训练了。我们会在每训练 5 个批次的数据后打印一次训练日志，每训练 25 个批次后打印一次验证日志。训练花费的时间取决于读者的计算机性能，作者这里训练时长是 15 分钟左右。

```python
# 设置迭代次数，8 次
epochs = 8
# 创建检查点保存对象
saver = tf.train.Saver()
# 创建一个 TensorFlow 会话
with tf.Session() as sess:
    # 初始化全局变量
    sess.run(tf.global_variables_initializer())

    iteration = 1
    # 开始迭代
    for e in range(epochs):
        # 首次计算初始化状态
        state = sess.run(initial_state)

        # 将所有的数据都进行训练，get_batches()函数会获取数据生成器，然后进行迭代
```

```python
        for ii, (x, y) in enumerate(get_batches(train_x, train_y, batch_size), 1):
            feed = {inputs_: x,
                    labels_: y[:, None],
                    keep_prob: 0.5,
                    initial_state: state}
            loss, state, _ = sess.run([cost, final_state, optimizer], feed_dict=feed)

            # 每训练 5 次,打印一次训练日志
            if iteration%5==0:
                print("Epoch: {}/{}".format(e, epochs),
                      "Iteration: {}".format(iteration),
                      "Train loss: {:.3f}".format(loss))
            # 每训练 25 次,打印一次验证日志
            if iteration%25==0:
                val_acc = []
                val_state = sess.run(cell.zero_state(batch_size, tf.float32))
                # 对验证集的所有数据计算分值
                for x, y in get_batches(val_x, val_y, batch_size):
                    feed = {inputs_: x,
                            labels_: y[:, None],
                            keep_prob: 1,
                            initial_state: val_state}
                    batch_acc, val_state = \
                        sess.run([accuracy, final_state], feed_dict=feed)
                    # 每 25 次训练后,完全验证一次,得到的验证分值,保存在数组 val_acc 里
                    val_acc.append(batch_acc)
                # 每 25 次训练,打印验证的均值
                print("Val acc: {:.3f}".format(np.mean(val_acc)))
            iteration +=1

            # 每批次时都记录检查点
            saver.save(sess, "checkpoints/sentiment.ckpt")
    # 当所有的数据迭代训练完毕后,最后记录一次检查点
    saver.save(sess, "checkpoints/sentiment.ckpt")
```

训练时输出的日志信息如下。

```
Epoch: 0/8 Iteration: 5 Train loss: 0.286
Epoch: 0/8 Iteration: 10 Train loss: 0.244
Epoch: 0/8 Iteration: 15 Train loss: 0.227
Epoch: 0/8 Iteration: 20 Train loss: 0.249
Epoch: 0/8 Iteration: 25 Train loss: 0.232
Val acc: 0.603
……
Epoch: 7/8 Iteration: 295 Train loss: 0.000
Epoch: 7/8 Iteration: 300 Train loss: 0.000
Val acc: 0.699
Epoch: 7/8 Iteration: 305 Train loss: 0.000
Epoch: 7/8 Iteration: 310 Train loss: 0.000
```

7.2.4 模型测试

评估测试集数据，首先要读取训练时保存模型的检查点，然后将测试集数据传入网络模型中计算，而计算出来的分值就是预测的得分，我们这里就是计算每次得分的最终均值。

```
test_acc = []
with tf.Session() as sess:
    # 从检查点恢复已训练的模型
    saver.restore(sess, "checkpoints/sentiment.ckpt")
    # 在计算测试集数据前，先创建一个空的状态
    test_state = sess.run(cell.zero_state(batch_size, tf.float32))
    # 获取测试集数据生成器
    for ii, (x, y) in enumerate(get_batches(test_x, test_y, batch_size), 1):
        feed = {inputs_: x,
                labels_: y[:, None],
                keep_prob: 1,
                initial_state: test_state}
        # 开始分批次计算测试集数据
        batch_acc, test_state = sess.run([accuracy, final_state], feed_dict=feed)
        # 将每个批次的得分保存到数组
        test_acc.append(batch_acc)
    # 最后输出测试得分均值，即精确度
    print("Test accuracy: {:.3f}".format(np.mean(test_acc)))
```

输出精确度如下。

```
Test accuracy: 0.731
```

7.3 基于 Keras 的长短期记忆网络实现影评的情感分析

我们将使用 Keras 构建一个循环神经网络模型来预测 IMDb 的影评是积极的还是消极的。通过 Keras 的数据集模块可以直接获取数据集。

7.3.1 数据预处理

首先下载 Keras 的 datasets 模块下的 IMDb 数据集。我们获取到数据集后，把它分成训练集和测试集。最后对训练集和测试集数据进行截断和填充的预处理，因为我们要保证每个影评的长度是一样的，这里设置的最大长度是 500。

```
import numpy
from keras.datasets import imdb
from keras.models import Sequential
from keras.layers import Dense, LSTM
from keras.layers.embeddings import Embedding
from keras.preprocessing import sequence
# 为了确保可复现性，我们设置一个随机种子
numpy.random.seed(7)
```

```
# 设置 5000 的意思是,只保留前面 5000 个常见的单词,其他的都为 0
top_words = 5000
# 加载数据集
(X_train, y_train), (X_test, y_test) = imdb.load_data(num_words=top_words)
# 设置单个影评的最大长度是 500
review_max_length = 500
# 影评长度不够 500 的用 0 填充,超过 500 的截断
X_train = sequence.pad_sequences(X_train, maxlen=review_max_length)
X_test = sequence.pad_sequences(X_test, maxlen=review_max_length)
```

7.3.2 创建模型

我们创建一个 Sequential 模型,添加输入嵌入层,传入输入维度、输出维度和输入长度,并添加 100 个 LSTM 的隐藏层,最后添加输出层。这是一个非常简单的长短期记忆网络模型结构。

```
embedding_vecor_length = 32
model = Sequential()
# 添加输入嵌入层
model.add(Embedding(top_words, embedding_vecor_length, input_length=review_max_length))
# 添加 LSTM 隐藏层
model.add(LSTM(100))
# 添加输出层(全连接层),二分类问题,使用 sigmoid 激活函数
model.add(Dense(1, activation='sigmoid'))
# 编译模型,二分类问题,使用二进制交叉熵来计算损失
model.compile(loss='binary_crossentropy', optimizer='adam', metrics=['accuracy'])
```

7.3.3 预览模型架构

模型创建好后,我们通过 summary()函数就能预览模型的架构,代码如下。

```
model.summary()
```

输出模型架构图,如图 7.3 所示。

```
Layer (type)                 Output Shape              Param #
=================================================================
embedding_1 (Embedding)      (None, 500, 32)           160000
_____
lstm_1 (LSTM)                (None, 100)               53200
_____
dense_1 (Dense)              (None, 1)                 101
=================================================================
Total params: 213,301
Trainable params: 213,301
Non-trainable params: 0
```

图 7.3 预览 LSTM 模型架构图

7.3.4 训练模型

通过 fit()函数我们就能训练模型。这里将 epochs 设置为 3,表示所有的数据都会被训练 3 次;而 batch_size 设置为 64,表示每次训练的批次大小是 64 个样本。训练的过程大约需要 10

分钟,这取决于读者计算机的性能,越好越快。

```
model.fit(X_train, y_train, epochs=3, batch_size=64)
```

输出日志信息如下,从结果可以看到训练的精确度为 0.8938。

```
Epoch 1/3
25000/25000 [==============================] - 358s 14ms/step - loss: 0.4492 - acc: 0.7843
Epoch 2/3
25000/25000 [==============================] - 357s 14ms/step - loss: 0.3452 - acc: 0.8598
Epoch 3/3
25000/25000 [==============================] - 357s 14ms/step - loss: 0.2667 - acc: 0.8938
```

7.3.5 模型评估

通过 evaluate() 函数我们就能根据测试集数据来评估模型,代码如下。

```
scores = model.evaluate(X_test, y_test, verbose=0)
print("Accuracy: {}".format((scores[1]*100)))
```

输出如下,可见得到的分值约为 85.63,满分是 100。

```
Accuracy: 85.63199999999999
```

通过以上的 TensorFlow 和 Keras 的运行,我们看到了 TensorFlow 获得了 0.731 的分数,而 Keras 获得了约 85.63 的分数。当然我们也可以尝试通过调整神经网络和超参数以获得更高的分数。

7.4 小结

本章使用 IMDb 的影评数据,对它进行分析、预测及数据集的分割,然后分别使用 TensorFlow 和 Keras 来搭建 LSTM 模型对之进行训练和预测。其中在 Keras 下,我们用的是 Keras 的 datasets 模块下的数据进行加载的,最后两个模型的训练效果差不多。

第 8 章

语言翻译

语言作为人类十分重要的交流工具和思维工具，在我们的生活中无处不在、不可分离。有语言学家将世界上近百种的语言分成九大语系，他们将语音、词汇和语法规则之间相似程度较高和存在对应关系的语言划分为一类，称为同族语言。

互联网科技公司 Google、微软、Facebook、阿里巴巴、百度等先后在发布会上说他们的 AI Lab 已经研发出来近百种语言的机器翻译引擎。基于神经机器翻译（Neural Machine Translation）的开源库也比较多，有的比较复杂，有的比较容易。本章就为大家讲解将法语翻译成英语的神经网络模型的训练。读者也可以微调代码，进行将英文翻译成法文的神经网络模型的训练，甚至也可用来训练一些其他的印欧语系的机器翻译模型。

8.1 数据准备

作为机器翻译（Machine Translation）的数据集，它和一般的数据不同的是，它只有一个 txt 文本，里面的文本内容的语句中，前面是英文，后面的是对应的法文，然后非常多的这样的语句组成一个超大的数据集，我们称之为平行语料库（Parallel Corpora）。平行语料库的数据对（Data Pairs）越多，对神经网络模型的训练就越有好处。平行语料库的数据集下载地址请到本代码仓库中找到链接并下载。

8.1.1 环境准备

- numpy = 1.14.6。
- keras = 2.2.3。

8.1.2 数据准备

数据集已经准备好，读者可通过本书前言中提供的 Github 地址下载。这是一个 fra-eng.zip 里的 fra.txt 文件，共计有 160872 个法英句子对，读者自行解压。读取平行语料库的代码如下。

```
with open("fra.txt", 'rt', encoding="utf-8") as f:
    text = f.read()
# 查看文本前 80 个字符
text[:80]
```

输出如下。

'Go.\tVa !\nHi.\tSalut !\nRun!\tCours\u202f!\nRun!\tCourez\u202f!\nWow!\tÇa alors\u202f!\nFire!\tAu feu !\nH'

从输出的文本中我们看到每一行语句的前面是英文，后面就是对应的法文，中间用\t（制表符）符号分割两段语句，结尾是以\n 换行符来显示的。然后我们将此文本内容读取到数组中以方便预处理，并输出前 15 行语句对查看数据。

```
lines = text.strip().split('\n')
lines_pairs = [line.split('\t') for line in  lines]
lines_pairs[:15]
```

输出如下。

```
[['Go.', 'Va !'],
 ['Hi.', 'Salut !'],
 ['Run!', 'Cours\u202f!'],
 ['Run!', 'Courez\u202f!'],
 ['Wow!', 'Ça alors\u202f!'],
 ['Fire!', 'Au feu !'],
 ['Help!', "À l'aide\u202f!"],
 ['Jump.', 'Saute.'],
 ['Stop!', 'Ça suffit\u202f!'],
 ['Stop!', 'Stop\u202f!'],
 ['Stop!', 'Arrête-toi !'],
 ['Wait!', 'Attends !'],
 ['Wait!', 'Attendez !'],
 ['Go on.', 'Poursuis.'],
 ['Go on.', 'Continuez.']]
```

查看这些平行语句对的长度，代码如下。

```
pairs_len = len(lines_pairs)
eng_pair_lens = [len(line_pair[0]) for line_pair in lines_pairs]
fra_pair_lens = [len(line_pair[1]) for line_pair in lines_pairs]
print("一共有{}个英法文数据对；".format(pairs_len))
print("其中法文语句最短的长度为{}，最长的长度为{}；".format(min(fra_pair_lens), max(fra_pair_lens)))
print("其中英文语句最短的长度为{}，最长的长度为{}。".format(min(eng_pair_lens), max(eng_pair_lens)))
```

输出如下。

一共有 160872 个英法文数据对；
其中法文语句最短的长度为 4，最长的长度为 349；
其中英文语句最短的长度为 3，最长的长度为 286。

8.1.3 数据预处理

接下来我们要移除非英文字母的字母发音符号，比如在法语中，一些发音符号在字母上面,也有些在字母下面。如果读者不理解发音符号,作者就拿中文的拼音来举例子,中文中的拼音是我们对汉语汉字发音的基础,拼音的字母来源于拉丁字母,但是只是由字母组成,读起来没有轻重音,所以就引入了四个声调,这四个声调就是发音符号,更详细的解说请读者自行查阅资料。

这里我们用正则表达式来匹配指定的标点符号,符合的话就将之移除。并对每个语句文本进行 unicode 字符串的标准化处理,使用到了 unicodedata 模块,通过该模块可访问所有的 Unicode Character 数据库。

```python
import re
import numpy as np
from unicodedata import normalize
import string
# 生成正则表达式做字符匹配
re_print = re.compile('[^{}]'.format(re.escape(string.printable)))
# 构建英文标点符号转换表格,用来移除文本字符串里包含的标点符号
english_table = str.maketrans('', '', string.punctuation)
cleaned_pairs = list()
# 遍历每一行的英法文本平行语句
for pair in lines_pairs:
    clean_pair = list()
    for i, line in enumerate(pair):
        # 将语句文本做 Unicode 字符串的标准化处理
        line = normalize('NFD', line).encode('ascii', 'ignore')
        line = line.decode('UTF-8')
        # 对包含空格的文本进行分组
        line = line.split()
        # 然后全部转小写
        line = [word.lower() for word in line]
        # 移除文本里包含的指定的标点符号字符
        line = [word.translate(english_table) for word in line]
        # 移除每个字母上面或者下面的发音符号
        line = [re_print.sub('', w) for w in line]
        # 如果文本里包含的不是字母,就移除掉
        line = [word for word in line if word.isalpha()]
        # 把处理完后的文本添加到数组
        clean_pair.append(' '.join(line))
    cleaned_pairs.append(clean_pair)
# 转换成 Numpy 数组
cleaned_pairs = np.array(cleaned_pairs)
# 查看前 15 个元素
cleaned_pairs[:15]
```

输出如下。

```
array([['go', 'va'],
       ['hi', 'salut'],
       ['run', 'cours'],
       ['run', 'courez'],
       ['wow', 'ca alors'],
       ['fire', 'au feu'],
```

```
        ['help', 'a laide'],
        ['jump', 'saute'],
        ['stop', 'ca suffit'],
        ['stop', 'stop'],
        ['stop', 'arretetoi'],
        ['wait', 'attends'],
        ['wait', 'attendez'],
        ['go on', 'poursuis'],
        ['go on', 'continuez']], dtype='<U339')
```

我们看下 string 模块下的 printable 和 punctuation 对象有哪些字符。

```
print(string.printable)
print(string.punctuation)
```

输出如下。

```
0123456789abcdefghijklmnopqrstuvwxyzABCDEFGHIJKLMNOPQRSTUVWXYZ!"#$%&'()*+,-./:;<=>?@[\]^_`{|}~
!"#$%&'()*+,-./:;<=>?@[\]^_`{|}~
```

保存已经处理过的平行语句文本数据到本地，这是为了以后再次使用时，可以直接加载数据，而不用再去运行以上步骤的代码；使用 pickle 保存数据。

```
import pickle
with open("french_to_english.pkl", "wb") as f:
    pickle.dump(cleaned_pairs, f)
```

先加载刚刚保存的数据，然后分割数据。为了让训练的速度较快，我们使用前 10000 行平行语句对来训练模型，代码如下。

```
with open("french_to_english.pkl", "rb") as f:
    raw_dataset = pickle.load(f)
# 截取前 10,000 个英法文数据对进行训练
sequence_length = 10000
dataset = raw_dataset[:sequence_length]
# 随机打乱数据
np.random.shuffle(dataset)
# 输出前 15 行语句
dataset[:15]
```

输出的英法文数据对每次都不一样，这是因为通过 shuffle()函数打乱了顺序，输出如下。

```
array([['he has a video', 'il detient une video'],
       ['can you help me', 'pourraistu maider'],
       ['be respectful', 'soyez respectueuses'],
       ['youre naive', 'vous etes naif'],
       ['he hardly works', 'il travaille a peine'],
       ['show me', 'montrezmoi'],
       ['i must go', 'je dois y aller'],
       ['what a team', 'quelle equipe'],
       ['we are late', 'nous sommes en retard'],
       ['call security', 'appelle la securite'],
       ['is that love', 'estce de lamour'],
```

```
['he lay face up', 'il etait etendu le visage visible'],
['she helps us', 'elle nous aide'],
['she had twins', 'elle a eu des jumeaux'],
['you needed me', 'vous aviez besoin de moi']], dtype='<U339')
```

分割训练集和测试集,测试集占总数据的 15%;总数是 10000,那 15%的测试集就是 1500 行。

```
train_len = sequence_length - 1500
train, test = dataset[:train_len], dataset[train_len:]
# 保存数据
def save_dataset(sentences, filename):
    with open(filename, 'wb') as f:
        pickle.dump(sentences, f)
# 保存 10,000 行的数据集,含法文和英文
save_dataset(dataset, "french_to_english_dataset_top10000.pkl")
# 保存训练集数据
save_dataset(train, "french_to_english_train.pkl")
# 保存测试集数据
save_dataset(test, "french_to_english_test.pkl")
# 打印输出
print("train.shape={}, test.shape={}".format(train.shape, test.shape))
```

输出如下。

```
train.shape=(8500, 2), test.shape=(1500, 2)
```

8.2 基于 Keras 的长短期记忆网络实现语言翻译

选择长短期记忆网络来实现语言翻译是非常合适的,这是因为 LSTM 是有状态的,这意味着每个序列在神经网络的训练过程中,可以通过反向传播带着上一次的状态到下一次的计算中去。如果没有状态,那计算的结果可能就有很大的偏差。

8.2.1 Tokenize 文本数据

在对平行语料数据对(Data Pairs)进行 Tokenize 前,我们先做一个小实验:给出一些自定义的数据文本,Tokenizer 会如何处理呢?

```
from keras.preprocessing.text import Tokenizer
texts = ['I love AI in China', '特拉字节', 'AI 人工智能']
tokenizer = Tokenizer()
tokenizer.fit_on_texts(texts)
print("tokenizer.word_index={}.".format(tokenizer.word_index))
print("tokenizer.texts_to_sequences={}.".format(tokenizer.texts_to_sequences(texts)))
```

输出如下。

```
tokenizer.word_index={'ai': 1, 'i': 2, 'love': 3, 'in': 4, 'china': 5, '特拉字节': 6, '人工智能': 7}.
tokenizer.texts_to_sequences=[[2, 3, 1, 4, 5], [6], [1, 7]].
```

其中 fit_on_texts()方法会将输入的内容做向量化处理，就是将每个单词映射成一个索引数值，索引是唯一的，但是单词会重复，所以最终索引的个数和所有的单词个数长度不一致。这也能看出来，不管本次数据文本中的使用到的单词有多少个，转换后的单词也全部都小写了。然后通过 texts_to_sequences()方法将每个单词对应的索引组成的数组一并返回。

对英法文本数据进行 Tokenization。数据集变量 dataset[:, 0]表示对整个数据集中的第 0 列取出来并作为新的数组返回；对应的，dataset[:, 1]表示对整个数据集中的第 1 列取出来并作为新的数组返回。索引为 0 的列表示英文，索引为 1 的列表示法义。

```
from keras.preprocessing.text import Tokenizer
def create_tokenizer(lines):
    tokenizer = Tokenizer()
    tokenizer.fit_on_texts(lines)
    return tokenizer
def max_length(lines):
    return max(len(line.split()) for line in lines)
# 准备法文的 tokenizer
fra_tokenizer = create_tokenizer(dataset[:, 1])
fra_vocab_size = len(fra_tokenizer.word_index) + 1
fra_length = max_length(dataset[:, 1])
print('法文序列单词最大个数{}，单词有{}个。'.format(fra_length, fra_vocab_size))
# 准备英文的 tokenizer
eng_tokenizer = create_tokenizer(dataset[:, 0])
eng_vocab_size = len(eng_tokenizer.word_index) + 1
eng_length = max_length(dataset[:, 0])
print('英文序列单词最大个数{}，单词有{}个。'.format(eng_length, eng_vocab_size))
```

输出如下。

法文序列单词最大个数 10，单词有 4397 个。
英文序列单词最大个数 5，单词有 2125 个。

8.2.2 数据编码和填充

我们知道了每个序列的最大长度，现在可以通过 pad_sequences()方法对不够最大长度的数据用 0 进行填充。然后对 target 数据进行独热编码。

```
from keras.preprocessing import sequence
from keras import utils
def encode_sequences(tokenizer, length, lines):
    X = tokenizer.texts_to_sequences(lines)
    # 填充序列，如果原文语句的序列不够长，就用 0 填充
    # maxlen 表示序列的最大长度
    # padding 的值有 pre 和 post，分别表示在前面填充、在后面填充
    X = sequence.pad_sequences(X, maxlen=length, padding='post')
    return X
def encode_output(sequences, vocab_size):
    ylist = list()
    # 对序列数组进行遍历
    for sequence in sequences:
        # 进行 one-hot encoding 编码
```

```
        encoded = utils.to_categorical(sequence, num_classes=vocab_size)
        ylist.append(encoded)
    # 将编码后的数组转换成 Numpy 数组
    y = np.array(ylist)
    y = y.reshape(sequences.shape[0], sequences.shape[1], vocab_size)
    return y
# 准备训练数据
X_train = encode_sequences(fra_tokenizer, fra_length, train[:, 1])
y_train = encode_sequences(eng_tokenizer, eng_length, train[:, 0])
y_train = encode_output(y_train, eng_vocab_size)
# 准备测试数据
X_test = encode_sequences(fra_tokenizer, fra_length, test[:, 1])
y_test = encode_sequences(eng_tokenizer, eng_length, test[:, 0])
y_test = encode_output(y_test, eng_vocab_size)
```

8.2.3 创建模型

我们定义了一个创建模型的函数 create_model()，然后传入参数。其中添加 Embedding 输入层时，参数解释如下。

- 参数 1：输入维度。因为我们是用法语翻译英语，所以输入维度就是法语词汇表的长度。
- 参数 2：输出维度。自身的输出维度。
- 参数 3：输入的序列长度。
- 参数 4：表示输入值含 0 的是否要被屏蔽。因为 0 是通过 pad_sequences()函数填充的，且要求输入的词汇表的大小必须+1。

RepeatVector 层是自编码器，它会将 LSTM 编码器的输入序列转换为包含整个序列中的时间步数，然后重复这个向量 n 次（其中 n 就是输出序列中的时间步数，用 tar_timesteps 表示），最后再利用 LSTM 解码器将这个常量序列转换为目标序列。

```
from keras import Sequential
from keras.layers import Dense, Embedding, LSTM, RepeatVector, TimeDistributed
def create_model(src_vocab, tar_vocab, src_timesteps, tar_timesteps, n_units):
    model = Sequential()
    model.add(Embedding(src_vocab, n_units, input_length=src_timesteps, mask_zero=True))
    model.add(LSTM(n_units))
    model.add(RepeatVector(tar_timesteps))
    model.add(LSTM(n_units, return_sequences=True))
    # 添加输出层，输出层的大小就是英文词汇表的大小
    model.add(TimeDistributed(Dense(tar_vocab, activation='softmax')))
    return model
# 创建模型
model = create_model(fra_vocab_size, eng_vocab_size, fra_length, eng_length, 256)
# 编译模型
model.compile(optimizer='adam', loss='categorical_crossentropy')
# 预览定义的模型架构
model.summary()
```

输出模型预览架构图如下。

```
Layer (type)                      Output Shape              Param #
=================================================================
embedding_1 (Embedding)           (None, 10, 256)           1125632
_____
lstm_1 (LSTM)                     (None, 256)               525312
_____
repeat_vector_1 (RepeatVecto      (None, 5, 256)            0
_____
lstm_2 (LSTM)                     (None, 5, 256)            525312
_____
time_distributed_1 (TimeDist      (None, 5, 2125)           546125
=================================================================
Total params: 2,722,381
Trainable params: 2,722,381
Non-trainable params: 0
```

8.2.4 训练模型

在训练模型时,我们为其添加了两个回调,一个是检查点,另一个是早期停止。分别解释如下。

(1) 检查点 (ModelCheckPoint),表示模型检查点保存。
- 参数 1:表示模型的文件名路径。
- 参数 2:表示在保存模型时的监听对象,val_loss 就是表示验证损失。
- 参数 3:日志打印类型。
- 参数 4:只保存最佳的模型检查点。
- 参数 5:当 monitor 等于 val_loss 时,就设置 min,表示验证损失越小越好;当 monitor 等于 val_acc 时,就设置 max,表示验证精确度越高越好。

(2) 早期停止 (EarlyStopping),模型训练时的验证回调,在给定的 3 个周期内,如果验证损失值没有降低,就自动停止训练。

设置 50 个 epochs 用来训练,每次训练的最佳位置可能不一样。

```python
from keras.callbacks import ModelCheckpoint, EarlyStopping
model_filename = 'translator_weights_model.h5'
checkpoint_ModelCheckpoint = ModelCheckpoint(model_filename,
                              monitor='val_loss',
                              verbose=1,
                              save_best_only=True,
                              mode='min')
callbacks_EarlyStopping = EarlyStopping(monitor='val_loss',
                              patience=3)
# 训练模型
history = model.fit(X_train,
                    y_train,
                    epochs=50,
                    batch_size=64,
```

```
                          validation_data=(X_test, y_test),
                          callbacks=[checkpoint_ModelCheckpoint, callbacks_EarlyStopping],
                          verbose=2)
```

训练输出如下。

```
Train on 8500 samples, validate on 1500 samples
Epoch 1/50
8500/8500 [==========================] - 34s 4ms/step - loss: 4.3055 - val_loss: 3.3909
Epoch 00001: val_loss improved from inf to 3.39091, saving model to translator_weights_model.h5
Epoch 2/50
8500/8500 [==========================] - 32s 4ms/step - loss: 3.3292 - val_loss: 3.2612
……
Epoch 00032: val_loss did not improve from 1.80462
Epoch 33/50
8500/8500 [==========================] - 32s 4ms/step - loss: 0.2715 - val_loss: 1.8113
Epoch 00033: val_loss did not improve from 1.80462
```

如果读者在调试训练代码时，需要重新构建神经网络模型和再次训练模型，则要先清空 Keras 会话，代码如下。

```
from keras import backend as K
K.clear_session()
```

8.2.5　测试模型

先加载数据集，然后对训练集和测试集进行序列编码，再加载已训练的模型，最后来测试模型。我们会打印 10 组数据用来测试效果，并用 NLTK 的 corpus_bleu()方法来计算 BLEU 分数。BLEU 是一种在两种语言之间用于评估文本质量的算法，在业界已经是自动化且廉价的度量标准之一。它的值在 0 到 1 之间，越接近 1 表示越相似的文本，但是一般没有必要是 1。

```
def load_clean_sentences(filename):
    return pickle.load(open(filename, 'rb'))
# 加载数据集
dataset = load_clean_sentences('french_to_english_dataset_top10000.pkl')
# 加载训练集
train_ds = load_clean_sentences('french_to_english_train.pkl')
# 加载测试集
test_ds = load_clean_sentences('french_to_english_test.pkl')
# 准备英文的 tokenizer
eng_tokenizer = create_tokenizer(dataset[:, 0])
# 准备法文的 tokenizer
fra_tokenizer = create_tokenizer(dataset[:, 1])
# 法文序列单词最大个数
fra_length = max_length(dataset[:, 1])
# 对训练集序列进行编码
X_train = encode_sequences(fra_tokenizer, fra_length, train_ds[:, 1])
# 对测试集序列进行编码
X_test = encode_sequences(fra_tokenizer, fra_length, test_ds[:, 1])
```

```python
from keras import models
# 加载已训练的模型
model = models.load_model('translator_weights_model.h5')
```

测试模型,去 tokenizer 的字典中寻找每次每个序列预测到的最大概率值对应的单词,最后拼接起来就是预测到的句子。

```python
def word_for_id(integer, tokenizer):
    # 根据索引,去 tokenizer 的字典里找到该索引对应的单词,并返回
    for word, index in tokenizer.word_index.items():
        if index == integer:
            return word
    return None
def predict_sequence(model, tokenizer, source):
    # 根据法文预测它的英文
    prediction = model.predict(source, verbose=0)[0]
    # 根据预测的结果在每个数组里选择最大概率值的索引
    # 最后返回所有的最大概率的值
    integers = [np.argmax(vector) for vector in prediction]
    target = list()
    # 寻找这些最大概率值对应的单词
    for i in integers:
        word = word_for_id(i, tokenizer)
        if word is None:
            break
        target.append(word)
    return ' '.join(target)
from nltk.translate.bleu_score import corpus_bleu
def test_model(model, tokenizer, sources, raw_dataset):
    actual, predicted = list(), list()
    # 对数据进行遍历
    for i, source in enumerate(sources):
        # 将法文的编码数据进行预测
        source = source.reshape((1, source.shape[0]))
        translation = predict_sequence(model, tokenizer, source)
        raw_target, raw_src = raw_dataset[i]
        if i < 10:
            print('源语句=[{}], 目标语句=[{}], 预测语句=[{}]'.format(raw_src, raw_target, translation))
        actual.append(raw_target.split())
        predicted.append(translation.split())
    # 计算 BLEU 分数
    print('BLEU-1: {}'.format(corpus_bleu(actual, predicted, weights=(1.0, 0, 0, 0))))
    print('BLEU-2: {}'.format(corpus_bleu(actual, predicted, weights=(0.5, 0.5, 0, 0))))
    print('BLEU-3: {}'.format(corpus_bleu(actual, predicted, weights=(0.3, 0.3, 0.3, 0))))
    print('BLEU-4: {}'.format(corpus_bleu(actual, predicted, weights=(0.25, 0.25, 0.25, 0.25))))
# 测试训练集序列
print('训练集: ')
test_model(model, eng_tokenizer, X_train, train_ds)
# 测试测试集序列
print('测试集: ')
test_model(model, eng_tokenizer, X_test, test_ds)
```

输出如下。

训练集：
源语句=[il detient une video]，目标语句=[he has a video]，预测语句=[he has a video]
源语句=[pourraistu maider]，目标语句=[can you help me]，预测语句=[can you help me]
源语句=[soyez respectueuses]，目标语句=[be respectful]，预测语句=[be respectful]
源语句=[vous etes naif]，目标语句=[youre naive]，预测语句=[youre naive]
源语句=[il travaille a peine]，目标语句=[he hardly works]，预测语句=[he hardly works]
源语句=[montrezmoi]，目标语句=[show me]，预测语句=[show me]
源语句=[je dois y aller]，目标语句=[i must go]，预测语句=[i must to go]
源语句=[quelle equipe]，目标语句=[what a team]，预测语句=[what a team]
源语句=[nous sommes en retard]，目标语句=[we are late]，预测语句=[were late]
源语句=[appelle la securite]，目标语句=[call security]，预测语句=[call security]
/usr/local/lib/python3.6/dist-packages/nltk/translate/bleu_score.py:490: UserWarning:
Corpus/Sentence contains 0 counts of 2-gram overlaps.
BLEU scores might be undesirable; use SmoothingFunction().
　warnings.warn(_msg)
BLEU-1: 0.08884628871460039
BLEU-2: 0.28372048975984243
BLEU-3: 0.4514308812293625
BLEU-4: 0.5070098604719333
测试集：
源语句=[venez la]，目标语句=[come here]，预测语句=[come back]
源语句=[laissemoi sortir]，目标语句=[let me out]，预测语句=[let me out]
源语句=[ne soyez pas mesquin]，目标语句=[dont be mean]，预测语句=[dont be mean]
源语句=[vous devez le faire]，目标语句=[you must do it]，预测语句=[you need do it]
源语句=[ty rendrastu]，目标语句=[will you go]，预测语句=[shut you serious]
源语句=[elles ont refuse]，目标语句=[they refused]，预测语句=[they refused]
源语句=[ils nous ont trouvees]，目标语句=[they found us]，预测语句=[they found us]
源语句=[jadore cuisiner]，目标语句=[i love cooking]，预测语=[i love baking]
源语句=[tom semble perdu]，目标语句=[tom seems lost]，预测语句=[tom lost]
源语句=[avaisje tort]，目标语句=[was i wrong]，预测语句=[am was i]
BLEU-1: 0.07970902428748444
BLEU-2: 0.26403639977318066
BLEU-3: 0.42631079123622802
BLEU-4: 0.48055373499039605

8.3　小结

　　本章把从开源的网站上下载的语言语句平行语料库，用作神经机器翻译的数据对，也就是把句子 A 翻译成句子 B，然后对句子进行 Tokenizer 等一系列处理，下载的原始数据集有 16 万多个语句数据对，如果完全训练，会得到更好的模型表现效果，但是为了读者更快地学习，作者截取了 1 万个数据对来训练，这虽然可以加快训练速度，但是会降低一些模型表现效果，不过最终构建了 Keras 的长短期记忆网络（LSTM）模型，经过训练模型，使用 corpus_bleu() 来测试 BLEU 的得分的表现还是不错的。

　　如果读者想训练更好的、更高精确度的模型，那么可以在上面截取数据对的代码中，将 1 万个数据对，修改成全部的数据对即可。

第二篇

识别类项目实战

- 第 9 章　MNIST 手写数字识别
- 第 10 章　狗的品种识别
- 第 11 章　人脸识别
- 第 12 章　人脸面部表情识别
- 第 13 章　人体姿态识别
- 第 14 章　皮肤癌分类
- 第 15 章　对象检测

第 9 章

MNIST 手写数字识别

MNIST 是深度学习图像识别方向的一个简单的"Hello, World!"应用。MNIST 是一个简单的计算机视觉识别数据集,全部由手写数字图像组成。本章将会为大家讲解如何通过多层感知器(MLP)和卷积神经网络(CNN)进行 MNIST 手写数字识别。

在之前的章节中我们已经用到过 TensorFlow,那么对于 Tensor(张量)在 TensorFlow 中的使用是这样的概念和表示。

- 标量(Scalar):一个具体的数字或者数值,也称为零阶张量。
- 向量(Vector):一个一维数组,也称为一阶张量。
- 矩阵(Matrix):一个二维数组,也称为二阶张量。

一个三维数组,就是三阶张量。依次类推,多维数组就是多阶张量。

9.1 MNIST 数据集

在深度学习视觉领域中,MNIST 数据集是非常适合作为训练入门模型的数据集,不管是使用 TensorFlow 还是 Keras,MNIST 数据集都已集成到开源库中,我们可以直接拿来使用。该数据集是公开的,请到本代码仓库中找到并下载。

9.1.1 简介

MNIST 数据集的所有图像都是由 0~9 的数字组成,共 10 个阿拉伯数字,图像都是灰度的,且图像中的数字都位于图像的中间位置。我们先随机预览几个数字在图像中的呈现,如图 9.1 所示。

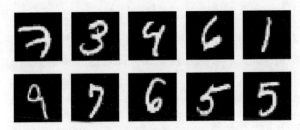

图 9.1　数字在图像中的呈现

在 MNIST 数据集中，一个图像对应着一个数字标签，表示该图像是数字几。该数据集包含 60000 个训练集样本，10000 个测试集样本。我们再来看数据集文件的组成部分，如表 9.1 所示。

表 9.1　MNIST 数据集文件组成部分

文件名称	文件大小	使用目的
train-images-idx3-ubyte.gz	大约 9.45MB	训练集图像数据
train-labels-idx1-ubyte.gz	大约 0.03MB	训练集图像标签
t10k-images-idx3-ubyte.gz	大约 1.57MB	测试集图像数据
t10k-labels-idx1-ubyte.gz	大约 0.0043MB	测试集图像标签

文件 train-images-idx3-ubyte.gz 和文件 train-labels-idx1-ubyte.gz 是一对训练样本集和标签，而文件 t10k-images-idx3-ubyte.gz 和文件 t10k-labels-idx1-ubyte.gz 是一对测试样本集和标签。它们的关系是一一对应的，一个图像对应着一个数字标签。在后面的代码中我们将详细使用该样本来训练模型。

9.1.2　数据下载

环境准备：

- numpy = 1.14.5。
- tensorflow = 1.10.1。
- keras = 2.1.6。
- matplotlib = 2.1.2。

我们可以通过 9.1 节中的链接直接下载，也可以通过代码下载。通过 TensorFlow 代码下载时，需要导入 TensorFlow 的 examples 里的相应数据集组件。

```
from tensorflow.examples.tutorials.mnist import input_data
mnist = input_data.read_data_sets("MNIST_data/", one_hot=True)
```

其中 MNIST_data 会自动在当前目录下创建一个目录，叫作 MNIST_data；在该目录下会存放如表 9.1 中所示的文件。one_hot=True 表示对训练集和测试集的标签进行独热编码，可以理解为一个有效编码。

比如，在一个有 0~9 的数字的 10 个标量的向量中，假设要表示 6 这个标签，那么向量的值应该是[0, 0, 0, 0, 0, 0, 1, 0, 0, 0]。也就是说，自己是 1，其他位都是 0。然后，变量 mnist 会存储训练集、验证集和测试集的数据，我们来打印看这三个数据集的大小。

```
print("训练集图像大小: {}".format(mnist.train.images.shape))
print("训练集标签大小: {}".format(mnist.train.labels.shape))
print("验证集图像大小: {}".format(mnist.validation.images.shape))
print("验证集标签大小: {}".format(mnist.validation.labels.shape))
print("测试集图像大小: {}".format(mnist.test.images.shape))
print("测试集标签大小: {}".format(mnist.test.labels.shape))
```

输出如下。

```
训练集图像大小: (55000, 784)
训练集标签大小: (55000, 10)
验证集图像大小: (5000, 784)
验证集标签大小: (5000, 10)
测试集图像大小: (10000, 784)
测试集标签大小: (10000, 10)
```

细心的读者会发现，这里的训练集少了 5000 个样本，这是因为，这 5000 个样本被分配给了验证集，而测试集大小保持不变。图像是二维呈现的，为什么这里的图像大小是 784 呢？这是因为所有的 MNIST 数据集的图像大小都是 28×28，将图像转换成向量后，就是 784 了。

9.1.3 可视化数据

为了方便读取，在可视化数据之前，我们先把数据各自用一个变量存储起来。

```
x_train, y_train = mnist.train.images, mnist.train.labels
x_valid, y_valid = mnist.validation.images, mnist.validation.labels
x_test, y_test = mnist.test.images, mnist.test.labels
```

现在，我们绘制上面 10 个数字的图像。

```
import matplotlib.pyplot as plt
# 这行代码表示如果读者使用的是 Jupyter Notebook, 在 Jupyter Notebook 中显示 matplotlib 绘制的图
%matplotlib inline
import numpy as np
# figsize 表示每个图像绘制后的大小，单位是英寸
fig = plt.figure(figsize=(10, 10))
# 绘制和显示前 5 个训练集图像
for i in range(5):
    ax = fig.add_subplot(1, 5, i+1, xticks=[], yticks=[])
    ax.imshow(np.reshape(x_train[i:i+1], (28, 28)), cmap='gray')
```

得到的结果如图 9.1 的第一行图像所示。

```
# 绘制和显示前（2*12）个之后的 5 个训练集图像
fig = plt.figure(figsize=(10, 10))
for i in range(5):
    ax = fig.add_subplot(1, 5, i+1, xticks=[], yticks=[])
    ax.imshow(np.reshape(x_train[i+2*12:i+1+2*12], (28, 28)), cmap='gray')
```

输出的结果如图 9.1 的第二行图像所示。然后,我们来分析计算机是如何查看图像的。首先,我们通过以下代码输出图像。

```python
# 定义可视化图像的函数,传入一个图像向量和 figure 对象
def visualize_input(img, ax):
    # 绘制并输出图像
    ax.imshow(img, cmap='gray')

    # 对于该图像的宽和高,我们输出具体的数值
    # 便于我们更清晰地知道计算机是如何查看图像的
    width, height = img.shape

    # 将图像中的具体数值转换成 0~1 之间的值
    thresh = img.max()/2.5
    # 遍历行
    for x in range(width):
        # 遍历列
        for y in range(height):
            # 将图像的数值在它对应的位置上标出,且水平垂直居中
            ax.annotate(str(round(img[x][y],2)), xy=(y,x),
                        horizontalalignment='center',
                        verticalalignment='center',
                        color='white' if img[x][y]<thresh else 'black')
fig = plt.figure(figsize=(10, 10))
ax = fig.add_subplot(111)
# 假设我们就取出下标为 5 的样本来作为例子
visualize_input(np.reshape(x_train[5:6], (28, 28)), ax)
```

输出结果如图 9.2 所示。

图 9.2 数字 8 的具体数值图像

RGB 颜色模式中的纯黑色色值为（0,0,0），纯白色色值为（255, 255, 255），这里我们将 0 到 255 的色值转换成 0 到 1 之间的数值。然后我们可以清晰地看到，在一张纯黑色的背景中，大于 0 的数值会显示具体的颜色，这些数值在不同的地方组合起来就是一个具体的图像了。通过这个图像，我们可以看到这上面显示的是什么。

9.2 基于多层感知器的 TensorFlow 实现 MNIST 识别

在深度学习训练模型中，多层感知器（Multi-Layer Perceptron）模型是最基本的模型。即在神经网络模型中，每个节点就是一个感知器，每个感知器的基本公式是$(x \times W) + b$。然后再通过激活函数输出两个类别或多个类别，这些类别输出结果的概率值都是 0 到 1 之间的数值，加起来的和是 1；在此基础上，添加多层，就是多层感知器。

9.2.1 参数准备

在训练任何神经网络模型时，我们都需要预先设定参数，然后根据训练的实际情况来调整参数的值，模型的参数如下。

```
# 图像大小
img_size = 28 * 28
# 要预测的类别有多少个
num_classes = 10
# 学习率，也叫作梯度下降值
learning_rate = 0.1
# 迭代次数
epochs = 100
# 每批次大小
batch_size = 128
```

9.2.2 创建模型

本模型使用 softmax 多类别分类激活函数，配合交叉熵来计算损失值，然后通过梯度下降来定义优化器，感知器的算法公式一般为$(x \times W) + b$。

```
# x 表示输入，创建输入占位符，该占位符在训练时，会对每次迭代的数据进行填充
# None 表示在训练时传入的图像数量，每张大小是 img_size
x = tf.placeholder(tf.float32, [None, img_size])
# W 表示weight，创建权重，初始值都是 0，它的大小是(图像的向量大小，图像的总类别个数)
W = tf.Variable(tf.zeros([img_size, num_classes]))
# b 表示bias，创建偏置项，初始值都是 0
b = tf.Variable(tf.zeros([num_classes]))
# y 表示计算输出结果，softmax 表示激活函数是多类别分类的输出
# 计算公式：softmax((x * W) + b)
y = tf.nn.softmax(tf.matmul(x, W) + b)
# 定义输出预测占位符 y_
y_ = tf.placeholder(tf.float32, [None, 10])
# 创建给 TensorFlow 的训练模型时的参数数据字典
```

```
valid_feed_dict = { x: x_valid, y_: y_valid }
test_feed_dict = { x: x_test, y_: y_test }
# 通过激活函数 softmax 的交叉熵来定义损失函数
cost = tf.reduce_mean(tf.nn.softmax_cross_entropy_with_logits(labels=y_, logits=y))
# 定义梯度下降优化器，根据学习率来梯度下降，并且下降过程中，损失值也越来越小
optimizer = tf.train.GradientDescentOptimizer(learning_rate).minimize(cost)
# 比较正确的预测结果
correct_prediction = tf.equal(tf.argmax(y, 1), tf.argmax(y_, 1))
# 计算预测准确率
accuracy = tf.reduce_mean(tf.cast(correct_prediction, tf.float32))
```

其中，tf.argmax()函数返回的是最大值的索引，第一个参数是一个 Tensor（张量）。模型预测的变量 y 的形状是(N, 10)，实际标签的变量 y_的形状是(N, 10)，N 表示输入模型的样本个数。然后通过 tf.equal()函数来比较它们是否相等，将结果保存到变量 correct_prediction 中，该结果里的值都是 True 或者 False。最后再通过 tf.cast()函数将 True 或者 False 转换成 1 或者 0。

9.2.3 训练模型

在本模型训练结束后，我们需要通过模型来预测测试集数据。所以我们使用 tf.train.Saver()类来保存模型的最佳检查点，模型会保存在 checkpoints 的目录下。

```
import math
iteration = 0
# 定义训练时的检查点
saver = tf.train.Saver()
# 创建一个 TensorFlow 的会话
with tf.Session() as sess:
    # 初始化全局变量
    sess.run(tf.global_variables_initializer())
    
    # 根据每批次训练 128 个样本，计算出一共需要迭代多少次
    batch_count = int(math.ceil(mnist.train.labels.shape[0] / 128.0))
    
    # 开始迭代训练样本
    for e in range(epochs):
        
        # 每个样本都需要在 TensorFlow 的会话里进行运算和训练
        for batch_i in range(batch_count):
            
            # 样本的索引，间隔是 128 个
            batch_start = batch_i * batch_size
            # 取出图像样本
            batch_x = mnist.train.images[batch_start:batch_start+batch_size]
            # 取出图像对应的标签
            batch_y = mnist.train.labels[batch_start:batch_start+batch_size]
            # 训练模型
            loss, _ = sess.run([cost, optimizer], feed_dict={x: batch_x, y_: batch_y})
```

```
        # 每训练 20 次图像后打印一次日志信息
        if batch_i % 20 == 0:
            print("Epoch: {}/{}".format(e+1, epochs),
                  "Iteration: {}".format(iteration),
                  "Training loss: {:.5f}".format(loss))
        iteration += 1
        # 每训练 128 个样本，验证一下训练的模型效果如何，并输出日志信息
        if iteration % batch_size == 0:
            valid_acc = sess.run(accuracy, feed_dict=valid_feed_dict)
            print("Epoch: {}/{}".format(e, epochs),
                  "Iteration: {}".format(iteration),
                  "Validation Accuracy: {:.5f}".format(valid_acc))
    # 保存训练模型的检查点
    saver.save(sess, "checkpoints/mnist_mlp_tf.ckpt")
```

该训练会在几分钟内就完成，我们可以看它起始和结束的训练日志信息，输出如下。

```
Epoch: 1/60 Iteration: 0 Training loss: 2.30259
Epoch: 1/60 Iteration: 20 Training loss: 2.27273
Epoch: 1/60 Iteration: 40 Training loss: 2.19583
Epoch: 1/60 Iteration: 60 Training loss: 2.15586
Epoch: 1/60 Iteration: 80 Training loss: 2.07523
Epoch: 1/60 Iteration: 100 Training loss: 2.08399
Epoch: 1/60 Iteration: 120 Training loss: 1.94980
Epoch: 0/60 Iteration: 128 Validation Accuracy: 0.63220
Epoch: 99/100 Iteration: 42880 Validation Accuracy: 0.92800
Epoch: 100/100 Iteration: 42890 Training loss: 1.57692
Epoch: 100/100 Iteration: 42910 Training loss: 1.54435
Epoch: 100/100 Iteration: 42930 Training loss: 1.53651
Epoch: 100/100 Iteration: 42950 Training loss: 1.56185
Epoch: 100/100 Iteration: 42970 Training loss: 1.57890
Epoch: 100/100 Iteration: 42990 Training loss: 1.53769
```

最终模型的训练损失值从约 2.3 降低到了约 1.5，验证集的验证精确度达到了约 0.92。

9.2.4 模型预测

最后，我们对测试数据集的样本进行测试，得到精确度。在预测前，先把训练时的 checkpoint 读取到，也是通过 tf.train.Saver()。

```
saver = tf.train.Saver()
with tf.Session() as sess:
    # 从训练模型的检查点恢复
    saver.restore(sess, tf.train.latest_checkpoint('checkpoints'))
    
    # 预测测试数据集样本的精确度
    test_acc = sess.run(accuracy, feed_dict=test_feed_dict)
    print("test accuracy: {:.5f}".format(test_acc))
```

测试精确度的输出如下。

```
test accuracy: 0.92560
```

9.3 基于多层感知器的 Keras 实现 MNIST 识别

本节将使用 Keras 实现 MNIST 识别。使用 Keras 的代码和 TensorFlow 的代码相比存在不同之处，比如在创建模型、选择优化器、编译、训练和预测时，Keras 的代码要比 TensorFlow 少很多，而且也更简洁易懂。

9.3.1 数据准备

通过 Keras 实现 MNIST 识别需要先下载 MNIST 数据集，由于 load_data()函数只返回训练集和测试集，所以我们需要自己手动将训练集中的数据分割出 5000 个样本作为验证集。然后对训练集、验证集和测试集的数据进行归一化处理，并对它们的标签进行独热编码处理。

```
import keras
from keras.datasets import mnist
from keras.models import Sequential
from keras.layers import Dense, Dropout, Activation
from keras.optimizers import RMSprop
# 参数准备
# 每次训练样本的批次大小
batch_size = 128
# MNIST 只有 10 个类别，0 到 9 的数字
num_classes = 10
# 训练所有的样本，重复迭代训练 20 次
epochs = 20
# 图像宽和高
img_size = 28 * 28
# 下载并读取 MNIST 数据集的数据
(x_train, y_train), (x_test, y_test) = mnist.load_data()
# 分割 5000 个验证数据集样本数据
valid_len = 5000
x_len = x_train.shape[0]
train_len = x_len-valid_len
# 验证集数据
x_valid = x_train[train_len:]
y_valid = y_train[train_len:]
# 训练集数据
x_train = x_train[:train_len]
y_train = y_train[:train_len]
# 将训练集、验证集和测试集数据进行图像的向量转换
x_train = x_train.reshape(x_train.shape[0], img_size)
x_valid = x_valid.reshape(x_valid.shape[0], img_size)
x_test = x_test.reshape(x_test.shape[0], img_size)
# 将训练集、验证集和测试集数据都转换成 float32 类型
x_train = x_train.astype('float32')
x_valid = x_valid.astype('float32')
x_test = x_test.astype('float32')
# 将训练集、验证集和测试集数据都转换成 0 到 1 之间的数值，这就是归一化处理
```

```python
x_train /= 255
x_valid /= 255
x_test /= 255
# 通过 to_categorical()函数将训练集标签、验证集标签和测试集标签独热编码
y_train = keras.utils.to_categorical(y_train, num_classes)
y_valid = keras.utils.to_categorical(y_valid, num_classes)
y_test = keras.utils.to_categorical(y_test, num_classes)
# 我们来看下训练集、验证集和测试集的数据大小
print("训练集图像大小：{}".format(x_train.shape))
print("训练集标签大小：{}".format(y_train.shape))
print("验证集图像大小：{}".format(x_valid.shape))
print("验证集标签大小：{}".format(y_valid.shape))
print("测试集图像大小：{}".format(x_test.shape))
print("测试集标签大小：{}".format(y_test.shape))
```

输出如下。

```
训练集图像大小：(55000, 784)
训练集标签大小：(55000, 10)
验证集图像大小：(5000, 784)
验证集标签大小：(5000, 10)
测试集图像大小：(10000, 784)
测试集标签大小：(10000, 10)
```

9.3.2　创建模型

在 Keras 中，一般可以直接通过顺序模型（Sequential）类来创建模型，然后给这个模型添加输入层、激活层、Dropout 层、输出层等，最后还可以通过 summary()函数来预览模型架构。

Dropout 是在训练过程中一种降低神经网络过拟合的方式，顾名思义就是丢弃的意思，而传入的值就是丢弃的概率。这里的 Dropout(0.2)表示有 20%的感知器节点在训练过程中会被随机地暂时丢弃，待下一次反向传播时，可能就会丢弃其他的 20%感知器节点，那上一次被丢弃的感知器节点就不会被丢弃，而是去参与训练，依次类推。

```python
# 创建一个模型
model = Sequential()
# 添加输入层，输入层大小是图像大小，激活函数使用修正线性单元 ReLU
# input_shape 参数是必需的，它的值就是图像大小
model.add(Dense(512, activation='relu', input_shape=(img_size,)))
# 添加 Dropout 层
model.add(Dropout(0.2))
# 添加 512 个全连接层，并使用 ReLU 激活函数
model.add(Dense(512, activation='relu'))
# 添加 Dropout 层
model.add(Dropout(0.2))
# 添加输出层，输出个数是 10 个类别，使用 softmax 多类别分类激活函数
model.add(Dense(num_classes, activation='softmax'))
# 模型架构预览
model.summary()
```

模型架构预览，如图 9.3 所示。

```
Layer (type)                 Output Shape              Param #
=================================================================
dense_1 (Dense)              (None, 512)               401920
_____
dropout_1 (Dropout)          (None, 512)               0
_____
dense_2 (Dense)              (None, 512)               262656
_____
dropout_2 (Dropout)          (None, 512)               0
_____
dense_3 (Dense)              (None, 10)                5130
=================================================================
Total params: 669,706
Trainable params: 669,706
Non-trainable params: 0
```

图 9.3　多层感知器的 Keras 模型架构图

9.3.3　训练模型

在 Keras 中，训练模型前我们需要先编译模型，然后指定损失函数的类型。categorical_crossentropy 表示多类别分类交叉熵；优化器使用 RMSprop()，对于它的参数值，官方的建议是保留默认值，这里也保留默认值；metrics 衡量方式就用 accuracy 精确度。

训练模型时，指定它的训练样本和训练标签，验证样本和验证标签。如果希望在训练时输出打印进度和日志信息，就添加参数 verbose=1。

```
# 编译模型
model.compile(loss='categorical_crossentropy', optimizer=RMSprop(), metrics=['accuracy'])
# 训练模型
model.fit(x_train, y_train, epochs=epochs, batch_size=batch_size, verbose=1,
          validation_data=(x_valid, y_valid))
```

训练的输出打印部分日志信息如下。

```
Epoch 1/20
55000/55000 [==============================] - 9s 169us/step - loss: 0.2631 - acc: 0.9197 - val_loss: 0.0966 - val_acc: 0.9726
Epoch 2/20
55000/55000 [==============================] - 9s 158us/step - loss: 0.1086 - acc: 0.9666 - val_loss: 0.0688 - val_acc: 0.9818
Epoch 3/20
55000/55000 [==============================] - 9s 156us/step - loss: 0.0781 - acc: 0.9760 - val_loss: 0.0823 - val_acc: 0.9764
……
Epoch 19/20
55000/55000 [==============================] - 9s 157us/step - loss: 0.0178 - acc: 0.9950 - val_loss: 0.1058 - val_acc: 0.9832
Epoch 20/20
55000/55000 [==============================] - 9s 157us/step - loss: 0.0160 - acc: 0.9956 - val_loss: 0.1051 - val_acc: 0.9838
```

我们可以看到训练损失值从 0.26 降到了 0.01，精确度从 0.91 提升到了 0.99。

9.3.4 模型预测

通过 evaluate()函数，我们对测试集数据评估模型得分，输出打印日志信息。

```
score = model.evaluate(x_test, y_test, verbose=0)
print('Test accuracy:{}, Test loss: {}, {}'.format(score[1], score[0], score))
```

输出如下。

```
Test accuracy:0.9829, Test loss: 0.116766489373, [0.11676648937257796, 0.9829]
```

9.3.5 单个图像预测

取出测试数据集中的索引为 7 的图像，然后把该图像送入模型进行预测，返回结果为一个向量，向量里的各个元素表示模型预测到的数值，最后绘图展示。

```
import matplotlib.pyplot as plt
import numpy as np
# 取出测试数据集中的索引为 7 的图像向量数据
x_img = x_test[7:8]
# 预测单个图像的概率，结果是以 10 个类别的概率的向量返回
prediction = model.predict(x_img)
# 绘图预览
x_coordinates = np.arange(prediction.shape[1])
plt.bar(x_coordinates, prediction[0][:])
plt.xticks(x_coordinates, np.arange(10))
plt.show()
```

输出结果如图 9.4 所示。

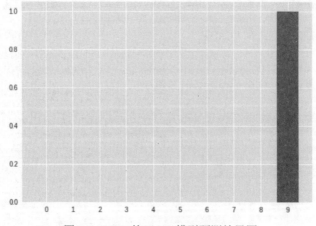

图 9.4　MLP 的 Keras 模型预测结果图

9.4　基于卷积神经网络的 TensorFlow 实现 MNIST 识别

卷积神经网络（CNN）是在 MLP 之后发展出来的，而且从精确度和效率来讲，都要优越

于 MLP。在卷积层中，每个感知器的节点算法公式就是卷积。卷积有一个滑动窗口，我们在写网络模型时主要就是对卷积的大小、通道数和每次移动的步长进行定义。

9.4.1 参数准备

在 9.1 节中已经介绍过如何下载和加载 MNIST 数据集，这里就不再展开。在训练任何神经网络模型时，我们都需要预先设定参数，然后根据训练的实际情况来调整参数的值。本模型的参数如下。

```
# 图像的宽和高
img_size = 28 * 28
# 图像有 10 个类别，数字 0 到 9
num_classes = 10
# 学习率，1e-4 是科学计数表示法，它的值是 0.0001
learning_rate = 1e-4
# 迭代次数
epochs = 10
# 每批次大小
batch_size = 50
```

9.4.2 创建模型

我们使用卷积来添加网络模型中的隐藏层，这里就添加两层卷积层，然后使用 softmax 多类别分类激活函数，配合交叉熵来计算损失值，通过 Adam 来定义优化器。

```
# 定义输入占位符，图像的输入形状大小是 [batch, image_vector]
x = tf.placeholder(tf.float32, shape=[None, img_size])
# 图像的输入形状大小在卷积中是 [batch, in_height, in_width, in_channels]
x_shaped = tf.reshape(x, [-1, 28, 28, 1])
# 定义输出占位符
y = tf.placeholder(tf.float32, shape=[None, num_classes])
# 定义卷积函数
def create_conv2d(input_data, num_input_channels, num_filters, filter_shape, pool_shape, name):
    # 卷积的过滤器形状大小是 [filter_height, filter_width, in_channels, out_channels]
    conv_filter_shape = [filter_shape[0], filter_shape[1], num_input_channels, num_filters]
    # 定义权重 Tensor 变量，初始化时是截断正态分布，标准差是 0.03
    weights = tf.Variable(tf.truncated_normal(conv_filter_shape, stddev=0.03), name=name+"_W")
    # 定义偏置项 Tensor 变量，初始化时是截断正态分布
    bias = tf.Variable(tf.truncated_normal([num_filters]), name=name+"_b")
    # 定义卷积层
    # 参数 1：输入图像；参数 2：权重；参数 3：步长；参数 4：填充
    # 参数 4 填充 padding，如果是 SAME 表示在移动窗口中，不够 filter 大小的数据就用 0 填充，
    # 相反，如果是 padding=VALID，则表示不够 filter 大小的那块数据就不要了
    out_layer = tf.nn.conv2d(input_data, weights, (1, 1, 1, 1), padding="SAME")
    out_layer += bias
    # 通过激活函数 ReLU 来计算输出
    out_layer = tf.nn.relu(out_layer)
    # 添加最大池化层，ksize 值的形状是 [batch, height, width, channels]，
    # 步长 strides 值的形状是 [batch, stride, stride, channels]
```

```python
        out_layer = tf.nn.max_pool(out_layer, ksize=(1, pool_shape[0], pool_shape[1], 1),
                                    strides=(1, 2, 2, 1), padding="SAME")
        return out_layer
# 添加第一层卷积层,深度为 32
layer1 = create_conv2d(x_shaped, 1, 32, (5, 5), (2, 2), name="layer1")
# 添加第二层卷积层,深度为 64
layer2 = create_conv2d(layer1, 32, 64, (5, 5), (2, 2), name="layer2")
# 添加扁平化层,扁平化为一个大向量
flattened = tf.reshape(layer2, (-1, 7 * 7 * 64))
# 添加全连接层
wd1 = tf.Variable(tf.truncated_normal((7 * 7 * 64, 1000), stddev=0.03), name="wd1")
bd1 = tf.Variable(tf.truncated_normal([1000], stddev=0.01), name="bd1")
dense_layer1 = tf.add(tf.matmul(flattened, wd1), bd1)
dense_layer1 = tf.nn.relu(dense_layer1)
# 添加输出全连接层,深度为 10,因为只需要十个类别
wd2 = tf.Variable(tf.truncated_normal((1000, num_classes), stddev=0.03), name="wd2")
bd2 = tf.Variable(tf.truncated_normal([num_classes], stddev=0.01), name="bd2")
dense_layer2 = tf.add(tf.matmul(dense_layer1, wd2), bd2)
# 添加激活函数的 softmax 输出层
y_ = tf.nn.softmax(dense_layer2)
# 通过 softmax 交叉熵定义计算损失值
cost = tf.reduce_mean(tf.nn.softmax_cross_entropy_with_logits(logits=y_, labels=y))
# 定义优化器是 Adam
optimizer = tf.train.AdamOptimizer(learning_rate=learning_rate).minimize(cost)
# 比较正确的预测结果
correct_prediction = tf.equal(tf.argmax(y, 1), tf.argmax(y_, 1))
# 计算预测的精确度
accuracy = tf.reduce_mean(tf.cast(correct_prediction, tf.float32))
```

变量 correct_prediction 保存的值都是 True 或者 False,然后我们通过 tf.cast()函数将 True 转换成 1,将 False 转换成 0,最后通过 tf.reduce_mean()函数计算元素的均值。

9.4.3 训练模型

训练完本模型后,需要通过模型来预测测试集数据,所以我们使用 tf.train.Saver()类来保存模型的最佳检查点,模型会保存在 checkpoints 的目录下,名称是 mnist_cnn_tf.ckpt。

```python
import math
# 定义要保存训练模型的变量
saver = tf.train.Saver()
# 创建 TensorFlow 会话
with tf.Session() as sess:
    # 初始化 TensorFlow 的全局变量
    sess.run(tf.global_variables_initializer())
    # 计算所有的训练集需要被训练多少次,当每批次是 batch_size 个时
    batch_count = int(math.ceil(x_train.shape[0] / float(batch_size)))
    # 要迭代 epochs 次训练
    for e in range(epochs):
        # 对每个图像进行训练
        for batch_i in range(batch_count):
```

```
# 每次取出 batch_size 个图像
batch_x, batch_y = mnist.train.next_batch(batch_size=batch_size)
# 训练模型，计算训练损失值
_, loss = sess.run([optimizer, cost], feed_dict={x: batch_x, y: batch_y})

# 每训练 20 次图像打印一次日志信息，也就是 20 次乘以 batch_size 个图像已经被训练了
if batch_i % 20 == 0:
    print("Epoch: {}/{}".format(e+1, epochs),
          "Iteration: {}".format(iteration),
          "Training loss: {:.5f}".format(loss))
iteration += 1
# 每迭代一次，做一次验证，并打印日志信息
if iteration % batch_size == 0:
    valid_acc = sess.run(accuracy, feed_dict={x: x_valid, y: y_valid})
    print("Epoch: {}/{}".format(e, epochs),
          "Iteration: {}".format(iteration),
          "Validation Accuracy: {:.5f}".format(valid_acc))
# 保存模型的检查点
saver.save(sess, "checkpoints/mnist_cnn_tf.ckpt")
```

大约 5 分钟训练完毕，训练损失值从一开始的 2.30 降到了 1.46，而精确度也从 0.11 提升到了 0.97。基本日志信息如下。

```
Epoch: 1/10 Iteration: 0 Training loss: 2.30495
Epoch: 1/10 Iteration: 20 Training loss: 2.28635
Epoch: 1/10 Iteration: 40 Training loss: 2.29930
Epoch: 0/10 Iteration: 50 Validation Accuracy: 0.11260
Epoch: 10/10 Iteration: 10960 Training loss: 1.46412
Epoch: 10/10 Iteration: 10980 Training loss: 1.46157
Epoch: 9/10 Iteration: 11000 Validation Accuracy: 0.97960
```

9.4.4 模型预测

最后，我们对测试数据集的样本进行预测，得到精确度。在预测前，先通过 tf.train.Saver() 读取训练时的模型检查点 checkpoint。

```
# 预测测试数据集
saver = tf.train.Saver()
with tf.Session() as sess:
    # 从 TensorFlow 会话中恢复之前保存的模型检查点
    saver.restore(sess, tf.train.latest_checkpoint('checkpoints/'))

    # 通过测试数据集预测精确度
    test_acc = sess.run(accuracy, feed_dict={x: x_test, y: y_test})
    print("test accuracy: {:.5f}".format(test_acc))
```

输出如下。

```
test accuracy: 0.98170
```

9.5 基于卷积神经网络的 Keras 实现 MNIST 识别

我们通过使用 Keras 的 CNN 实现 MNIST 识别，所需代码要比使用 TensorFlow 时的代码简洁得多。在数据预处理、创建模型、选择优化器、编译、训练和预测时，我们会使用到 Conv2D 和 MaxPooling2D 类，它们简洁易懂，但也有不同点。

9.5.1 数据准备

在 Keras 中，MNIST 数据集的下载和加载与 9.1 节介绍的方法不一样，但是在 Keras 中下载和加载数据也比较简单，我们可通过 load_data()函数来实现。在训练神经网络模型时，需要先设定参数，然后根据训练的实际情况来调整参数的值。

```
import numpy as np
import keras
from keras.datasets import mnist
from keras.models import Sequential
from keras.layers import Dense, Dropout, Flatten, Conv2D, MaxPooling2D
from keras import utils
# 参数准备
# 每批次大小
batch_size = 128
# 迭代次数
epochs = 15
# 类别总个数
num_classes = 10
# 图像的宽
img_width = 28
# 图像的高
img_height = 28
# 图像通道数
img_channels = 1
# 下载并读取MNIST数据集数据
(x_train, y_train), (x_test, y_test) = mnist.load_data()
# 分割5000个验证集样本数据
valid_len = 5000
x_len = x_train.shape[0]
train_len = x_len-valid_len
# 分割验证集数据
x_valid = x_train[train_len:]
y_valid = y_train[train_len:]
# 分割训练集数据
x_train = x_train[:train_len]
y_train = y_train[:train_len]
# 将训练集、验证集和测试集数据进行图像转换,
# 图像的形状大小是 [batch, height, width, channels]
```

```python
x_train = x_train.reshape(x_train.shape[0], img_height, img_width, img_channels)
x_valid = x_valid.reshape(x_valid.shape[0], img_height, img_width, img_channels)
x_test = x_test.reshape(x_test.shape[0], img_height, img_width, img_channels)
# 将训练集、验证集和测试集数据都转换成 float32 类型
x_train = x_train.astype(np.float32)
x_valid = x_valid.astype(np.float32)
x_test = x_test.astype(np.float32)
# 将训练集、验证集和测试集数据都转换成 0 到 1 之间的数值,这就是归一化处理
x_train /= 255
x_valid /= 255
x_test /= 255
# 通过 to_categorical()函数将训练集标签、验证集标签和测试集标签进行独热编码
y_train = keras.utils.to_categorical(y_train, num_classes)
y_valid = keras.utils.to_categorical(y_valid, num_classes)
y_test = keras.utils.to_categorical(y_test, num_classes)
```

我们对训练集、验证集和测试集数据预处理和归一化,然后通过 to_categorical()函数对它们的标签进行独热编码。

9.5.2 创建模型

在 Keras 中,我们直接使用顺序模型(Sequential)类创建模型,然后给这个模型添加输入层、卷积层、最大池化层、激活层、Dropout 层、输出层等,最后通过 summary()函数来预览模型架构。

```python
# 创建一个模型
model = Sequential()
# 添加输入层,深度是 32,内核大小 3×3,激活函数为 ReLU,input_shape 就是图像的大小和通道数
model.add(Conv2D(filters=32, kernel_size=(3, 3), activation='relu',
                 input_shape=(img_width, img_height, img_channels)))
# 添加卷积层,深度是 64,内核大小 3×3,激活函数为 ReLU
model.add(Conv2D(filters=64, kernel_size=(3, 3), activation='relu'))
# 添加最大池化层,每个池子大小 2×2
model.add(MaxPooling2D(pool_size=(2, 2)))
# 添加 dropout 层,每次保留 75%的节点,随机地暂时丢弃剩下的 25%的节点
model.add(Dropout(0.25))
# 添加扁平化层,扁平一个超大向量
model.add(Flatten())
# 添加全连接层,深度为 128,激活函数为 ReLU
model.add(Dense(128, activation='relu'))
# 添加 dropout 层,每次保留 50%的节点,随机地暂时丢弃剩下的 50%的节点
model.add(Dropout(0.5))
# 添加全连接层输出层,深度是 10,使用激活函数 softmax 多类别分类
model.add(Dense(num_classes, activation='softmax'))
# 模型架构预览
model.summary()
```

输出结果如图 9.5 所示。

```
Layer (type)                 Output Shape              Param #
=================================================================
conv2d_1 (Conv2D)            (None, 26, 26, 32)        320
conv2d_2 (Conv2D)            (None, 24, 24, 64)        18496
max_pooling2d_1 (MaxPooling2 (None, 12, 12, 64)        0
dropout_1 (Dropout)          (None, 12, 12, 64)        0
flatten_1 (Flatten)          (None, 9216)              0
dense_1 (Dense)              (None, 128)               1179776
dropout_2 (Dropout)          (None, 128)               0
dense_2 (Dense)              (None, 10)                1290
Total params: 1,199,882
Trainable params: 1,199,882
Non-trainable params: 0
```

图 9.5 Keras 卷积神经网络模型架构图

9.5.3 训练模型

在 Keras 中，训练模型前我们需要先编译模型，然后指定损失函数的类型。categorical_crossentropy 表示多类别分类交叉熵；优化器使用 Adadelta，对于它的参数值，官方的建议是保留默认值，这里也保留默认值；metrics 衡量方式就用 accuracy 精确度。

训练模型时，我们要指定它的训练样本和训练标签，验证样本和验证标签，如果希望在训练时输出打印进度和日志信息，就添加参数 verbose=1。

```
# 编译模型
model.compile(loss=keras.losses.categorical_crossentropy,
              optimizer=keras.optimizers.Adadelta(), metrics=['accuracy'])
# 训练模型
model.fit(x_train, y_train, batch_size=batch_size, epochs=epochs,
          verbose=1, validation_data=(x_valid, y_valid))
```

训练输出打印的部分日志信息如下。

```
Epoch 1/15
55000/55000 [==============================] - 11s 191us/step - loss: 0.2786 - acc: 0.9144 - val_loss: 0.0550 - val_acc: 0.9842
Epoch 2/15
55000/55000 [==============================] - 9s 156us/step - loss: 0.0939 - acc: 0.9723 - val_loss: 0.0421 - val_acc: 0.9888
……
55000/55000 [==============================] - 9s 157us/step - loss: 0.0240 - acc: 0.9928 - val_loss: 0.0311 - val_acc: 0.9910
Epoch 15/15
55000/55000 [==============================] - 9s 157us/step - loss: 0.0233 - acc: 0.9924 - val_loss: 0.0332 - val_acc: 0.9932
```

训练时的损失值从 0.27 降到了 0.02，而精确度从 0.91 提升到了 0.99。

9.5.4 模型预测

通过 evaluate()函数，我们对测试数据集评估模型预测得分，不输出打印日志信息。

```
score = model.evaluate(x_test, y_test, verbose=0)
print("Test Loss: {:.5f}, Test Accuracy: {:.5f}".format(score[0], score[1]))
```

输出如下。

```
Test Loss: 0.02434, Test Accuracy: 0.99250
```

9.5.5 单个图像预测

我们取出测试数据集中的索引为 7 的图像，然后把该图像送入模型进行预测，返回结果为一个向量，向量里的各个元素表示模型预测到的数值，最后绘图展示。

```
# 单个图像预测
import matplotlib.pyplot as plt
# 取出第一个图像
x_img = x_test[0:1]
# 通过模型预测
prediction = model.predict(x_img)
# 绘制图展示
x_coordinate = np.arange(prediction.shape[1])
plt.bar(x_coordinate, prediction[0][:])
plt.xticks(x_coordinate, np.arange(10))
plt.show()
```

输出结果如图 9.6 所示。预览图显示的是数字 7，我们可以通过 y_test[0:1]来查看这个图像是否是数字 7。

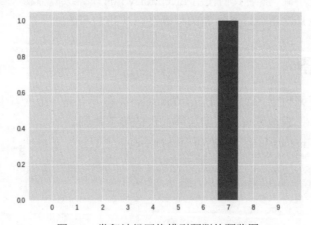

图 9.6　卷积神经网络模型预测的预览图

9.6　小结

本章通过使用 MNIST 手写图像数据集，分析图像在计算机中是如何被查看的。实际上，对于计算机来说，一个图像就是一个矩阵，矩阵里有许多数字，各个数字代表着不同的颜色。然后我们基于 TensorFlow 和 Keras 构建多层感知器和卷积神经网络模型来识别手写数字图像。

第10章

狗的品种识别

当你在街道或公园里散步时,有没有遇到过一只特别可爱或者外形极酷但不知道品种的狗呢?如果有,你想知道这个品种的狗叫什么吗?本章将介绍如何通过直接给狗拍一张照片,然后就可以识别它的品种。最后还有一个小测试,测试作者被识别得最像120个品种的狗中的哪一种。

本章会通过卷积神经网络(CNN)和迁移学习(Transfer Learning)对120种国内外的狗进行品种识别。一开始我们使用自己构建的卷积神经网络来训练并进行狗的品种识别,但是发现训练后的精确度不高。然后我们使用InceptionV3的预训练模型和MobileNet v2加以迁移学习,发现只要训练几次就能达到很高的精确度,那我们开始吧。

10.1 数据准备

狗的图片数据集来自于Stanford Dogs Dataset,由世界各地120个品种的狗的图片组成,共计20580张图片。下载地址可以在前言所提供链接下找到。

下载并解压后,Images目录下会有每个品种的狗的子目录,子目录里包含狗的一张张图片。

```
Images/
        n02085620-Chihuahua/
        n02085620_7.jpg
        n02085620_199.jpg
        ……
        n02085782-Japanese_spaniel/
            n02085782_2.jpg
            n02085782_17.jpg
            ……
        n02085936-Maltese_dog/
            n02085936_37.jpg
            n02085936_66.jpg
            ……
        ……
```

每张狗的图片大小可能不一样,我们在下面的代码中会分析图片的大小,最后归一化处理图片数据以便于神经网络模型的训练。

10.1.1 环境准备

- numpy = 1.14.6。
- matplotlib = 2.1.2。
- sklearn = 0.19.2。
- keras = 2.1.6。
- tqdm = 4.27.0。
- tensorflow = 1.11.0。
- tensorflow_hub = 0.2.0。

10.1.2 数据可视化

首先,使用 sklearn 库下加载文件的函数来加载狗的完整数据集,并且对狗的 120 个品种进行独热编码处理。

```python
from sklearn.datasets import load_files
from keras.utils import np_utils
import numpy as np
from glob import glob
# 共有 120 个品种的狗
num_classes = 120
# 定义加载数据集的函数
def load_dataset(path):
    # 通过 sklearn 提供的 load_files() 方法加载文件
    # 返回一个类字典对象,包含文件相对路径和文件所属编号
    data = load_files(path)
    # 将文件路径转变成 NumPy 对象
    dog_files = np.array(data['filenames'])
    # 狗的每张图片都按照顺序排成列表
    raw_targets = np.array(data['target'])
    # 通过 to_categorical() 方法将文件所属编号转换成二进制类别矩阵(就是 one-hot encoding)
    dog_targets = np_utils.to_categorical(raw_targets, num_classes)
    # 返回所有图片文件路径,图片文件编号和图片文件的二进制类别矩阵
    return dog_files, raw_targets, dog_targets

# 加载数据集
dog_filepaths, dog_raw_targets, dog_targets = load_dataset('Images/')
# 加载狗的品种名称列表
# glob 是一个文件操作相关的模块,通过指定的匹配模式,返回相应的文件或文件夹路径
# 这里的操作就是返回 Images 目录下的所有文件夹
# 最后通过列表推导式遍历每个文件路径字符串,并截取狗品种名称字符串
```

```
dogpath_prefix_len = len('Images/n02085620-')
dog_names = [item[dogpath_prefix_len:] for item in sorted(glob("Images/*"))]
print('狗的品种有{}种。'.format(len(dog_names)))
print('狗的图片一共有{}张。\n'.format(len(dog_filepaths)))
```

输出如下。

狗的品种有 120 种。
狗的图片一共有 20580 张。

查看 dog_names 里的前 5 种狗的名称,代码如下。

```
dog_names[:5]
```

输出如下。

```
['Chihuahua', 'Japanese_spaniel', 'Maltese_dog', 'Pekinese', 'Shih-Tzu']
```

查看 dog_filepaths 里的前 5 条数据,代码如下。

```
dog_filepaths[:5]
```

输出信息就是每张图片的相对路径,如下所示。

```
array(['Images/n02098105-soft-coated_wheaten_terrier/n02098105_2842.jpg',
       'Images/n02091467-Norwegian_elkhound/n02091467_4110.jpg',
       'Images/n02111129-Leonberg/n02111129_2617.jpg',
       'Images/n02088238-basset/n02088238_11281.jpg',
       'Images/n02089867-Walker_hound/n02089867_1243.jpg'], dtype='<U67')
```

查看 dog_raw_targets 里的前 10 条数据,代码如下。

```
dog_raw_targets[:10]
```

输出信息是数组里的每个元素的狗编号,如下。

```
array([ 51,  23, 103,  10,  15, 105,  49,  11,  58, 105])
```

最后,再来看 dog_targets 里的前 3 个狗品种的独热编码的数据,代码如下。

```
dog_targets[:3]
```

从如下输出中,我们可以看到在这 3 个狗的品种数据中,如果当前这个品种是代表自己,值就是 1;如果不是,值就是 0。这就是独热编码后的数据形态。我们也看到这个[3, 120]的矩阵中,每个矩阵代表一个狗的编号,就是数值是 1 的那个。

```
array([[0., 0., 0., 0., 0., 0., 0., 0., 0., 0., 0., 0., 0., 0., 0.,
        0., 0., 0., 0., 0., 0., 0., 0., 0., 0., 0., 0., 0., 0., 0.,
        0., 0., 0., 0., 0., 0., 0., 0., 0., 0., 0., 0., 0., 0., 0.,
        0., 0., 0., 1., 0., 0., 0., 0., 0., 0., 0., 0., 0., 0., 0.,
        0., 0., 0., 0., 0., 0., 0., 0., 0., 0., 0., 0., 0., 0., 0.,
        0., 0., 0., 0., 0., 0., 0., 0., 0., 0., 0., 0., 0., 0., 0.,
        0., 0., 0., 0., 0., 0., 0., 0., 0., 0., 0., 0., 0., 0., 0.,
        0., 0., 0., 0., 0., 0., 0.],
       [0., 0., 0., 0., 0., 0., 0., 0., 0., 0., 0., 0., 0., 0., 0.,
```

```
             0., 0., 0., 0., 0., 0., 0., 1., 0., 0., 0., 0., 0., 0.,
             0., 0., 0., 0., 0., 0., 0., 0., 0., 0., 0., 0., 0., 0.,
             0., 0., 0., 0., 0., 0., 0., 0., 0., 0., 0., 0., 0., 0.,
             0., 0., 0., 0., 0., 0., 0., 0., 0., 0., 0., 0., 0., 0.,
             0., 0., 0., 0., 0., 0., 0., 0., 0., 0., 0., 0., 0., 0.,
             0., 0., 0., 0., 0., 0., 0.],
            [0., 0., 0., 0., 0., 0., 0., 0., 0., 0., 0., 0., 0., 0.,
             0., 0., 0., 0., 0., 0., 0., 0., 0., 0., 0., 0., 0., 0.,
             0., 0., 0., 0., 0., 0., 0., 0., 0., 0., 0., 0., 0., 0.,
             0., 0., 0., 0., 0., 0., 0., 0., 0., 0., 0., 0., 0., 0.,
             0., 0., 0., 0., 0., 0., 1., 0., 0., 0., 0., 0., 0., 0.,
             0., 0., 0., 0., 0., 0., 0., 0.]], dtype=float32)
```

分割训练集、验证集和测试集数据，代码如下。

```
from sklearn.model_selection import train_test_split
# 为了训练得更快些，也考虑到一些读者的计算机性能不高，我们就用前 9000 张狗的图片吧
# 如果读者的计算机性能还不错，那就注释这两行，直接训练所有的图片数据
dog_filepaths = dog_filepaths[:9000]
dog_targets = dog_targets[:9000]
# 分割训练数据集和测试数据集
X_train, X_test, y_train, y_test = train_test_split(dog_filepaths, dog_targets, test_size=0.2)
# 将测试数据集数据分割一半给验证数据集
half_test_count = int(len(X_test) / 2)
X_valid = X_test[:half_test_count]
y_valid = y_test[:half_test_count]
X_test = X_test[half_test_count:]
y_test = y_test[half_test_count:]
print("X_train.shape={}, y_train.shape={}.".format(X_train.shape, y_train.shape))
print("X_valid.shape={}, y_valid.shape={}.".format(X_valid.shape, y_valid.shape))
print("X_test.shape={}, y_test.shape={}.".format(X_test.shape, y_test.shape))
```

输出如下。

```
X_train.shape=(7200,), y_train.shape=(7200, 120).
X_valid.shape=(900,), y_valid.shape=(900, 120).
X_test.shape=(900,), y_test.shape=(900, 120).
```

接着，我们随机从训练数据集中抽取 9 张狗的图片预览。随机从某个范围中选取一些数字使用 NumPy 下 random 模块的 choice()函数。因为要显示图片，所以我们使用 matplotlib 包下 image 模块的 imread()函数读取，然后在 Axes 对象中显示。最后看到输出的每张狗图片大小不一。

```
import matplotlib.pyplot as plt
# 设置 matplotlib 在绘图时的默认样式
plt.style.use('default')
```

```python
from matplotlib import image
# 查看随机 9 张狗的图片
def draw_random_9_dog_images():
    # 创建 9 个绘图对象,3 行 3 列
    fig, axes = plt.subplots(nrows=3, ncols=3)
    # 设置绘图的总容器大小
    fig.set_size_inches(10, 9)
    # 随机选择 9 个数,也就是 9 个品种的狗(可能重复,且每次都不一样)
    random_9_nums = np.random.choice(len(X_train), 9)
    # 从训练集中选出 9 张图
    random_9_imgs = X_train[random_9_nums]
    print(random_9_imgs)
    # 根据这随机的 9 张图片路径,截取狗相应的品种名称
    imgname_list = []
    for imgpath in random_9_imgs:
        imgname = imgpath[dogpath_prefix_len:]
        imgname = imgname[:imgname.find('/')]
        imgname_list.append(imgname)
    index = 0
    for row_index in range(3): # 行
        for col_index in range(3): # 列
            # 读取图片的数值内容
            img = image.imread(random_9_imgs[index])
            # 获取绘图 Axes 对象,根据[行索引, 列索引]
            ax = axes[row_index, col_index]
            # 在 Axes 对象上显示图片
            ax.imshow(img)
            # 在绘图对象上设置狗品种名称
            ax.set_xlabel(imgname_list[index])
            # 索引加 1
            index += 1

draw_random_9_dog_images()
```

输出的文字信息如下,输出的 9 张狗的图片如图 10.1 所示。

```
['Images/n02089078-black-and-tan_coonhound/n02089078_3893.jpg'
 'Images/n02090622-borzoi/n02090622_3210.jpg'
 'Images/n02096294-Australian_terrier/n02096294_4295.jpg'
 'Images/n02092339-Weimaraner/n02092339_3698.jpg'
 'Images/n02088466-bloodhound/n02088466_10724.jpg'
 'Images/n02108089-boxer/n02108089_8739.jpg'
 'Images/n02102040-English_springer/n02102040_1519.jpg'
 'Images/n02086079-Pekinese/n02086079_19690.jpg'
 'Images/n02094433-Yorkshire_terrier/n02094433_9618.jpg']
```

图 10.1 随机 9 张狗的图片预览

接着，我们来分析狗图片的宽和高都在什么范围。首先，我们来读取所有狗品种的图片的 shape。我们这里要对数据集数据进行分析，所以就对变量 dog_filepaths 进行遍历，再通过 matplotlib 包下的 image 模块的 imread()函数读取图片，并获取它的大小。

```
# 对数据集进行遍历，读取每张图片，并获取它的大小，
# 最后返回的图片 shape 存储在变量 dogs_shape_list 列表里
dogs_shape_list = []
for filepath in dog_filepaths:
    shape = image.imread(filepath).shape
    if len(shape) == 3:
        dogs_shape_list.append(shape)
dogs_shapes = np.asarray(dogs_shape_list)
print("总共{}张。".format(len(dogs_shapes)))
print("随机抽取3张图片的维度是{}。".format(dogs_shapes[np.random.choice(len(dogs_shapes), 3)]))
```

输出如下。

```
总共 9000 张。
随机抽取 3 张图片的维度是[[500 361   3]
 [179 237   3]
 [375 500   3]]。
```

然后,看图片的平均宽度和高度。mean()函数是 NumPy 包下求平均值的。切片[:,0]表示冒号前后没有值,就是所有的数据;冒号后面有个逗号,并跟着值,就是说该数组里的元素是有长度的。我们这里想要取出宽和高,那宽就是索引 0 的位置,高就是索引 1 的位置。

```
dogs_mean_width = np.mean(dogs_shapes[:,0])
dogs_mean_height = np.mean(dogs_shapes[:,1])
print("狗的图片的平均宽: {:.1f} * 平均高: {:.1f}。".format(dogs_mean_width, dogs_mean_height))
```

输出如下。

狗的图片的平均宽: 384.5 * 平均高: 439.1。

然后我们通过散点图看图片的大小分布。

```
# 显示图片的宽和高的范围
# 参数 1: 图片的宽
# 参数 2: 图片的高
# 参数 3: 每个数据点用"o"标记
plt.plot(dogs_shapes[:, 0], dogs_shapes[:, 1], "o")
# 设置图题
plt.title("The Image Sizes of All Categories of Dog")
# 设置图的 X 轴标题
plt.xlabel("Image Width")
# 设置图的 Y 轴标题
plt.ylabel("Image Height")
```

输出结果如图 10.2 所示。

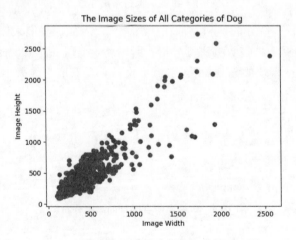

图 10.2 数据集所有图片的宽和高的分布图

我们再通过直方图看图片的宽度分布。

```
plt.hist(dogs_shapes[:, 0], bins=100)
plt.title("Dog Images Width Distribution")
```

输出结果如图 10.3 所示。

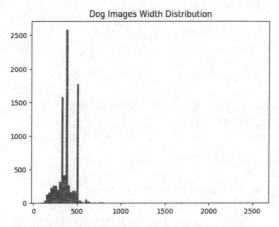

图 10.3　数据集所有图片的宽度分布图

我们继续通过直方图看图片的高度分布，如下代码。

```
plt.hist(dogs_shapes[:, 1], bins=100)
plt.title("Dog Images Height Distribution")
```

输出结果如图 10.4 所示。

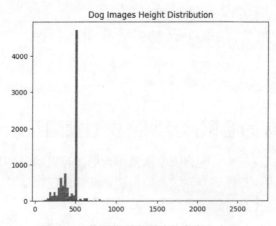

图 10.4　数据集所有图片的高度分布图

10.1.3　预处理数据

在训练模型前，我们需要对图片数据进行预处理，将每张图片都转换成标准的(1, 224, 224, 3)的四维张量；RGB 色彩模式的图片有红、绿、蓝 3 个颜色通道，所以这里的 3 就代表（红、绿和蓝）3 层；我们对图像的具体数值都除以 255 以进行归一化处理。这里我们用到了 tqdm 包，这是一个显示进度条的工具包，在对下面的所有数据进行归一化处理时，可能会花费几分钟,如果不加上进度条可能会给读者造成程序假死的表面现象。

```python
from keras.preprocessing import image
from tqdm import tqdm
# 定义一个函数,将每张图片都转换成标准大小(224, 224, 3)
def path_to_tensor(img_path):
    # 加载图片
    # 用的是 PIL 库加载图片对象,通过 load_img()方法返回的就是一个 PIL 对象
    img = image.load_img(img_path, target_size=(224, 224, 3))
    # 将 PIL 图片对象类型转化为格式(224, 224, 3)的三维张量
    x = image.img_to_array(img)
    # 将 3 维张量转化格式为(1, 224, 224, 3)的四维张量并返回
    return np.expand_dims(x, axis=0)
# 定义一个函数,将数组里所有路径的图片都转换成图像数值类型并返回
def paths_to_tensor(img_paths):
    # tqdm 模块表示使用进度条显示,传入一个所有图片的数组对象
    # 将所有图片的对象都转换成 numpy 数值对象张量后,并返回成数组
    list_of_tensors = [path_to_tensor(img_path) for img_path in tqdm(img_paths)]
    # 将对象垂直堆砌排序摆放
    return np.vstack(list_of_tensors)
from PIL import ImageFile
# 为了防止 PIL 读取图片对象时出现 IO 错误,设置截断图片为 True
ImageFile.LOAD_TRUNCATED_IMAGES = True
# 将所有图片都转换成标准大小的数值图像对象,然后除以 255,进行归一化处理
# RGB 的颜色值,最大为 255,最小为 0
# 对训练集数据进行处理
train_tensors = paths_to_tensor(X_train).astype(np.float32) / 255
# 对验证集数据进行处理
valid_tensors = paths_to_tensor(X_valid).astype(np.float32) / 255
# 对测试集数据进行处理
test_tensors = paths_to_tensor(X_test).astype(np.float32) / 255
```

等待几分钟,训练集、验证集和测试集数据就能归一化处理完毕。

10.2 基于 Keras 的卷积神经网络模型预测

我们通过自己设计的卷积神经网络模型来预测狗的品种识别。此神经网络结构比较简单,最终的结果可能会不精准,但是值得尝试,因为这才能体现迁移学习(Transfer Learning)的优秀表现能力。在下一节中会讲解通过迁移学习对狗进行品种识别。

10.2.1 创建模型

我们通过 Sequential 对象创建模型,添加一个输入层、一个中间卷积层、一个输出层。其中卷积核都设置为 2×2 的大小,步长都是 1。除了输出层的激活函数是 softmax,其他层的激活函数都是 ReLU。最后输出层是 120 个品种输出。

```python
from keras.layers import Conv2D, MaxPooling2D, GlobalAveragePooling2D
from keras.layers import Dropout, Flatten, Dense
from keras.models import Sequential
```

```python
# 创建 Sequential 模型
model = Sequential()
# 创建输入层，输入层必须传入 input_shape 参数以表示图像大小，深度是 16
model.add(Conv2D(filters=16, kernel_size=(2, 2), strides=(1, 1), padding='same',
                 activation='relu', input_shape=train_tensors.shape[1:]))
# 添加最大池化层，大小为 2×2，有效范围默认是 valid，就是说，不够 2×2 的大小的空间数据就丢弃
model.add(MaxPooling2D(pool_size=(2, 2)))
# 添加 Dropout 层，每次丢弃 20% 的网络节点，防止过拟合
model.add(Dropout(0.2))
# 添加卷积层，深度是 32，内核大小是 2×2，跨步是 1×1，有效范围是 same 则表示不够数据范围的就用 0 填充
model.add(Conv2D(filters=32, kernel_size=(2, 2), strides=(1, 1), padding='same',
activation='relu'))
# 添加卷积层，深度是 64
model.add(Conv2D(filters=64, kernel_size=(2, 2), strides=(1, 1), padding='same',
activation='relu'))
model.add(MaxPooling2D(pool_size=(2, 2)))
model.add(Dropout(0.2))
# 添加全局平均池化层
model.add(GlobalAveragePooling2D())
# 添加 Dropout，每次丢弃 50%
model.add(Dropout(0.5))
# 添加输出层，120 个品种输出
model.add(Dense(num_classes, activation="softmax"))

# 打印输出网络模型架构
model.summary()
```

输出结果如图 10.5 所示。

```
Layer (type)                 Output Shape              Param #
=================================================================
conv2d_1 (Conv2D)            (None, 224, 224, 16)      208
_____
max_pooling2d_1 (MaxPooling2 (None, 112, 112, 16)      0
_____
dropout_1 (Dropout)          (None, 112, 112, 16)      0
_____
conv2d_2 (Conv2D)            (None, 112, 112, 32)      2080
_____
max_pooling2d_2 (MaxPooling2 (None, 56, 56, 32)        0
_____
dropout_2 (Dropout)          (None, 56, 56, 32)        0
_____
conv2d_3 (Conv2D)            (None, 56, 56, 64)        8256
_____
max_pooling2d_3 (MaxPooling2 (None, 28, 28, 64)        0
_____
dropout_3 (Dropout)          (None, 28, 28, 64)        0
_____
global_average_pooling2d_1 ( (None, 64)                0
_____
dropout_4 (Dropout)          (None, 64)                0
_____
dense_1 (Dense)              (None, 120)               7800
=================================================================
Total params: 18,344
Trainable params: 18,344
Non-trainable params: 0
```

图 10.5　卷积神经网络模型架构图

10.2.2 训练模型

我们先使用多类别的交叉熵编译模型。然后通过 ModelCheckpoint 对象来保存每次训练时的最佳检查点，将检查点文件保存到 saved_models 目录下；一共要训练 20 个批次，每个批次训练大小是 20。

```
# 编译模型
model.compile(optimizer='rmsprop', loss='categorical_crossentropy', metrics=['accuracy'])
# 导入检查点模块
from keras.callbacks import ModelCheckpoint
# 创建一个检查点对象，指定存储路径，手动创建一个 saved_models 目录
# 参数 save_best_only 表示只存储最佳的训练
checkpointer = ModelCheckpoint(filepath='saved_models/weights.best.from_scratch.hdf5',
                               verbose=1,
                               save_best_only=True)
# 训练模型
epochs = 20
model.fit(train_tensors,
          y_train,
          validation_data=(valid_tensors, y_valid),
          epochs=epochs,
          batch_size=20,
          callbacks=[checkpointer],
          verbose=1)
```

我们只输出训练开始和结尾的日志。

```
Train on 7200 samples, validate on 900 samples
Epoch 1/20
7200/7200 [==============================] - 20s 3ms/step - loss: 4.7886 - acc: 0.0093 - val_loss: 4.7861 - val_acc: 0.0122
Epoch 00001: val_loss improved from inf to 4.78609, saving model to saved_models/weights.best.from_scratch.hdf5
Epoch 2/20
7200/7200 [==============================] - 16s 2ms/step - loss: 4.7824 - acc: 0.0117 - val_loss: 4.7833 - val_acc: 0.0078
……
Epoch 00019: val_loss improved from 4.65432 to 4.64252, saving model to saved_models/weights.best.from_scratch.hdf5
Epoch 20/20
7200/7200 [==============================] - 16s 2ms/step - loss: 4.6085 - acc: 0.0296 - val_loss: 4.6347 - val_acc: 0.0356
Epoch 00020: val_loss improved from 4.64252 to 4.63467, saving model to saved_models/weights.best.from_scratch.hdf5
<keras.callbacks.History at 0x7f0a0840b358>
```

10.2.3 模型评估

我们先加载训练时保存的模型检查点，然后通过预测的分值来计算准确度。

```
## 加载具有最好验证权重的模型
model.load_weights('saved_models/weights.best.from_scratch.hdf5')
# 获取测试数据集中每一个图像所预测的狗品种的 index
dog_breed_predictions = [np.argmax(model.predict(np.expand_dims(tensor, axis=0))) for tensor in test_tensors]
# 测试准确率
test_accuracy = 100*np.sum(np.array(dog_breed_predictions)==np.argmax(y_test, axis=1))/len(dog_breed_predictions)
print('Test Accuracy: {:.4f}'.format(test_accuracy))
```

输出如下。

```
Test Accuracy: 3.2222
```

10.3　基于 Keras 的 InceptionV3 预训练模型实现预测

预训练模型有很多，InceptionV3 是其中一个，它由谷歌团队从 ImageNet 的 1000 个类别的超大型数据集训练而来，表现优异，经常用来做计算机视觉方面的迁移学习研究和应用。使用预训练模型的好处有两点：一是该模型本身就可以预测图像，因为它是基于 1000 个类别训练而来；二是迁移学习的技术可以让我们不需要从头开始训练分类模型，我们在一定数据量的情况下只需要微调就可以满足需求。

10.3.1　模型函数声明

我们先定义预测后的绘图对象函数 plot_bar()和加载图片到数值类型的函数 load_img()，模型对输入的图片要求的大小是 224×224，所以对加载图片的函数的参数 target_size 要指定输入图片大小。

```
from keras.preprocessing import image
def plot_bar(predictions):
    types = [pred[1] for pred in predictions]
    probs = [pred[2] for pred in predictions]
    plt.barh(np.arange(len(probs)), probs)
    _ = plt.yticks(np.arange(len(predictions)), types)
    plt.show()

def load_img(img_path):
    img = image.load_img(img_path, target_size=(224, 224))
    x = image.img_to_array(img)
    x = np.expand_dims(x, axis=0)
    return x
```

然后，创建 InceptionV3 模型，如果本地没有下载 InceptionV3 模型，那么 Keras 会自己主动下载，下载后的模型会保存在当前登录用户的根目录的一个扩展名为 keras 的隐藏文件里。最后加载狗的图片，直接预测狗的品种。

```
import matplotlib
from keras.applications.inception_v3 import InceptionV3, preprocess_input, decode_predictions
```

```python
def predicts_inceptionV3(img_path):
    """
    通过预训练模型 InceptionV3 直接预测
    """
    # 加载图片
    x = load_img(img_path)
    # 预处理图片数值
    x = preprocess_input(x)
    
    # 加载 InceptionV3 预训练模型
    # 如果本地不存在该模型, 就自动下载
    model = InceptionV3(weights='imagenet')
    # 预测狗的品种
    preds = model.predict(x)
    # 对预测的值解码, 只显示前 5 个最大的值
    predictions = decode_predictions(preds, top=5)[0]
    
    # 绘直方图显示
    plot_bar(predictions)
    
    # 绘制原始图和预测概率图
    # 创建一个绘图对象
    fig, ax = plt.subplots()
    # 设置绘图的总容器大小
    fig.set_size_inches(5, 5)
    # 取出预测的品种名称和它对应的概率值
    fig_title = "".join(["{}: {:.5f}%\n".format(pred[1], pred[2]*100) for pred in predictions])
    # 设置在图像旁边的预测注解文字
    ax.text(1.01, 0.7,
            fig_title,
            horizontalalignment='left',
            verticalalignment='bottom',
            transform=ax.transAxes)
    # 读取图片的数值内容
    img = matplotlib.image.imread(img_path)
    # 在 Axes 对象上显示图像
    ax.imshow(img)
```

10.3.2 预测单张图片

我们会先从图片数据集中随机选取一张图片进行测试预训练模型，这里就以 EntleBucher（恩特雷布赫山地犬）举例。

```python
# 从数据集中选一张图片来识别
img_path = "Images/n02108000-EntleBucher/n02108000_93.jpg"
predicts_inceptionV3(img_path)
```

输出结果如图 10.6 所示。

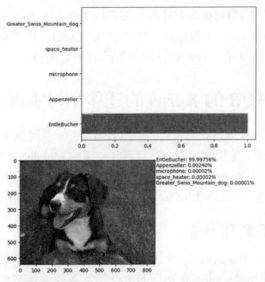

图 10.6 通过 InceptionV3 直接预测狗的品种

模型最终的预测结果也是 EntleBucher。在 ImageNet 数据集里的 1000 种图片类别中包含一些狗的品种，所以通过 InceptionV3 的预训练模型预测得很精确。假如，作者通过该模型来识别自己的图片，那它会把作者识别成哪些类别呢？

```
img_path = "victor_test.jpeg"  # 作者的图像，该图像在 dogImages 的同级目录
predicts_inceptionV3(img_path)
```

输出结果如图 10.7 所示。

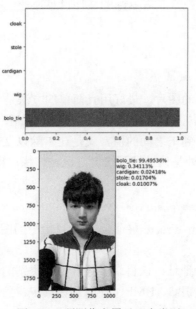

图 10.7 预测作者属于哪个类别

由于在 InceptionV3 模型训练时，并没有在原始数据集上加入作者的图片和类别，所以识别的结果完全不对。InceptionV3 模型尽可能去寻找和识别与作者图片相似的类别，最终得出 bolo_tie（波洛领带），这就是一种领带，跟作者的衣领有些像。

10.4 基于 TFHUB 的 Keras 的迁移学习实现预测

通过迁移学习我们可以训练我们自己的图片数据，训练速度很快，不用从头开始训练。因为我们是基于 MobileNet 预训练模型，因此在训练时会冻结 MobileNet 神经网络中的所有层，在训练模型前，要先去掉预训练模型的最后的分类层，然后添加我们自己的分类层，再进行训练模型。我们采用的是 TFHUB 模块中心的 MobileNet v2 模型来训练、预测和测试。

10.4.1 数据集下载和准备

我们可以通过 Keras 模块下的 utils 来下载文件，只需要指定一个文件 URL 即可。这里我们就是下载狗的图片数据集，其实在本章的一开始就下载过该文件，但是如果想用代码下载图片压缩包文件，可以使用 get_file()函数。其中参数 untar 的意思是文件下载完后，是否解压缩文件，传入 True 表示要解压缩。

```
dataset_path = \
tf.keras.utils.get_file("Images",
            "http://vision.stanford.edu/aditya86/ImageNetDogs/images.tar",
            untar=True)
```

等待几分钟，使文件下载和解压完成。默认情况下，该函数会把文件下载到 Keras 的默认目录下，我们可以通过该函数来查看返回的文件目录地址。

```
print(dataset_path)
```

输出如下。

```
'/root/.keras/datasets/Images'
```

现在，我们来加载目录中的图片，最好用且简单的方法是使用 Keras 的预处理模块下的 ImageDataGenerator 类，然后传入 rescale 参数，表示所有的图片的矩阵数值都被用做计算。我们定义图片的计算值是在 0 到 1 之间的值，就是让 1 除以 255（RGB 颜色值的范围最小是 0，最大是 255）。这样做的好处是让神经网络更好地被训练，也是 TensorFlow Hub 期待的图片的浮点值输入。

```
image_generator = tf.keras.preprocessing.image.ImageDataGenerator(rescale=1/255)
```

然后，我们调用 flow_from_directory()函数加载目录中的图片的增强数据，它的参数解释分别如下。

- 参数 1：图片的目录地址，该目录下包含每个类别的子目录，子目录下包含具体的图片，格式有 PNG、JPG、BMP、PPM 以及 TIF。我们这里的图片全部都是 JPG 格式的。

- 参数 2：所有的图片的输入大小最终会被裁剪为指定的大小。这里指定 target_size 为 224×224。
- 参数 3：数据的批次的大小，默认值是 32，也可以设定为 64、128、256 等，这取决于您的计算机的性能或模型表现。
- 参数 4：所有的图像要被转换的通道是什么。这里指定 color_mode 为 rgb，默认值也是 rgb；除了 rgb，还可以设置为 grayscale 或 rgba。
- 参数 5：输入的 Label（标签）的类型方式。这里指定 class_mode 为 categorical，意思是类别。它的类别类型还可以是 binary 或 sparse。

```
input_img_size = (224, 224)
image_data = image_generator.flow_from_directory("/root/.keras/datasets/Images",
                                                 target_size=input_img_size,
                                                 batch_size=32,
                                                 color_mode='rgb',
                                                 class_mode='categorical')
for image_batch,label_batch in image_data:
  print("Image batch shape: ", image_batch.shape)
  print("Label batch shape: ", label_batch.shape)
  break
steps_per_epoch = image_data.samples//image_data.batch_size
print("图片样本有{}张, batch size 定在{}。".format(image_data.samples, image_data.batch_size))
print("需要训练{}次就能做一次完整的训练。".format(steps_per_epoch))
```

输出如下。

```
Found 20580 images belonging to 120 classes.
Image batch shape:  (32, 224, 224, 3)
Label batch shape:  (32, 120)
```

图片样本有 20580 张，batch size 定在 32。
需要训练 643 次就能做一次完整的训练。

10.4.2 预训练模型下载

我们使用的是 mobilenet v2 预训练模型，所以需要先把它下载到本地目录。TensorFlow 的 Hub 中心提供了非常多的预训练模型，模型的分类也很清晰，并且做了顶层的删除处理，有的是到特征向量层（Feature Vector），有的是到分类层（Classification）。本次我们用到的模型是到特征向量层。

```
import tensorflow_hub as tf_hub
feature_extractor_url = "https://tfhub.dev/google/imagenet/mobilenet_v2_100_224/feature_vector/2"
# feature_extractor_url = "/tmp/tfhub_modules/adfe0cf8d843e3588bfb9602e32a718b12212904"
def feature_extractor(x):
  feature_extractor_module = tf_hub.Module(feature_extractor_url)
  return feature_extractor_module(x)
```

我们通过 tf_hub 的 Module 来导入一个预训练模型，导入时可以传入一个 URL，也可以是本地文件系统的磁盘目录地址。默认情况下，Module 会把指定的 URL 的模型下载到本地系统的临时目录下，所以这里在下载时也是到了系统的 /tmp/ 目录下。发现 TFHUB 自动创建了一个目录，叫作 tfhub_modules，然后在该目录下，就是下载的预训练模型。有时候我们需要多个预训练模型，在下载时，也都默认在这个目录下了。

在第一次执行此代码时，我们要去网络请求下载，第二次就是从本地计算机的临时目录下加载了。该模型已经放在本篇章的代码库的说明里，请看说明下载。

10.4.3 创建模型

我们使用 Keras 下的 Lambda 类来初始化，将任意表达式包装成一个 Keras 的 Layer 对象。传入两个参数，解释分别如下。

- 参数 1：是一个函数（function），该函数可以被评估计算。函数 feature_extractor 就是在上一小节中定义的。
- 参数 2：模型第一层的输入的图片的形状大小。这里指定 input_shape 是 3 个通道的图片，它的 shape 是 224×224×3。

```
features_extractor_layer = tf.keras.layers.Lambda(feature_extractor, input_shape=input_img_size + (3,))
```

然后将该预训练模型的所有的层都设置成不可训练。

```
features_extractor_layer.trainable = False
```

再通过 Keras 的 Sequential 类来创建模型，新模型的前面的层就是预训练模型的那些层，输出的是一个全连接层（Dense），输出的大小是狗的总品种数。总品种数可以通过变量 image_data 的 num_classes 来获取，这个属性是在加载图片数据集时的子目录数；然后多类别分类就用 softmax 激活函数。

```
model = tf.keras.Sequential([
  features_extractor_layer,
  tf.keras.layers.Dense(image_data.num_classes, activation='softmax')
])
model.summary()
```

输出如下。

```
Layer (type)                 Output Shape              Param #
=================================================================
lambda (Lambda)              (None, 1280)              0
_____
dense (Dense)                (None, 120)               153720
=================================================================
Total params: 153,720
Trainable params: 153,720
Non-trainable params: 0
_____
```

10.4.4 训练模型

当我们使用 TensorFlow 下的 Keras 时，TFHUB 的 modules 需要手动初始化。

```
import tensorflow.keras.backend as K
sess = K.get_session()
init = tf.global_variables_initializer()
sess.run(init)
```

然后我们编译模型，来配置训练的过程。优化器就使用 Adam，损失类型用多类别交叉熵，度量方式使用 accuracy。

```
model.compile(
  optimizer=tf.train.AdamOptimizer(),
  loss='categorical_crossentropy',
  metrics=['accuracy'])
```

训练模型。在调用 fit() 函数训练时，传入的参数解释分别如下。

- 参数 1：图片数据集，就是一批批形如（32, 224, 224, 3）的维度数值数据。
- 参数 2：迭代训练次数。这里设置 epochs 为 5。
- 参数 3：每个 epoch 训练的步数。
- 参数 4：用于记录训练时的 loss 和 accuracy。所以我们需要自定义一个 Collect BatchStats 类，每结束一个 batch 训练，就记录一次。

```
class CollectBatchStats(tf.keras.callbacks.Callback):
  def __init__(self):
    self.batch_losses = []
    self.batch_acc = []
  def on_batch_end(self, batch, logs=None):
    self.batch_losses.append(logs['loss'])
    self.batch_acc.append(logs['acc'])

batch_stats = CollectBatchStats()
model.fit((item for item in image_data),
          epochs=5,
          steps_per_epoch=steps_per_epoch,
          callbacks = [batch_stats]
          )
```

训练模型时的输出信息如下。

```
Epoch 1/5
643/643 [==============================] - 128s 199ms/step - loss: 1.0509 - acc: 0.7234
Epoch 2/5
643/643 [==============================] - 127s 198ms/step - loss: 0.4209 - acc: 0.8676
Epoch 3/5
643/643 [==============================] - 127s 198ms/step - loss: 0.2879 - acc: 0.9124
Epoch 4/5
643/643 [==============================] - 126s 196ms/step - loss: 0.2057 - acc: 0.9417
Epoch 5/5
643/643 [==============================] - 125s 195ms/step - loss: 0.1532 - acc: 0.9610
```

对训练时记录的 Loss 进行绘图显示；如果希望 y 轴显示最低和最高的数值，那就设置

plt.ylim()函数的参数。

```
import matplotlib.pylab as plt
plt.figure()
plt.ylabel("Loss")
plt.xlabel("Training Steps")
# plt.ylim([0,2])
plt.plot(batch_stats.batch_losses)
```

输出结果如图 10.8 所示。

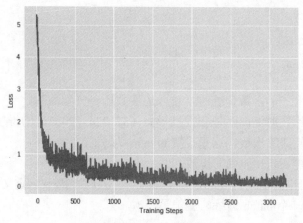

图 10.8　Loss 曲线图

对训练时记录的 Accuracy 进行绘图显示。

```
plt.figure()
plt.ylabel("Accuracy")
plt.xlabel("Training Steps")
# plt.ylim([0,1])
plt.plot(batch_stats.batch_acc)
```

输出结果如图 10.9 所示。

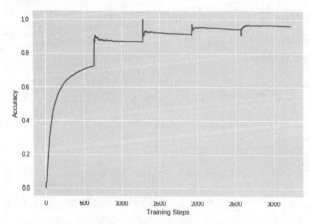

图 10.9　Accuracy 的曲线图

10.4.5 测试模型

我们用 9 张狗的图片数据测试模型的表现。首先我们获取所有狗品种的名称。

```
# 获取狗品种名称的顺序列表
label_names = sorted(image_data.class_indices.items(), key=lambda pair:pair[1])
# 仅返回识别出的狗品种的名称
label_names = np.array([key.title() for key, value in label_names])
# 查看前 15 个名称
label_names[:15]
```

输出如下。

```
array(['N02085620-Chihuahua', 'N02085782-Japanese_Spaniel',
       'N02085936-Maltese_Dog', 'N02086079-Pekinese',
       'N02086240-Shih-Tzu', 'N02086646-Blenheim_Spaniel',
       'N02086910-Papillon', 'N02087046-Toy_Terrier',
       'N02087394-Rhodesian_Ridgeback', 'N02088094-Afghan_Hound',
       'N02088238-Basset', 'N02088364-Beagle', 'N02088466-Bloodhound',
       'N02088632-Bluetick', 'N02089078-Black-And-Tan_Coonhound'],
      dtype='<U40')
```

然后，我们预测狗的图片所对应的类别的概率。

```
result_batch = model.predict(image_batch)
```

将识别到的最大概率数组取出与之对应的狗品种的名称。

```
# 取出最大的概率
max_index = np.argmax(result_batch, axis=-1)
labels_batch = label_names[max_index]
```

将识别到最大概率数组取出与之对应的狗品种的名称的概率值。

```
result_batch_probs = [l[max_index[i]] for i, l in enumerate(result_batch)]
labels_batch[:10]  # 预览前 10 个
```

输出如下。

```
array(['N02098413-Lhasa', 'N02098105-Soft-Coated_Wheaten_Terrier',
       'N02092002-Scottish_Deerhound',
       'N02095314-Wire-Haired_Fox_Terrier', 'N02091635-Otterhound',
       'N02086910-Papillon', 'N02096585-Boston_Bull',
       'N02106550-Rottweiler', 'N02088632-Bluetick',
       'N02099712-Labrador_Retriever'], dtype='<U40')
```

然后输出前 10 个预测品种的概率值。

```
result_batch_probs[:10]  # 预览前 10 个
```

输出如下。

```
[0.58318573,
 0.58909327,
 0.5498832,
 0.99628145,
```

```
0.9984805,
0.99941444,
0.946167,
0.94342893,
0.9836289,
0.5137576]
```

绘图显示狗的图片,并对应名称和概率。

```python
# 绘图的大小
plt.figure(figsize=(10, 10))
# 绘图的标题
plt.suptitle("Model predictions")
# 循环 9 次,显示 9 张
for n in range(9):
    # 取出图片的第 n 张
    plt.subplot(3,3,n+1)
    # 显示图片
    plt.imshow(image_batch[n])
    # 设置图片的标题,是"品种名称:概率"
    name_prefix = len("N02085620-")
    # 概率值转换成百分比的值后,保留两位小数
    prob = round(result_batch_probs[n] * 100, 2)
    plt.title(labels_batch[n][name_prefix:] + ":" + str(prob) + "%")
    # 不显示网格
    plt.axis('off')
```

输出结果如图 10.10 所示。

图 10.10　9 张预测狗的图像

10.4.6　模型预测单张图片

现在，我们来看测试模型对单张图片的表现，我们随机取一张狗的图片，这里用 Maltese（马尔他犬）的图片作为例子来测试模型对它的识别会得出多大的概率。然后我们又拿作者的图片作为例子，又会被识别成哪种狗呢？

```
import matplotlib
from tensorflow.keras.preprocessing import image
from tensorflow.keras.applications.mobilenet import preprocess_input, decode_predictions

# 加载图片，返回图片数组
def load_img(img_path):
    # 加载进来的图片都以 224×224 的表示
    img = image.load_img(img_path, target_size=(224, 224))
    x = image.img_to_array(img)
    x = np.expand_dims(x, axis=0)
    return x

# 预测狗的品种
def predict_dog_breed(model, dog_names, img_path):
    # 加载图片
    x = load_img(img_path)
    # 图片预处理
    x = preprocess_input(x)
    # 模型预测
    predictions = model.predict(x)
    # 取出预测数值
    prediction_list = predictions[0]
    # 取出最大值索引和最大值
    def get_max_arg_value(prediction_list):
        arg_max = np.argmax(prediction_list)
        max_val = prediction_list[arg_max]
        preds = np.delete(prediction_list, arg_max)
        return preds, arg_max, max_val
    # 取出前 3 个预测值的最大值索引和最大值
    def get_list_of_max_arg_value(prediction_list):
        preds, argmax1, max1val = get_max_arg_value(prediction_list)
        preds, argmax2, max2val = get_max_arg_value(preds)
        preds, argmax3, max3val = get_max_arg_value(preds)
        top_3_argmax = np.array([argmax1, argmax2, argmax3])
        top_3_max_val = np.array([max1val, max2val, max3val])
        return top_3_argmax, top_3_max_val
    top_3_argmax, top_3_max_val = get_list_of_max_arg_value(prediction_list)
    title_prefix = len("N02098413-")
    dog_titles = [dog_names[index][title_prefix:] for index in top_3_argmax]
    print('前 3 个最大值: {}'.format(top_3_max_val))
    # 如果希望显示直方图，可以取消注释这 3 行代码
    plt.barh(np.arange(3), top_3_max_val)
```

```
plt.yticks(np.arange(3), dog_titles)
plt.show()
# 创建绘图对象
fig, ax = plt.subplots()
# 设置绘图的总容器大小
fig.set_size_inches(5, 5)
# 将最大值乘以 100 就是百分比
top_3_max_val *= 100
# 拼接前 3 个最大值的字符串
dog_title = "{}: {:.2f}%\n".format(dog_titles[0], top_3_max_val[0]) + \
            "{}: {:.2f}%\n".format(dog_titles[1], top_3_max_val[1]) + \
            "{}: {:.2f}%\n".format(dog_titles[2], top_3_max_val[2])
# 在绘图的右上角显示加上识别的值字符串
ax.text(1.01, 0.8,
        dog_title,
        horizontalalignment='left',
        verticalalignment='bottom',
        transform=ax.transAxes)
# 读取图片的数值内容
img = plt.image.imread(img_path)
# 在 Axes 对象上显示图像
ax.imshow(img)
plt.grid(False)
```

然后，先识别狗的品种。这张狗的图片既没有参与模型的训练也没有参与验证或测试。读者也可以从其他地方下载一张狗的图片，拿来测试模型的好坏。

```
dog_img_path = "Maltese_test.jpg"
predict_dog_breed(model, label_names, dog_img_path)
```

输出结果如图 10.11 所示。

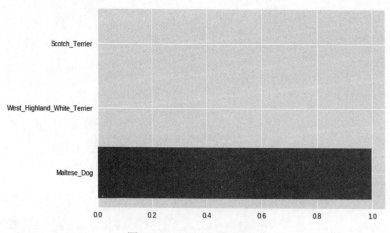

图 10.11　Maltese 狗的识别

图 10.11　Maltese 狗的识别（续）

这张图片中的狗的品种确实是 Maltese（马耳他犬），预测精确度达到 99.75%！最后，我们拿这个模型来判断作者最像这 120 个品种中的哪一种狗。

```
dog_img_path = "victor_test.jpeg"
predict_dog_breed(model, label_names, dog_img_path)
```

输出结果如图 10.12 所示。

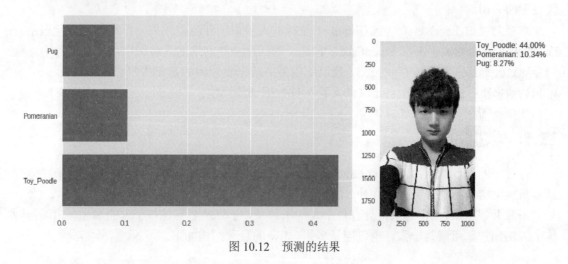

图 10.12　预测的结果

事实是，作者最像 Toy_Poodle（玩具贵宾犬），精确度是 44%，这概率也不低了。

10.5　小结

本章先对开源的国内外 120 个品种的狗的图片数据集进行预处理，使其统一大小；然后通过预训练模型识别狗的品种；最后基于 InceptionV3 来测试识别狗的品种，用 MobileNet v2 的预训练模型来做狗的图片数据集的迁移学习训练模型，训练后的模型识别效果良好。

第 11 章

人脸识别

人脸识别在当下的生活中越来越普遍了。比如去公司上班,刷脸打卡;乘坐高铁,刷脸进站;食堂就餐,刷脸支付;银行办卡,刷脸认证等。笔者认为,未来我们的生活将从出门不带现金、只带一个智能手机,慢慢转变成出门什么都不用带,因为人脸(您的面容)可以代表你的一切信息。

本章会通过 FaceNet(一个 Github 上流行的人脸识别开源库)来分析和使用人脸识别。FaceNet 开源库的灵感来自 OpenFace,它是一个深度神经网络的人脸识别库。在项目中,我们会通过预训练模型来评估和预测,预训练模型的数据集使用的是 LFW(Labeled Faces in the Wild)数据集;接着会使用自定义的人脸图片数据训练模型。

11.1 数据准备

已标记的人脸数据库 LFW(Labeled Faces in the Wild)由美国马萨诸塞州阿默斯特分校计算机视觉实验室整理完成,共计 5700 张不同的人脸,13000 多张图片。

下载并解压后,lfw 目录下会有每个人的子目录,子目录名称是每个人的名字,里面包含这个人的图片,可能是一张,也可能是多张。lfw 的目录结构如下。

```
lfw/
    Aaron_Eckhart/
        Aaron_Eckhart_0001.jpg
    Aaron_Guiel/
        Aaron_Guiel_0001.jpg
    ……
    Michael_Capellas/
        Michael_Capellas_0001.jpg
        ……
    Vin_Diesel/
```

```
Vin_Diesel_0001.jpg
......
```
......

每张人脸图片的大小是一样的,我们会在下面的代码中证明。

11.1.1 环境准备

- numpy = 1.14.6。
- scipy = 0.19.1。
- matplotlib = 2.1.2。
- pillow = 4.0.0。
- opencv-python = 3.4.3。
- tensorflow = 1.11.0。

11.1.2 数据下载和分析

我们使用 Linux 系统下常用的文件传输工具 curl 下载文件,它可以通过一个 HTTP 或者 HTTPS 地址下载文件到本地。如果是在 Jupyter Notebook 中下载文件的话,需要在命令前面加一个感叹号,表示是用命令执行,而不是代码执行;Jupyter Notebook 的单元格默认可以直接执行 Python 代码,下载 lfw 的数据集文件代码如下。

```
curl -O 请指定 lfw.tgz 下载链接
```

下载完人脸数据集后,我们可以通过 tar 命令把 lfw.tgz 解压到当前目录,参数-xvf 表示在当前目录下解压每个文件,且每解压一个文件就打印一个日志,代码如下。

```
tar -xvf lfw.tgz
```

文件几分钟就可以解压完毕。我们先通过 glob 模块来加载所有的人脸图片文件,它是一个文件处理模块,给定一个路径(可以包含通配符),来查找所有匹配的文件。

```
import random
from glob import glob
import numpy as np
# 加载所有人脸图片,返回每张图片的路径,形成一个数组,最后通过 np.array()将数组转换成 NumPy 数组
human_filepaths = np.array(glob("lfw/*/*"))
# 通过 shuffle()函数将 human_filepaths 数组变量里的数据打乱混洗
random.shuffle(human_filepaths)
# 查看前 10 个图片的路径
human_filepaths[:10]
```

输出如下。

```
array(['lfw/David_Collenette/David_Collenette_0001.jpg',
       'lfw/Rainer_Schuettler/Rainer_Schuettler_0001.jpg',
       'lfw/Silvan_Shalom/Silvan_Shalom_0004.jpg',
       'lfw/Jennifer_Aniston/Jennifer_Aniston_0014.jpg',
       'lfw/Tommy_Franks/Tommy_Franks_0012.jpg',
```

```
                'lfw/Albert_Pujols/Albert_Pujols_0001.jpg',
                'lfw/Elin_Nordegren/Elin_Nordegren_0001.jpg',
                'lfw/Lleyton_Hewitt/Lleyton_Hewitt_0035.jpg',
                'lfw/Colin_Powell/Colin_Powell_0075.jpg',
                'lfw/Pascal_Quignard/Pascal_Quignard_0002.jpg'], dtype='<U84')
```

可以看到数组里的每个元素都是一个图片路径。我们来看有多少条数据。

```
print("human_files.shape={}".format(human_filepaths.shape))
```

输出如下。

```
human_files.shape=(13233,)
```

11.1.3 人脸图片数据预览

我们首先检测图片中的人脸，这可以通过 OpenCV 的人脸检测模型做到。作者已经把这个人脸检测模型提前下载到了这个项目中，读者可以直接使用，也可以自己去 Github 上下载人脸检测模型。

这里用的人脸检测模型的文件是 haarcascade_frontalface_alt.xml。

```
import matplotlib.pyplot as plt
# 设置 matplotlib 在绘图时使用默认样式
plt.style.use('default')
from matplotlib import image
import cv2
# 提取 OpenCV 的人脸检测模型
face_cascade = cv2.CascadeClassifier('haarcascade_frontalface_alt.xml')
# 加载彩色（通道顺序为 BGR）图片
# 我们随机选择一张图片作为检测的图片
img = cv2.imread('test0.jpg')
# 将 BGR 图片进行灰度处理
gray = cv2.cvtColor(img, cv2.COLOR_BGR2GRAY)
# 在图像中找出人脸
faces = face_cascade.detectMultiScale(gray)
# 打印图片中检测到的人脸的个数
print('Number of faces detected:', len(faces))
# 获取每一个检测到的人脸的识别框
for (x,y,w,h) in faces:
    # 在人脸图片中以矩形形式绘制出识别框
    # 参数 1：目标图片
    # 参数 2：(x, y) 起始坐标
    # 参数 3：(x, y) 检测到的人脸最大坐标
    # 参数 4：框线的颜色，因为顺序是 BGR，所以第三个数据 255 表示 red，就是红色
    # 参数 5：框线的宽度
    cv2.rectangle(img, (x, y), (x + w, y + h), (0, 0, 255), 2)

# 将 BGR 图片转变为 RGB 图片以打印
cv_rgb = cv2.cvtColor(img, cv2.COLOR_BGR2RGB)
```

```
# 展示含有识别框的图片
plt.imshow(cv_rgb)
plt.show()
```

输出在图片中检测到的人脸,如图 11.1 所示。

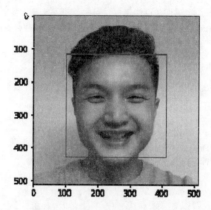

图 11.1　图片中检测到的人脸

我们来看该数据集有多少张人脸。

```
from glob import glob
facepath_prefix_len = len('lfw/')
# 先加载 lfw 目录下的所有目录,然后把每个目录的名称前缀 lfw/去掉,
# 通过列表推导式将裁切的目录名称装起来,然后返回一个列表
face_names = [item[facepath_prefix_len:] for item in sorted(glob("lfw/*"))]
print("共计有{}张人脸".format(len(face_names)))
```

输出如下。

共计有 5749 张人脸

我们再看前 10 个人名。

```
face_names[:10]
```

输出结果如下。

```
['AJ_Cook',
 'AJ_Lamas',
 'Aaron_Eckhart',
 'Aaron_Guiel',
 'Aaron_Patterson',
 'Aaron_Peirsol',
 'Aaron_Pena',
 'Aaron_Sorkin',
 'Aaron_Tippin',
 'Abba_Eban']
```

接下来，我们随机抽取 9 张人脸的图片进行预览。

```python
# 创建 9 个绘图对象，3 行 3 列
fig, axes = plt.subplots(nrows=3, ncols=3)
# 设置绘图的总容器大小
fig.set_size_inches(10, 10)
# 随机选择 9 个数，也就是 9 张人脸图片（可能重复，且每次都不一样）
random_9_nums = np.random.choice(len(human_filepaths), 9)
# 从数据集中选出 9 张图片
random_9_imgs = human_filepaths[random_9_nums]
print(random_9_imgs)
# 根据这随机的 9 张图片路径，取得相应的人脸所属人物的名字
imgname_list = []
for imgpath in random_9_imgs:
    imgname = imgpath[facepath_prefix_len:]
    imgname = imgname[:imgname.find('/')]
    imgname_list.append(imgname)
index = 0
for row_index in range(3):  # 行
    for col_index in range(3):  # 列
        # 读取图片的数值内容
        img = image.imread(random_9_imgs[index])
        # 根据[行索引，列索引]获得绘图 Axes 对象
        ax = axes[row_index, col_index]
        # 在 Axes 对象上显示图像
        ax.imshow(img)
        # 在绘图对象上设置人的名字
        ax.set_xlabel(imgname_list[index])
        # 索引加 1
        index += 1
```

输出 9 张随机图片路径的日志信息如下。

```
['lfw/Ronnie_Jagday/Ronnie_Jagday_0001.jpg'
 'lfw/Jean_Charest/Jean_Charest_0015.jpg'
 'lfw/Maha_Habib/Maha_Habib_0001.jpg'
 'lfw/Philip_Zalewski/Philip_Zalewski_0001.jpg'
 'lfw/Hamid_Karzai/Hamid_Karzai_0008.jpg'
 'lfw/Heidi_Klum/Heidi_Klum_0004.jpg' 'lfw/Ernie_Els/Ernie_Els_0002.jpg'
 'lfw/Joe_Dicaro/Joe_Dicaro_0001.jpg'
 'lfw/Russell_Simmons/Russell_Simmons_0002.jpg']
```

读者在运行代码后就可以看到 9 张图片，这里就不展示了。最后，我们来看每张图片的大小。先把所有的图片都读取出来，取出它们的 shape，然后比较大小。

```python
# 对数据集进行遍历，读取每张图片，并获取它的大小，对比每张图片大小是否一致
faces_shape_list = []
for filepath in human_filepaths:
    shape = image.imread(filepath).shape
    if len(shape) == 3 and shape[0] == 250 and shape[1] == 250 and shape[2] == 3:
        faces_shape_list.append(shape)
```

```
        else:
            print("找到一张异样大小的人脸图片。路径是：{}".format(filepath))

faces_shapes = np.asarray(faces_shape_list)
print("总共{}张。".format(len(faces_shapes)))
print("随机抽取3张图片的维度是{}。".format(faces_shapes[np.random.choice(len(faces_shapes), 3)]))
```

输出如下。

```
总共13233张。
随机抽取3张图片的维度是[[250 250   3]
 [250 250   3]
 [250 250   3]]。
```

11.2　基于 FaceNet 的人脸对齐和验证

FaceNet 是一个在 Github 上非常火的开源项目，它由《FaceNet：人脸识别和聚类的统一嵌入》论文而来，是利用 TensorFlow 实现的。我们接下来将会讲解如何使用 MTCNN 来进行人脸对齐，如何使用预训练模型来评估 LFW 人脸图片数据集。

11.2.1　下载和对齐图片

我们先把 FaceNet 仓库下载下来，通过 git 命令来克隆一份远程 Github 的仓库。这里作者提供的是自己的仓库作为讲解教程，作者从原始 FaceNet 仓库克隆出来一个副本，并对该 FaceNet 仓库做了一些修改，下载命令如下。

```
git clone 请指定本代码仓库地址
```

等待几分钟，待文件下载完后，设置环境变量。

```
import os
# 切换目录到 facenet
os.chdir('facenet')
# 输出当前目录地址
print(os.getcwd())
# 设置环境变量，地址是刚下载的 facenet 目录下的 src 子目录
os.environ["PYTHONPATH"] = "/content/facenet/src"
# 查看环境变量
print(os.environ["PYTHONPATH"])
```

输出如下。

```
/content/facenet
/content/facenet/src
```

接下来将我们将上面下载的 LFW 数据集挪到 facenet/data 目录下，为了方便一会儿做 MTCNN（Multi-Task Cascaded Convolutional Neural Networks）多任务级联卷积网络处理。简单地说，就是把检测到的人脸用一张只显示人脸大小的图片存储起来，该图片大小由传入参

数决定。拷贝 lfw 目录的代码如下。

```
cp -r ../lfw data/lfw
```

当前，我们所处在的目录是 facenet，然后开始使用 MTCNN 处理图片。

```
python src/align/align_dataset_mtcnn.py data/lfw/ data/lfw_output/ --image_size 160
 --margin 32 --random_order --gpu_memory_fraction 0.25
```

通过 python 去执行路径下 src/align/align_dataset_mtcnn.py 文件，它有一些参数，分别表示如下意思。

- 参数 1：在 data/lfw 目录下的原始图片数据。这就是刚移动到这个目录的原始人脸图片文件。
- 参数 2：在 data/目录下，新建一个 lfw_output 文件夹，用来放置已经处理过的 MTCNN 图片。处理后的图片大小全部一致。
- 参数 3：以 160（像素）的宽和高来处理每张图片。
- 参数 4：对已检测到的人脸矩形框以指定的边距裁切。
- 参数 5：使用多个进程来启用对齐进行图片的混洗处理。
- 参数 6：进程将使用 GPU 的内存量的上限的百分比。

由于图片数据集有 13000 多张图片，所以处理的过程会慢一点儿，10 到 20 分钟可以执行完毕。待执行完后，我们可以去 data/lfw_output/目录下查看任意一个人的图片，大小都是 182×182 的。

然后，我们下载一个预训练模型 20180402-114759，该预训练模型下载地址在本代码仓库内。滑动到页面底部，就能看到。由于该预训练模型的大小超出了 Github 上传文件的大小，所以我将其上传到了百度云盘，下载地址也在本代码仓库内。

下载并解压，里面包含扩展名为 meta、ckpt 和 pb 文件，然后将整个目录拷贝到 data/models/目录下。这个预训练模型在 LFW 数据集上预测，可达到的识别精确度是 0.99650，它的架构基于 Inception ResNet V1，训练的数据集是 VGGFace2。

11.2.2 在 LFW 上验证

我们通过使用 src/validate_on_lfw.py 脚本文件来执行在 LFW 数据集上的验证。

```
python src/validate_on_lfw.py \
    data/lfw_output/ \
    data/models/20180402-114759 \
    --distance_metric 1 \
    --use_flipped_images \
    --subtract_mean \
    --use_fixed_image_standardization
```

相关参数解释如下。

- data/lfw_output：刚才已经对齐的人脸图片数据目录。
- data/models/20180402-114759：刚才下载的预训练模型目录。

- --distance_metric：距离度量，0 表示欧几里得距离，1 表示余弦相似度距离。
- --use_flipped_images：连接图片的嵌入和水平翻转的对应物。
- --subtract_mean：在计算距离前减去特征均值。
- --use_fixed_image_standardization：执行固定的图片标准化。

验证 LFW 人脸对齐图片数据集大约会花费 15 分钟，也可能会更快。输出的信息部分如下。

```
Instructions for updating:
To construct input pipelines, use the `tf.data` module.
Runnning forward pass on LFW images
..................
Accuracy: 0.99550+-0.00342
Validation rate: 0.98600+-0.00975 @ FAR=0.00100
Area Under Curve (AUC): 1.000
Equal Error Rate (EER): 0.004
```

我们可以看到精确度很高，它的表现是非常不错的！

11.3 训练自己的人脸识别模型

在上面的代码中，我们已经通过一个预训练模型对 LFW 图片数据集进行人脸识别，并达到很高的精确度，那我们是否可以训练自己的人脸识别模型呢？当然是可以的，作者已经在 mydata/my_own_datasets 目录下提供了一个 5 张人脸的数据集。按照逻辑，我们还是需要先对齐人脸图片，然后训练模型，最后评估模型。

11.3.1 图片数据准备和对齐

我们需要先准备好图片数据，在 mydata/my_own_datasets 目录下有 5 张人脸图片，这将是我们的训练数据。我们首先要做的就是对齐图片。

```
python src/align/align_dataset_mtcnn.py \
    mydata/my_own_datasets/ \
    mydata/my_own_datasets_output/ \
    --image_size 160 \
    --random_order \
    --gpu_memory_fraction 0.25
```

通过 src/align/align_dataset_mtcnn.py 脚本文件去对齐人脸图片，相关参数解释如下。

- mydata/my_own_datasets：人脸图片数据集。
- mydata/my_own_datasets_output：人脸图片被对齐后输出的目录。
- --image_size 160：对齐图片的宽和高。
- --random_order：使用多个进程来启用对齐进行图片的混洗处理。
- --gpu_memory_fraction 0.25：进程将使用 GPU 的内存量的上限的比例。

1~2 分钟就对齐完毕。

11.3.2 训练模型

我们使用该项目提供的 src/classifier.py 脚本来训练模型,虽然最终的识别结果并不好,但是这个流程值得大家去关注。

```
python src/classifier.py \
  TRAIN \
  mydata/my_own_datasets_output \
  data/models/20180402-114759/ \
  data/models/20180402-114759/myown_classifier.pkl \
  --image_size 160
```

通过 src/classifier.py 脚本文件去训练模型,相关参数解释如下。

- TRAIN:训练。
- mydata/my_own_datasets_output:人脸图片被对齐后输出目录。
- data/models/20180402-114759:模型的路径,这里是预训练模型的路径。
- data/models/20180402-114759/myown_classifier.pkl:训练后的模型保存路径名。
- --image_size 160:每张图片以 160(像素)的宽和高来处理。

11.3.3 验证模型

我们现在要验证刚才训练的模型,还是使用 src/classifier.py 脚本文件。

```
python src/classifier.py \
  CLASSIFY \
  mydata/my_own_datasets_output \
  data/models/20180402-114759/ \
  data/models/20180402-114759/myown_classifier.pkl \
  --image_size 160
```

通过 src/classifier.py 脚本文件去验证模型,相关参数解释如下。

- CLASSIFY:验证。
- mydata/my_own_datasets_output:人脸图片被对齐后输出的目录。
- data/models/20180402-114759:模型的路径,这里是预训练模型的路径。
- data/models/20180402-114759/myown_classifier.pkl:训练后的模型保存路径名。
- --image_size 160:每张图片以 160(像素)的宽和高来处理。

输出信息如下。

```
Loaded classifier model from file "data/models/20180402-114759/myown_classifier.pkl"
   0  Bill Gates: 0.214
   1  Bill Gates: 0.234
   2  Zhang Qiang: 0.223
   3  Bill Gates: 0.211
   4  Zhang Qiang: 0.224
   5  Bill Gates: 0.218
   6  Zhang Qiang: 0.232
```

```
 7  Zhang Qiang: 0.217
 8  Zhang Qiang: 0.226
 9  Zhang Qiang: 0.236
10  Zhang Qiang: 0.244
11  Zhang Qiang: 0.221
12  Zhang Qiang: 0.215
13  Zhang Qiang: 0.243
14  Zhang Qiang: 0.221
15  Bill Gates: 0.221
16  Bill Gates: 0.225
17  Zhang Qiang: 0.220
18  Zhang Qiang: 0.212
19  Zhang Qiang: 0.239
20  Bill Gates: 0.220
21  Zhang Qiang: 0.232
22  Zhang Qiang: 0.234
23  Bill Gates: 0.218
24  Zhang Qiang: 0.225
25  Bill Gates: 0.223
26  Zhang Qiang: 0.243
27  Zhang Qiang: 0.220
28  Zhang Qiang: 0.227
29  Bill Gates: 0.241
30  Bill Gates: 0.231
31  Bill Gates: 0.231
Accuracy: 0.188
```

识别结果的精确度是非常低的。

11.3.4 再训练模型

再次训练模型前，我们需要先将 mydata/my_own_datasets 目录下的 31 张图片进行处理，让其跟 LFW 的 data/pairs.txt 文件格式一样。pairs.txt 格式的文件用来表述图片数据集，这里作者编写了一个 generate_pairs.py 脚本，读者可自行调用。

```
# 进入到 mydata 目录
os.chdir('mydata')
# 生成 pairs.txt 文件，用于我们自定义的模型
!python generate_pairs.py
# 返回上一层目录
os.chdir('../')
```

训练模型使用 src/train_softmax.py 脚本文件，它是一个基于 TensorFlow 实现的以 softmax 交叉熵损失训练得来的人脸识别器。我们通过该脚本就可以训练我们自己的分类器。

```
python src/train_softmax.py \
    --logs_base_dir my_lfw_train/logs/facenet/ \
    --models_base_dir my_lfw_train/models/facenet/ \
    --data_dir mydata/my_own_datasets_output/ \
    --image_size 160 \
    --model_def models.inception_resnet_v1 \
```

```
--lfw_dir mydata/my_own_datasets_output/ \
--optimizer ADAM \
--learning_rate -1 \
--max_nrof_epochs 1 \
--keep_probability 0.8 \
--random_crop \
--random_flip \
--use_fixed_image_standardization \
--learning_rate_schedule_file data/learning_rate_schedule_classifier_casia.txt \
--weight_decay 5e-4 \
--embedding_size 512 \
--lfw_distance_metric 1 \
--lfw_use_flipped_images \
--lfw_subtract_mean \
--validation_set_split_ratio 0.05 \
--validate_every_n_epochs 5 \
--prelogits_norm_loss_factor 5e-4 \
--epoch_size 8 \
--lfw_batch_size 16 \
--lfw_pairs mydata/pairs.txt
```

相关参数解释如下。

- --logs_base_dir：一个用于写入事件日志的目录。
- --models_base_dir：一个用于存储训练模型和 CheckPoints 的目录。
- --data_dir：人脸图片被对齐后的输出目录。
- --image_size：每张图片按照指定的像素来处理宽和高。
- --model_def：模型定义，指的是包含推理计算图（Inference Graph）的定义。
- --lfw_dir：验证时人脸图片对齐的输出目录。
- --optimizer：要使用的模型优化算法，常见的有 AdaGrad、Adam、RMSProp 和 AdaDelta。
- --learning_rate：学习率，如果设置负值的话，则使用 learning_rate_schedule.txt 文件中指定的学习率。
- --max_nrof_epochs：运行多少个迭代次数。
- --keep_probability：对于全连接层的 Dropout 保留率。
- --random_crop：执行随机对训练图片的中心像素位置裁剪。
- --random_flip：执行随机水平翻转训练集图片。
- --use_fixed_image_standardization：执行固定图片的标准化。
- --learning_rate_schedule_file：指定学习率的文件。
- --weight_decay：L2 权重正则化。
- --embedding_size：嵌入层的维度。
- --lfw_distance_metric：使用距离度量的类型，0 表示欧几里得距离，1 表示余弦相似度距离。
- --lfw_use_flipped_images：连接图片的嵌入和它的水平翻转对应物。

- --lfw_subtract_mean:在计算距离前减去特征均值。
- --validation_set_split_ratio:整个数据集用于验证时的比例。
- --validate_every_n_epochs:在验证时的迭代次数。
- --prelogits_norm_loss_factor:基于预先分对数(pre-logits)层中激活的正则的损失因素值。
- --epoch_size:每次迭代有多少个批次。
- --lfw_batch_size:在 LFW 的测试集中每个 batch 处理的图片数量。
- --lfw_pairs:用于描述数据的格式文件,就是我们上一步生成的 pairs.txt 文件。

训练时输出的主要日志信息如下。

```
Running training
Epoch: [1][1/8]    Time 51.569    Loss 9.296    Xent 2.140    RegLoss 7.157
    Accuracy 0.222    Lr 0.05000    Cl 0.999
Epoch: [1][2/8]    Time 46.551    Loss 8.517    Xent 1.361    RegLoss 7.157
    Accuracy 0.500    Lr 0.05000    Cl 1.042
Epoch: [1][3/8]    Time 26.371    Loss 8.311    Xent 1.155    RegLoss 7.156
    Accuracy 0.600    Lr 0.05000    Cl 1.138
Epoch: [1][4/8]    Time 27.670    Loss 7.884    Xent 0.727    RegLoss 7.156
    Accuracy 0.778    Lr 0.05000    Cl 1.454
Epoch: [1][5/8]    Time 28.782    Loss 7.700    Xent 0.545    RegLoss 7.155
    Accuracy 0.856    Lr 0.05000    Cl 2.313
Epoch: [1][6/8]    Time 27.969    Loss 7.563    Xent 0.407    RegLoss 7.156
    Accuracy 0.867    Lr 0.05000    Cl 4.496
Epoch: [1][7/8]    Time 32.971    Loss 7.469    Xent 0.314    RegLoss 7.155
    Accuracy 0.933    Lr 0.05000    Cl 11.849
Epoch: [1][8/8]    Time 28.774    Loss 7.292    Xent 0.136    RegLoss 7.155
    Accuracy 0.978    Lr 0.05000    Cl 32.629
Running forward pass on validation set
```

11.3.5 再评估模型

在再次评估模型时,我们需要使用 mydata/pairs.txt 文件,这是我们通过自己的图片数据集生成的文件。评估时使用的模型,就是刚才训练的模型的 CheckPoint 的文件保存的目录。

```
python src/validate_on_lfw.py \
  mydata/my_own_datasets_output \
  my_lfw_train/models/facenet/20181107-034642 \
  --lfw_pairs mydata/pairs.txt \
  --lfw_batch_size 2 \
  --image_size 160 \
  --distance_metric 1 \
  --use_flipped_images \
  --subtract_mean \
  --use_fixed_image_standardization
```

相关参数解释如下。

- mydata/my_own_datasets_output:刚才已经对齐的人脸图片数据目录。

- my_lfw_train/models/facenet/20181107-034642：刚才下载的预训练模型目录。目录名称会随着日期而不同，如 20181107-034642 表示 2018 年 11 月 7 日训练的模型。
- --lfw_pairs：我们通过自己的人脸图片数据生成的 pairs.txt 文件。
- --lfw_batch_size：在 LFW 测试集上每个 batch 处理图片的数量。
- --image_size：每张图片按照指定像素的宽和高来处理。
- --distance_metric：距离度量，0 表示欧几里得的距离，1 表示余弦相似度距离。
- --use_flipped_images：连接图片的嵌入和水平翻转的对应物。
- --subtract_mean：在计算距离前减去特征均值。
- --use_fixed_image_standardization：执行固定的图片标准化。

输出的主要日志信息如下。

```
Accuracy: 1.00000+-0.00000
```

11.3.6 将模型 CheckPoints 文件转换成 pb 文件

在转换前，我们先分析每次模型训练完毕后，在 CheckPoints 目录保存的那些文件分别是什么。解释如下。

checkpoint 文件是文本文件，用来保存最新的一次检查点文件。大多只被用来保存和记录训练时最佳的检查点。

扩展名为 ckpt 的文件，是二进制文件，保存着所有权重、偏置项、梯度和一些其他变量的值。现在最新的 TensorFlow 版本保存 ckpt 文件分别由扩展名为 ckpt-1.data 和扩展名为 ckpt-1.index 的文件组成，1 表示最佳检查点，它会随着在训练时的 ckpt 文件每个记录最佳检查点的增多而增多。

扩展名为 meta 的文件，是 Protocol Buffer 文件，表示存储着 TensorFlow 的计算图（Graph）、所有的变量（Variables）、操作（Operations）和集合（Collections）等。

转换扩展名为 pb 的文件的脚本是 src/freeze_graph.py。该文件将 TensorFlow 模型的计算图进行序列化，包括一个或者多个的模型的图形定义，以及模型的元数据。

```
python src/freeze_graph.py \
    my_lfw_train/models/facenet/20181107-034642 \
    my_lfw_train/models/facenet/20181107-034642/20181107-034642.pb
```

名称 20181107-034642 表示我们训练 TensorFlow 模型的检查点目录。在读者的实际操作中，名称肯定不一样，读者将之修改成自己训练的模型检查点目录即可。命令执行完毕后，就可以在目录 my_lfw_train/models/facenet/20181107-034642 下找到 20181107-034642.pb 文件。

11.4 基于 FaceRecognition 的人脸识别

开源项目 FaceRecognition 在 LFW 人脸图片数据集上的表现非常不错，精确度在 99.38%。FaceRecognition 是在 dlib 开源项目上实现的。dlib 是一个 C++的工具包，包含了机器学习和深度学习的算法和工具，广泛应用于工业界、学术界、机器人、嵌入式和高性能的计算环境。

我们会先实现人脸检测，然后进行人脸实时识别。人脸识别的实现基于卷积神经网络算法模型。

11.4.1 配置环境

对于这里的配置环境要说明的是，dlib 是一个 C++ 的库，我们需要安装编译 C++ 代码的工具。这里提供了两种方式，如果读者以方式 1 安装失败，考虑以方式 2 手动安装。按照方式 1 安装，依次执行下面的 4 条命令。

```
apt-get install build-essential cmake
apt-get install libgtk-3-dev
apt-get install libboost-all-dev
pip install dlib
```

按照方式 2，手动安装，要下载 dlib.git。

```
git clone 请指定 dlib.git 代码仓库地址，在本代码仓库内查阅
```

下载完毕后，切换到 dlib 仓库目录下。

```
cd dlib
```

依次执行以下命令来安装 dlib，可能会花费 10 分钟左右。

```
mkdir build; cd build; cmake ..; cmake --build .
cd ..
python setup.py install
```

最后，不管是通过方式 1 还是方式 2 成功安装后，就可以安装 face_recognition 包了。

```
pip install face_recognition
```

11.4.2 人脸检测

作者准备了一张图片，名称是 victor_test.jpeg。先将这张图片加载，再进行人脸面部特征检测，包括下巴、左眉毛、右眉毛、鼻梁、鼻尖、左眼睛、右眼睛、上嘴唇、下嘴唇和脸轮廓。

```
# 检测人脸面部特征
from PIL import Image, ImageDraw
import face_recognition
# 将图片文件加载到 NumPy 数组中
image = face_recognition.load_image_file("victor_test.jpeg")
# 查找图片中所有的面部特征
face_landmarks_list = face_recognition.face_landmarks(image)
print("找到{}张人脸。".format(len(face_landmarks_list)))
for face_landmarks in face_landmarks_list:
    # 图片中每个面部特征的 key
    facial_features = [
        'chin',
```

```
            'left_eyebrow',
            'right_eyebrow',
            'nose_bridge',
            'nose_tip',
            'left_eye',
            'right_eye',
            'top_lip',
            'bottom_lip'
        ]
        # 将 NumPy 数组对象转换为 PIL.Image.Image 对象
        pil_image = Image.fromarray(image)
        # 再将 PIL.Image.Image 对象转换成 PIL.ImageDraw.ImageDraw，可供绘图
        new_img = ImageDraw.Draw(pil_image)
        # 根据找到的面部特征值
        for facial_feature in facial_features:
            # 绘制线条，线条宽度是 5
            new_img.line(face_landmarks[facial_feature], width=5)
        # 显示已经绘制过特征线条的图片
        pil_image.show()
```

输出日志信息如下，如图 11.2 所示。

找到 1 张人脸。

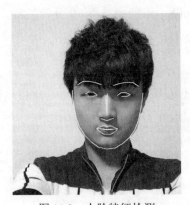

图 11.2　人脸特征检测

11.4.3　实时人脸识别

我们可以通过 OpenCV 获取网络摄像头，并将摄像头拍摄到的画面进行人脸检测和识别。首先，我们将人物的人脸进行编码，作为已知人脸对象保存起来；然后再通过摄像头把拍摄到的画面进行人脸检测和编码；最后和已知的人脸对象作比较，看看是否匹配。如果有匹配对象，则识别成功；反之，若无匹配对象，我们就定义其为未知。

```
import face_recognition
import cv2
# 获取网络摄像头 webcam 实例对象
video_capture = cv2.VideoCapture(0)
```

11.4 基于 FaceRecognition 的人脸识别

```python
# 加载人脸图片函数
def load_person_image(person_filename):
    # 将图片文件加载到内存,以数组呈现
    person_image = face_recognition.load_image_file(person_filename)
    # 将图片数据编码为 128 维度的人脸矩阵数据,以 NumPy 数组返回
    person_face_encoding = face_recognition.face_encodings(person_image)[0]
    return person_face_encoding
# 加载已知人脸对象
victor_face_encoding = load_person_image("victor_test.jpeg")
# 创建已知人脸编码数组
known_face_encodings = [
    victor_face_encoding
]
# 创建已知人脸对象名字数组
known_face_names = [
    "Zhang Qiang"
]
# 初始化一些变量
face_locations = []
face_encodings = []
face_names = []
process_this_frame = True
# 创建一个无限循环的环境,以便于让程序一直在 webcam 摄像头中读取图片内容
while True:
    # 获取摄像头的单帧 frame 图片
    ret, frame = video_capture.read()

    # 为了让检测和识别人脸更快些,我们将摄像头拍摄到的图片 frame 重新定义大小,是原大小的 1/4
    small_frame = cv2.resize(frame, (0, 0), fx=0.25, fy=0.25)

    # OpenCV 使用 BGR 颜色通道
    # face_recognition 使用 RGB 颜色通道
    # 这就将图片从 BGR 转换成 RGB 颜色通道
    rgb_small_frame = small_frame[:, :, ::-1]
    # 只处理每一帧视频,以节省时间
    if process_this_frame:
        # 在当前的视频帧中找出所有的人脸位置和人脸编码
        face_locations = face_recognition.face_locations(rgb_small_frame)
        face_encodings = face_recognition.face_encodings(rgb_small_frame, face_locations)
        face_names = []
        # 遍历人脸图片编码,因为图片中可能会有多张人脸
        for face_encoding in face_encodings:
            # 将已知的人脸编码和检测到的人脸编码进行对比,就知道它们是否是同一个人了
            # 已知的人脸编码和检测到的人脸编码,可能在数值上有些差异,但是,只要在容错率范围内就可以,默认值是 0.6
            matches = face_recognition.compare_faces(known_face_encodings, face_encoding)
            name = "Unknown"
            # 如果人脸编码匹配上了,则将人脸对应的人名字取出来,反之,如果没有匹配上,就显示未知人脸
            if True in matches:
```

```
                first_match_index = matches.index(True)
                name = known_face_names[first_match_index]

            face_names.append(name)
    process_this_frame = not process_this_frame
    # 遍历人物名字和找到的人脸特征位置,并把它们绘制到图片上显示
    for (top, right, bottom, left), name in zip(face_locations, face_names):
            # 上面在做人脸检测和识别时,我们缩小了图片到 1/4,现在我们要把图片在还原回去
            top *= 4
            right *= 4
            bottom *= 4
            left *= 4
            # 在人脸周围绘制一个矩形框
            cv2.rectangle(frame, (left, top), (right, bottom), (0, 0, 255), 2)
            # 在人脸矩形框的下面再绘制一个实心的矩形框,用来将人物的名字显示上去
            cv2.rectangle(frame, (left, bottom - 35), (right, bottom), (0, 0, 255), cv2.FILLED)
            font = cv2.FONT_HERSHEY_DUPLEX
            cv2.putText(frame, name, (left + 6, bottom - 6), font, 1.0, (255, 255, 255), 1)
    # 将图片显示到视频窗口中
    cv2.imshow('Video', frame)
    # 在视频窗口,允许通过按 q 键退出程序
    if cv2.waitKey(1) & 0xFF == ord('q'):
            break
# 将创建的 webcam 实例对象手动释放掉
video_capture.release()
# 销毁所有的 cv2 窗口
cv2.destroyAllWindows()
```

这段程序是实时人脸识别,但是因为书中只能放静态图,所以作者就截取了一张实时拍摄的人脸识别图,如图 11.3 所示。

图 11.3 实时人脸识别效果截图

11.5 小结

本章采用流行的 LFW 人脸图片数据集来训练人脸识别模型，使用流行的 FaceNet 和 FaceRecognition 开源库来训练、测试、检测和识别人脸的模型。最后，为了方便以后使用该模型，我们调用了一个将模型训练的检查点等文件转换成扩展名为 pb 的文件的代码，若需要再使用该模型，就可以直接加载 pb 文件。

第 12 章

人脸面部表情识别

不同的人脸面部表情形态,对于深度学习的神经网络来说,也是需要识别的。在我们现实生活中,它常见的应用场景有很多,比如,国内的 K12 在线教育公司研发的此类产品被应用在中小学教室内,用来检测学生上课时的状态:是认真听讲还是在打瞌睡等,这有助于老师有针对性地对学生进行单独辅导,以便于提高学生的知识水平。再比如,在一些公众场所,警方对高清的摄像头拍摄到的画面进行人脸面部表情识别,可找出一些有嫌疑的人,快速锁定关键人。

12.1 基于 Keras 的卷积神经网络实现人脸面部表情识别

我们将使用 fer2013(Facial Expression Recognition 2013)数据集并构建卷积神经网络(CNN)来训练人脸面部表情识别的模型。我们首先识别单张图片中的人脸面部表情,然后识别视频中的人脸面部表情,最后通过摄像头拍摄来实时识别人脸面部表情。

12.1.1 环境准备

- numpy = 1.15.4。
- matplotlib = 3.0.0。
- keras = 2.2.4。
- tensorflow = 1.12.0。
- PIL = 5.3.0。
- cv2 = 3.4.3。
- moviepy = 0.2.3.5。

12.1.2 数据准备

数据集由人脸的 48 像素×48 像素灰度图片组成，这些图片数据已经进行了处理，人脸占整个图片的大部分区域，而且大部分在图片正中间。该数据集将人脸的面部表情分为 7 个类别，分别是 angry=0、disgust=1、fear=2、happy=3、sad=4、surprise=5 和 neutral=6，对应的中文是愤怒=0、厌恶=1、恐惧=2、快乐=3、悲伤=4、惊喜=5 和中立=6。以下代码通过卷积神经网络对人脸面部表情的这 7 个类别进行模型训练和识别，包含静态的和动态的。

数据集是 fer2013.tar 压缩文件，解压后 fer2013 目录下有 fer2013.csv 文件，文件里面的内容包含训练集和测试集。数据集分三列，各列名称解释如下。

- 第一列是 emotion，表示人脸面部类别的序号，从 0 到 6。
- 第二列是 pixels，表示描述 48×48 的灰度图片矩阵数值。
- 第三列是 Usage，表示训练集或测试集，训练集用 Training 表示，测试集用 PublicTest 表示。

fer2013.csv 是一个逗号分隔符文件，可以通过 Excel 打开。我们打开后预览前几条数据，如图 12.1 所示。

	A	B	C
1	emotion	pixels	Usage
2	0	70 80 82 72 58 58 60 63 54 58 60 48 8	Training
3	0	151 150 147 155 148 133 111 140 170	Training
4	2	231 212 156 164 174 138 161 173 182	Training
5	4	24 32 36 30 32 23 19 20 30 41 21 22 3	Training
6	6	4 0 0 0 0 0 0 0 0 0 0 0 3 15 23 28 48 5	Training
7	2	55 55 55 55 55 54 60 68 54 85 151 163	Training
8	4	20 17 19 21 25 38 42 42 46 54 56 62 6	Training
9	3	77 78 79 79 78 75 60 55 47 48 58 73 7	Training
10	3	85 84 90 121 101 102 133 153 153 169	Training

图 12.1 fer2013.csv 数据集的前几条数据预览

然后，我们通过 open()函数将数据集的内容读取到内存中。

```
import numpy as np
# 通过 with 关键字读取 fer2013.csv 文件的内容，读取完毕后，它会自动调用文件的 close()函数
with open("fer2013/fer2013.csv") as f:
    # 读取所有的行
    content = f.readlines()
# 将数据内容装填到 NumPy 格式的数据矩阵
lines = np.array(content)
num_of_instances = lines.size
print("实例数量：{}。".format(num_of_instances))
print("实例长度：{}。".format(len(lines[1].split(",")[1].split(" "))))
```

输出如下。

```
实例数量：35888。
实例长度：2304。
```

35888 表示数据集的行数，2304 表示灰度图片的 48（宽）×48（高）的长度。

12.1.3　数据集分割

在 csv 文件中的数据是一行一行的，我们现在要将这个数据集分割成训练集、验证集和测试集。但是这个 csv 文件中的数据集并没有验证集，所以我们就对测试集进行折半处理，一半作验证集，一半作测试集。

```python
from keras import utils
# 定义面部表情的7个类别
class_list = ["angry", "disgust", "fear", "happy", "sad", "surprise", "neutral"]
num_classes = len(class_list)
# 定义训练集、验证集和测试集的数组
X_train, y_train, X_valid, y_valid, X_test, y_test = [], [], [], [], [], []
# 开始循环进行数据分割
for i in range(1, num_of_instances):
    try:
        # 读取一行数据
        emotion, img, usage = lines[i].split(",")
        # 用空格将图片的数值内容分割成数组
        val = img.split(" ")
        # 将图片转换成 NumPy 数组
        pixels = np.array(val, np.float32)
        # 对面部表情类别做 one-hot encoding
        emotion = utils.to_categorical(emotion, num_classes)
        # 如果是训练集数据，就添加到训练集数组
        if 'Training' in usage:
            y_train.append(emotion)
            X_train.append(pixels)
        elif 'PublicTest' in usage:
            # 如果是测试集数据，就添加到测试集数组
            y_test.append(emotion)
            X_test.append(pixels)
    except:
        print("", end="")
# 最后将测试集的数组数据分割一半给验证集，前半部分是验证集的，后半部分是测试集的
half_test_len = int(len(X_test) / 2)
X_valid = X_test[:half_test_len]
y_valid = y_test[:half_test_len]
X_test = X_test[half_test_len:]
y_test = y_test[half_test_len:]
```

12.1.4　数据集预处理

3 个数据集分割完后，我们开始将这些图片数据集进行预处理。我们都知道彩色图片是由 RGB（红、绿、蓝）3 种颜色组成的，但是这里的图片是灰度的，也就是只有一个颜色通道，颜色的数值范围是 0 到 255，所以我们会对图片数值进行归一化处理，即转换成 0 到 1 之间的数值。构建卷积神经网络时，要求输入的图片形状是（batch_size, height, width,

channels），所以我们会把三维图片的形状修改为四维数组的图片。

```
# 将训练集图片数值转换成float32，并且创建NumPy的数据格式表示
X_train = np.array(X_train, np.float32)
y_train = np.array(y_train, np.float32)
# 将验证集图片数值转换成float32，并且创建NumPy的数据格式表示
X_valid = np.array(X_valid, np.float32)
y_valid = np.array(y_valid, np.float32)
# 将测试集图片数值转换成float32，并且创建NumPy的数据格式表示
X_test = np.array(X_test, np.float32)
y_test = np.array(y_test, np.float32)
# 归一化处理输入的图片数值，将其转换成0到1之间的值
X_train /= 255
X_valid /= 255
X_test /= 255
# 定义图片的宽和高
img_width = 48
img_height = 48
# 将训练集图片的形状转换成（batch_size, height, width, channels）的四维数组
X_train = X_train.reshape(X_train.shape[0], img_width, img_height, 1)
X_train = X_train.astype(np.float32)
# 将验证集图片的形状转换成（batch_size, height, width, channels）的四维数组
X_valid = X_valid.reshape(X_valid.shape[0], img_width, img_height, 1)
X_valid = X_valid.astype(np.float32)
# 将测试集图片的形状转换成（batch_size, height, width, channels）的四维数组
X_test = X_test.reshape(X_test.shape[0], img_width, img_height, 1)
X_test = X_test.astype(np.float32)
# 打印输出
print("X_train.shape={}, y_train.shape={}.".format(X_train.shape, y_train.shape))
print("X_valid.shape={}, y_valid.shape={}.".format(X_valid.shape, y_valid.shape))
print("X_test.shape={}, y_test.shape={}.".format(X_test.shape, y_test.shape))
```

输出如下。

```
X_train.shape=(28709, 48, 48, 1), y_train.shape=(28709, 7).
X_valid.shape=(1794, 48, 48, 1), y_valid.shape=(1794, 7).
X_test.shape=(1795, 48, 48, 1), y_test.shape=(1795, 7).
```

12.1.5 构建 CNN 模型

使用 Keras 构建 CNN 模型，有 3 个卷积层，深度分别为 64、64 和 128。卷积窗大小从一开始的(5, 5)转换成后面的(3, 3)，最后添加 1024 个全连接层和 7 个类别的输出全连接层。7 个类别属于多分类问题，所以输出使用 softmax 激活函数。

```
from keras.models import Sequential
from keras.layers import Conv2D, MaxPooling2D, AveragePooling2D
from keras.layers import Dense, Activation, Dropout, Flatten
# 创建 Keras 的 Sequential 模型实例
model = Sequential()
```

```python
# 添加第一层的卷积层，需要传入图片的 input_shape 参数的值
model.add(Conv2D(64, (5, 5), activation='relu', input_shape=(img_width, img_height, 1)))
model.add(MaxPooling2D(pool_size=(5,5), strides=(2, 2)))
model.add(Dropout(0.5))
# 添加第二层卷积层
model.add(Conv2D(64, (3, 3), activation='relu'))
model.add(Conv2D(64, (3, 3), activation='relu'))
model.add(AveragePooling2D(pool_size=(3,3), strides=(2, 2)))
model.add(Dropout(0.5))
# 添加第三层卷积层
model.add(Conv2D(128, (3, 3), activation='relu'))
model.add(Conv2D(128, (3, 3), activation='relu'))
model.add(AveragePooling2D(pool_size=(3,3), strides=(2, 2)))
model.add(Dropout(0.5))
model.add(Flatten())
# 添加 1024 个全连接层
model.add(Dense(1024, activation='relu'))
model.add(Dropout(0.2))
model.add(Dense(1024, activation='relu'))
model.add(Dropout(0.2))
# 添加输出层
model.add(Dense(num_classes, activation='softmax'))
model.summary()
```

输出模型的网络架构信息如下。

Layer (type)	Output Shape	Param #
conv2d_1 (Conv2D)	(None, 44, 44, 64)	1664
max_pooling2d_1 (MaxPooling2	(None, 20, 20, 64)	0
dropout_1 (Dropout)	(None, 20, 20, 64)	0
conv2d_2 (Conv2D)	(None, 18, 18, 64)	36928
conv2d_3 (Conv2D)	(None, 16, 16, 64)	36928
average_pooling2d_1 (Average	(None, 7, 7, 64)	0
dropout_2 (Dropout)	(None, 7, 7, 64)	0
conv2d_4 (Conv2D)	(None, 5, 5, 128)	73856
conv2d_5 (Conv2D)	(None, 3, 3, 128)	147584
average_pooling2d_2 (Average	(None, 1, 1, 128)	0
dropout_3 (Dropout)	(None, 1, 1, 128)	0
flatten_1 (Flatten)	(None, 128)	0

```
dense_1 (Dense)                  (None, 1024)              132096
_____
dropout_4 (Dropout)              (None, 1024)              0
_____
dense_2 (Dense)                  (None, 1024)              1049600
_____
dropout_5 (Dropout)              (None, 1024)              0
_____
dense_3 (Dense)                  (None, 7)                 7175
=================================================================
Total params: 1,485,831
Trainable params: 1,485,831
Non-trainable params: 0
```

通过模型架构输出的信息我们可看见每一层的层数名称、输出形状和参数数量,这就是模型架构内部的样子,最终可训练的参数达到 148 万多个。而且构建的模型代码在当前的运行环境中,只能执行一次。假如执行了两次,那么每一层的名称和层数会在第一次运行的模型架构上进行追加,这是不可取的,会导致我们无法训练模型。所以当我们要重新运行这块构建模型的代码时,一定要清除当前运行环境的所有变量状态。

12.1.6 图片增强与训练模型

在训练模型前,我们通过 ImageDataGenerator 类对图片进行数据增强处理,它会遍历所有的图片以进行处理,最后返回一个由 yield 关键字产生的 iterator 对象,可通过 Sequential 对象的 fit_generator()函数来训练模型。

```
import keras
from keras.preprocessing.image import ImageDataGenerator
# 定义每批次大小
batch_size = 256
# 定义迭代训练次数
epochs = 20
# 创建图片数据增强生成器对象
imgGenerator = ImageDataGenerator()
# 增强图片数据后返回 iterator 对象
train_generator = imgGenerator.flow(X_train, y_train, batch_size=batch_size)
# 编译模型,使用类别交叉熵作为损失函数,以 Adam 为优化器,用 accuracy 来衡量效果
model.compile(loss='categorical_crossentropy'
    , optimizer=keras.optimizers.Adam()
    , metrics=['accuracy']
)
# 训练模型
history = model.fit_generator(train_generator,
                    steps_per_epoch=batch_size, epochs=epochs,
                    validation_data=(X_valid, y_valid),
                    verbose=1)
```

输出如下。

```
Epoch 1/20
256/256 [==============================] - 7s 27ms/step - loss: 1.8027 - acc: 0.2517 - val_loss: 1.7376 - val_acc: 0.2821
Epoch 2/20
256/256 [==============================] - 5s 21ms/step - loss: 1.6555 - acc: 0.3258 - val_loss: 1.5380 - val_acc: 0.4013
Epoch 3/20
……
Epoch 20/20
256/256 [==============================] - 6s 22ms/step - loss: 1.0733 - acc: 0.5940 - val_loss: 1.1167 - val_acc: 0.5674
```

训练模型结束，训练损失值一开始为 1.8027，最后降到了 1.0733；验证精确度从一开始的 0.2821，增加到了 0.5674。我们接下来看训练时的损失值和精确度的走势图。

12.1.7　评估模型

我们先通过测试集对模型进行评估，然后再通过 Keras 模型把在训练时保存的损失值数组和精确度数组进行绘图显示。评估模型的代码如下。

```
train_score = model.evaluate(X_train, y_train, verbose=0)
print('Train loss: {}.'.format(train_score[0]))
print('Train accuracy: {}.'.format(train_score[1]))
test_score = model.evaluate(X_test, y_test, verbose=0)
print('Test loss: {}.'.format(test_score[0]))
print('Test accuracy: {}.'.format(test_score[1]))
```

输出如下。

```
Train loss: 0.9199279783379052.
Train accuracy: 0.6604200773287053.
Test loss: 1.1018399342188927.
Test accuracy: 0.5855153203508648.
```

对模型在训练时和验证时保存的 loss 和 accuracy 的数组进行绘图显示，走势图清晰明了。history 变量在训练完模型后就返回了，该对象里包含哪些数值，可以通过 history.history.keys() 来查看，绘图代码如下。

```
import matplotlib.pyplot as plt
# 绘制训练时和验证时的精确度走势
plt.plot(history.history['acc'])
plt.plot(history.history['val_acc'])
plt.title('model accuracy')
plt.ylabel('accuracy')
plt.xlabel('epoch')
plt.legend(['train', 'test'], loc='upper left')
plt.show()
```

```
# 绘制训练时和验证时的损失值走势
plt.plot(history.history['loss'])
plt.plot(history.history['val_loss'])
plt.title('model loss')
plt.ylabel('loss')
plt.xlabel('epoch')
plt.legend(['train', 'test'], loc='upper left')
plt.show()
```

输出模型精确度走势如图 12.2 所示，模型损失值走势如图 12.3 所示。

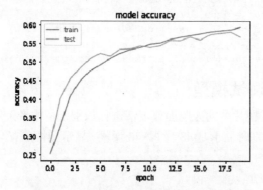

图 12.2　训练和验证模型时的精确度走势

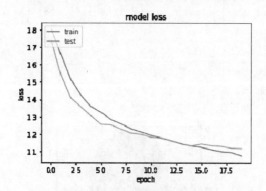

图 12.3　训练和验证模型时的损失值走势

12.1.8　保存与读取模型

如果我们希望下次使用该模型时不再重新运行以上代码、不再重新训练模型，那么最佳的做法就是将模型序列化保存到本地文件以便于下次直接读取和使用。模型的架构使用 to_json() 函数保存，模型的权重使用 save_weights() 函数保存。

```
# 序列化模型的架构保存到 JSON 对象
model_json = model.to_json()
# 保存到本地
with open("facial_expression_recog_model_architecture.json", "w") as json_file:
```

```
    json_file.write(model_json)
# 序列化模型的权重保存到 HDF5 文件
model.save_weights("facial_expression_recog_model_weights.h5")
```

然后我们从本地读取模型的架构文件和权重文件，就可以通过 loaded_model 变量来进行新图片的识别。

```
from keras.models import model_from_json
from keras.models import load_model
# 加载模型架构
with open('facial_expression_recog_model_architecture.json', 'r') as json_file:
    loaded_model_json = json_file.read()
    loaded_model = model_from_json(loaded_model_json)
# 加载模型权重
loaded_model.load_weights("facial_expression_recog_model_weights.h5")
```

12.1.9　单张图片测试模型

现在是时候来对单张图片中的人脸面部表情进行识别了。作者采用了一张 RGB 的图来作为识别图，原图在模型识别时显示是 48 像素×48 像素。显示图肯定有马赛克，因为像素太小了。

```
from keras.preprocessing import image
import matplotlib.pyplot as plt
def plot_src_image(img_path, grayscale=False):
    # 将图片从本地读取到内存中
    img = image.load_img(img_path, grayscale=grayscale, target_size=(48, 48, 3))
    # 显示图片
    plt.imshow(img)
    plt.show()
plot_src_image("test_imgs/victor_7.jpg")
```

输出结果如图 12.4 所示。

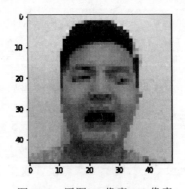

图 12.4　原图 48 像素×48 像素

显示这张图的灰度图。

```
plot_src_image("test_imgs/victor_7.jpg", grayscale=True)
```

输出结果如图 12.5 所示。

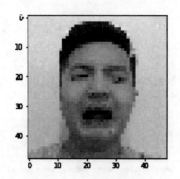

图 12.5　原图的灰度图

识别单张图片各个类别的概率图时，先用 load_img()函数读取图片并进行预处理，然后通过刚才加载的模型变量 loaded_model 对图片进行识别预测，最后通过直方图将产生的概率数组显示出来。

```python
import numpy as np
from keras.preprocessing import image
def load_img(img_path, width=48, height=48):
    # 以灰度的模式来加载指定的 RGB 图片，并且将其修改为 48 像素×48 像素
    img = image.load_img(img_path, grayscale=True, target_size=(width, height))
    # 将图片转换为数组
    x = image.img_to_array(img)
    # 扩展图片的数组维度为四维，这是 CNN 模型需要的
    x = np.expand_dims(x, axis=0)
    # 对图片的数值内容进行归一化处理
    x /= 255
    return x
def plot_analyzed_emotion(emotions_probs, class_list):
    # 绘制直方图显示各个类别的概率
    y_pos = np.arange(len(class_list))
    plt.bar(y_pos, emotions_probs, align='center', alpha=0.5)
    plt.xticks(y_pos, class_list)
    plt.ylabel('percentage')
    plt.title('emotion')
    plt.show()
# 加载图片
test_img = load_img("test_imgs/victor_7.jpg")
# 识别图片的概率
predicted_probs = loaded_model.predict(test_img)
# 定义类别
class_list = ["angry", "disgust", "fear", "happy", "sad", "surprise", "neutral"]
# 绘图显示
plot_analyzed_emotion(predicted_probs[0], class_list)
```

输出结果如图 12.6 所示。

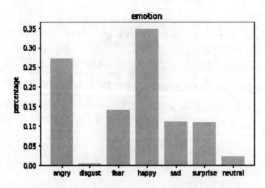

图 12.6　识别单张图片各个类别的概率图

在本章对应的目录 test_imgs 里有 8 张图片,我们将这些图片一一进行模型识别并绘图显示。

```python
from glob import glob
import matplotlib
import matplotlib.pyplot as plt
import cv2
import numpy as np
from PIL import ImageFont, ImageDraw, Image
import random
def plot_emotion_faces(filepaths):
    # 定义人脸矩阵排列的绘图函数
    # 将文件路径的数组随机打乱
    random.shuffle(filepaths)
    # 加载宋体字体
    fontpath = "Songti.ttc"
    font = ImageFont.truetype(fontpath, 120)
    # 创建 2 行 4 列的 figure 对象
    fig, axes = plt.subplots(nrows=2, ncols=4)
    # 设置整体宽和高
    fig.set_size_inches(12, 6)
    index = 0
    # 遍历 2 行
    for row_index in range(2):
        # 遍历 4 列
        for col_index in range(4):
            # 通过 load_img() 函数对加载的图片进行模型识别,返回类别概率
            predicted_probs = loaded_model.predict(load_img(filepaths[index]))
            # 将类别概率数组取出
            probs = predicted_probs[0]
            # 获取最大类别概率的数组索引
            max_index = np.argmax(probs)
            # 获取最大类别概率的值
            probs_val = probs[max_index]
            # 获取最大类别概率的具体中文名称
```

```
            emotion = class_list[max_index]
            # 拼接类别名称和概率值的字符串
            emotion_text = emotion + ":" + str(round(probs_val * 100, 2)) + "%"
            # 以下是将识别到的概率名称和值的字符串显示在图片左上角
            # 从文件路径读取图片
            img = matplotlib.image.imread(filepaths[index])
            # 将图片转换为 RGB 模式
            img_PIL = Image.fromarray(cv2.cvtColor(img, cv2.COLOR_BGR2RGB))
            # 创建绘图对象
            draw = ImageDraw.Draw(img_PIL)
            # 绘制概率的字符串到图片左上角 x=30 和 y=5 的位置
            draw.text((30, 5), emotion_text, font=font, fill=(0, 0, 255))
            # 再将图片从 RGB 转换到 BGR 模式
            final_img = cv2.cvtColor(np.asarray(img_PIL), cv2.COLOR_RGB2BGR)
            # 获取 matplotlib 的 Axes 对象
            ax = axes[row_index, col_index]
            # 显示图片到指定位置
            ax.imshow(final_img)
            index += 1
# 加载 test_imgs 目录下的所有 jpg 文件
test_img_filenames = glob("test_imgs/*.jpg")
# 绘图显示
plot_emotion_faces(test_img_filenames)
```

输出结果如图 12.7 所示。

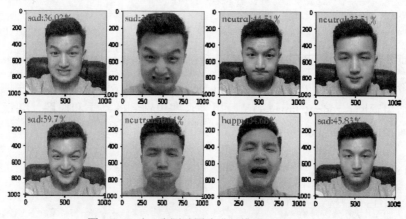

图 12.7 对 8 张测试图片进行模式识别的结果

12.2 视频中的人脸面部表情识别

我们做过了单张图片中的人脸面部表情识别，同样也可以进行视频中的人脸面部表情识别。视频是由一帧一帧的图片组成的，每一帧就是一张静态的图片，以每秒 25 帧到 60 帧的速度播放的视频是流畅的，不会卡顿。本节先对视频里的每一帧图片进行人脸面部表情识别，然后再将修改后的图片插入视频中的同步位置，最后导出视频。

12.2.1 读取模型

我们重新对模型进行读取，随后使用模型变量 loaded_model 来对视频中的每一帧进行识别。

```python
from keras.models import model_from_json
from keras.models import load_model
with open('facial_expression_recog_model_architecture.json', 'r') as json_file:
    loaded_model_json = json_file.read()
    loaded_model = model_from_json(loaded_model_json)
loaded_model.load_weights("facial_expression_recog_model_weights.h5")
```

12.2.2 模型参数定义

定义模型识别图片后显示到图片上的类别名称的字体，通过 cv2 模块的 haarcascade 文件来检测人脸。

```python
import cv2
import numpy as np
from PIL import ImageFont, ImageDraw, Image
# 加载字体
fontpath = "Songti.ttc"
font = ImageFont.truetype(fontpath, 32)
# 加载 cv2 的检测人脸的配置文件
cascPath = "haarcascade_frontalface_default.xml"
faceCascade = cv2.CascadeClassifier(cascPath)
# 定义人脸表情类别数组
class_list = ["angry", "disgust", "fear", "happy", "sad", "surprise", "neutral"]
```

12.2.3 视频的帧处理函数定义

定义 3 个用来处理图片的函数，其中：predict_class()函数用来识别图片中的人脸，然后把识别到的结果返回；redraw_image()函数用来合并识别到的结果和原图，并返回新图片；process_frame()函数在检测人脸和合成图片后，返回最终的图片。

```python
def predict_class(image_frame, x, y, w, h, target_class_list, model):
    # 通过模型识别人脸的对应类别
    # 只读取检测到的人脸矩阵区域
    face_crop = image_frame[y:y + h, x:x + w]
    # 将人脸图片大小修改为 48×48
    face_crop = cv2.resize(face_crop, (48, 48))
    # 将图片转换成灰度的
    face_crop = cv2.cvtColor(face_crop, cv2.COLOR_BGR2GRAY)
    # 将图片数值转换到 0 到 1 之间
    face_crop = face_crop.astype(np.float32) / 255
    face_crop = np.asarray(face_crop)
    # 转换图片的形状为 (batch_size, height, width, channels)
    face_crop = face_crop.reshape(1, face_crop.shape[0], face_crop.shape[1], 1)
    # 识别图片的表情类别
    emotion_result = target_class_list[np.argmax(model.predict(face_crop))]
```

```python
        return emotion_result
    def redraw_image(image_frame, emotion_result, x, y):
        # 将原图片和识别到的类别文字合并成新图片
        # 将该图片转换成 RGB 图片
        img_PIL = Image.fromarray(cv2.cvtColor(image_frame, cv2.COLOR_BGR2RGB))
        # 创建图片绘图对象
        draw = ImageDraw.Draw(img_PIL)
        # 绘制类别文字到图片上的(x, y)处
        draw.text((x, y - 50), emotion_result, font=font, fill=(32, 183, 228))
        # 图片合并后，将图片模式转换成 BGR
        img = cv2.cvtColor(np.asarray(img_PIL), cv2.COLOR_RGB2BGR)
        # 返回合并后的新图片
        return img
    def process_frame(frame):
        # 处理视频的每一帧图片
        global counter
        if counter % 1 == 0:
            # 将图片进行灰度处理
            gray = cv2.cvtColor(frame, cv2.COLOR_BGR2GRAY)
            # 通过 cv2 检测人脸，一张图片上可能会有多张人脸
            faces = faceCascade.detectMultiScale(
                gray,
                scaleFactor=1.1,
                minNeighbors=5,
                minSize=(30, 30)
            )
            # 对检测到的人脸数据数组进行循环
            for (x, y, w, h) in faces:
                # 将检测到人脸矩形数据绘制到原图片上
                cv2.rectangle(frame, (x, y), (x+w, y+h), (228, 32, 183), 2)
                # 识别图片中的人脸结果
                emotion_result = predict_class(frame, x, y, w, h, class_list, loaded_model)
                # 合并图片
                final_img = redraw_image(frame, emotion_result, x, y)
        counter += 1
        # 返回新图片
        return final_img
```

12.2.4 识别与转换视频

通过 moviepy 模块对视频做编辑处理，我们需要对视频中每一帧图片进行修改，然后再将新图片插入视频中同样的位置。原视频是 video.mp4 文件，识别人脸面部表情后导出的视频文件是 video_clipped.mp4。

```python
from moviepy.editor import VideoFileClip
# 输入的原视频文件路径
input_filename = 'test_imgs/video.mp4'
# 输出的视频文件路径
output_filename = 'test_imgs/video_clipped.mp4'
```

```
counter = 0
# 指定视频地址,通过函数 suclip(t_start, t_end)指定剪切的起始时间和终止时间
clip_playing = VideoFileClip(input_filename).subclip(0,10)
# 通过 process_frame 函数修改每一帧图片,将修改后的图片再更新到视频帧中
white_clip = clip_playing.fl_image(process_frame)
# 将视频导出到指定的文件中,并且带上百分比进度
%time white_clip.write_videofile(output_filename)
```

输出视频文件到本地 test_imgs 目录下的 video_clipped.mp4。因为纸质书籍无法播放视频,所以这里仅展示视频中的截图,如图 12.8 所示。

图 12.8　左图是原视频的截图,右图是识别人脸面部表情后的视频截图

12.3　实时人脸面部表情识别

我们对单张图片和视频中的人脸面部表情都做过识别了,但是,在现实生活和工作的场景中是要实时摄像头识别的。本节会用到上面的识别模型和图片处理函数。想象一下,视频是一帧一帧组成的,那摄像头拍摄到的画面呢?它每读取到一帧时,我们就对这一帧进行识别和图片修改,然后将这张图片显示到窗口中,连续的拍摄、连续的识别和连续的窗口显示,是不是就构成了基于摄像头的实时识别呢?

12.3.1　模型参数定义

我们在上面的代码中用到了中文字体(宋体),却没有使用中文。那么在这个实时识别的过程中,我们就用中文来显示识别的类别文字。

```
import cv2
import numpy as np
from PIL import ImageFont, ImageDraw, Image
# 开启计算机的第一个摄像头
video_capture = cv2.VideoCapture(0)
# 创建宋体的字体对象
fontpath = "Songti.ttc"
font = ImageFont.truetype(fontpath, 32)
```

```python
# 创建检测人脸 cv2 对象
cascPath = "haarcascade_frontalface_default.xml"
faceCascade = cv2.CascadeClassifier(cascPath)
# 面部表情类别对应的中文
# class_list = ["angry", "disgust", "fear", "happy", "sad", "surprise", "neutral"]
cn_class_list = ["生气", "厌恶", "恐惧", "高兴", "伤心", "惊喜", "中立"]
```

12.3.2　启动摄像头和识别处理

识别模型的模型变量在 12.2 节中已经加载过，这里不再重复代码。我们主要关注对摄像头拍摄到的画面进行人脸面部表情的实时识别。这可以通过 read() 函数来读取一帧一帧的画面图片，然后检测人脸，将检测到的人脸绘制矩形框显示到画面中，再去识别人脸面部表情类别。将识别到的面部表情类别文字显示到图片上，最终将新图片显示到窗口中。

```python
while True:
    # 捕获视频一帧一帧的画面
    ret, frame = video_capture.read()
    # 将这一帧图片转换成灰度图
    gray = cv2.cvtColor(frame, cv2.COLOR_BGR2GRAY)
    # 检测图片中的面部人脸
    faces = faceCascade.detectMultiScale(
        gray,
        scaleFactor=1.1,
        minNeighbors=5,
        minSize=(30, 30)
    )
    # 将检测到的人脸数组进行遍历
    for (x, y, w, h) in faces:
        # 绘制人脸矩形框
        cv2.rectangle(frame, (x, y), (x+w, y+h), (228, 32, 183), 2)
        # 识别这张人脸的面部表情
        emotion_result = predict_class(frame, x, y, w, h, cn_class_list, loaded_model)
        # 将原图片和表情类别文字合并成新图片
        img = redraw_image(frame, emotion_result, x, y)
        # 将新图片显示到窗口上，并将图片大小修改为 800×500
        cv2.imshow('Chapter 12 Facial Expression Recognition - by Zhang Qiang',
                   cv2.resize(img, (800, 500)))
    # 按下 q 键，即退出程序
    if cv2.waitKey(1) & 0xFF == ord('q'):
        break
# 记得释放视频和窗口对象占用的内存
video_capture.release()
cv2.destroyAllWindows()
```

对实时拍摄到的画面进行人脸面部表情识别时的截图如图 12.9 所示。

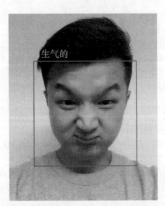

图 12.9 实时拍摄到的画面进行人脸面部表情识别的截图

12.4 小结

我们使用 fer2013 开源的人脸图片数据集，共有 7 类人脸面部表情。首先我们构建了 CNN 模型，然后对人脸图片做了增强处理。训练 CNN 模型后，我们测试了图片中包含的人脸的面部表情。然后我们对单张图片、视频和摄像头中的人脸进行面部表情识别。

第 13 章

人体姿态识别

试问,机器人如何识别人的动作行为呢?人类可以通过主观意识来判断一个人的动作是什么,但机器人只有眼睛(摄像头)拍摄到的一张张静态画面。而且机器人并不知道如果画面里有人,他在做什么、有什么行为。但是通过人体姿态识别这种视觉技术机器人就可以做到。此外,在安防和监控系统中,可以在人机交互上对老年人和小孩有更多的辅助照顾。当然了,人体姿态识别在机器人所具有的功能中只占了很小的一部分。本章,我们将通过此技术实现单张图片和视频中的人体姿态识别,以及摄像头拍摄到的人体姿态实时识别。

13.1 基于 TensorFlow 实现人体姿态识别

人体姿势识别是让计算机能够理解图像中的人体关键组成部分。为了检测图像中多人的 2D(二维)姿势,研究人员提出了一种非参数表示方法,称作部分亲和场(Part Affinity Fields,PAF)。PAF 学习了人身体的部位与图像中的个体的关联,不管图像中有多少人,它都自上而下地实现精度和实时性地识别。

13.1.1 环境准备

- numpy = 1.14.6。
- matplotlib = 3.0.2。
- PIL = 4.0.0。
- tensorflow = 1.12.0。
- cv2 = 3.4.3。
- slidingwindow = 0.0.13。
- moviepy = 0.2.3.5。

13.1.2 下载与安装

本次使用的 TensorFlow 代码库基于 OpenPose 基础代码库。OpenPose 是由美国卡内基梅隆大学（CMU）感知计算实验室的研究人员研究出来，并开源到 Github 上的，OpenPose 的实现基于 Caffe 深度学习框架。使用前，我们先通过 git 克隆该 TensorFlow 的实现版本的代码库。

```
git clone 请指定本tf-pose-estimation.git 代码仓库地址
```

克隆完后，进入 tf-pose-estimation 代码库目录。

```
import os
os.chdir("tf-pose-estimation")
```

主要的文件解释分别如下。

- setup.py：安装文件。
- run.py：单张图片测试推理计算人体姿态。
- run_video.py：视频中的人体姿态识别。
- run_webcam.py：实时摄像头拍摄的人体姿态识别。
- tf-pose：该目录下是主要实现文件。

我们可以手动安装 tf-pose 模块，也可以直接使用 tf-pose 目录下的文件。如果是手动安装就调用以下代码。

```
python setup.py install
```

该库使用了一个 slidingwindow 模块，因此我们要通过 pip 来下载它。下载安装完后，如果是在 Jupyter Notebook 环境中，我们需要重新启动 Jupyter Notebook 运行环境。

```
pip install slidingwindow
```

这个库提供了两个预训练的人体姿态识别模型，一个是 cmu，另一个 mobilenet_thin，下载模型的地址在前言中提到的库的 README.md 文件中。下载完毕后，将 cmu 的 graph_opt.pb 文件拷贝到目录 models/graph/cmu/下。将 mobilenet_thin 的 graph_opt.pb 文件拷贝到目录 models/graph/mobilenet_thin/下。

13.1.3 单张图片识别

我们先通过 TfPoseEstimator 类来加载预训练模型，并且指定目标的显示窗口大小；然后通过 PIL 模块来加载图片到内存中，以 NumPy 数组的数据格式显示；再通过 TfPoseEstimator 的实例对象来推理计算图片中的人的映射点和线；最后将这些点和线绘制到原始图片上。

```
import time
import numpy as np
import matplotlib.pyplot as plt
from PIL import Image
# 导入tf_pose模块
from tf_pose.estimator import TfPoseEstimator
from tf_pose.networks import get_graph_path
```

```python
# 指定预训练模型的名称，可以是 cmu，也可以是 mobilenet_thin
model = "cmu" #"mobilenet_thin"
# 设置图片输出大小的比率
resize_out_ratio = 4.0
# 设置目标窗口的宽和高
window_width = 375
window_height = 667
# 通过 get_graph_path()函数来获取模型地址
# 使用 TfPoseEstimator 类来初始化预训练模型和目标的窗口显示大小
estimator = TfPoseEstimator(get_graph_path(model), target_size=(window_width, window_height))
# 通过 PIL 模块来加载图片
victor_test_img = Image.open('images/Victor_test_1.jpg')
# 将图片的数据格式转换成 NumPy 的数组数据格式
victor_test_img = np.asarray(victor_test_img)
# 推理计算图片，返回已获知的人的关键部位的数据
humans = estimator.inference(victor_test_img,
                             resize_to_default=(window_width > 0 and window_height > 0),
                             upsample_size=resize_out_ratio)
# 将关键部位的数据绘制到原图上
image = TfPoseEstimator.draw_humans(victor_test_img, humans, imgcopy=False)
# 初始化一个 7×12 大小的窗口
fig, ax = plt.subplots(figsize=(7, 12))
# 显示图片
ax.imshow(image)
# 图片上不显示网格
plt.grid(False)
plt.show()
```

输出结果如图 13.1 所示。

图 13.1　左图是原始图片，右图是人体姿态识别后的图片

然后我们来看测试推理计算图片特征过程中 pafmap 和 heatmap 的 4 张图片，计算出点和连接线，显示向量图（VectorMap）的 X 和 Y 的方式。4 张图片是 2 行 2 列的排列方式，所以 Matplotlib 中 Figure 对象的 add_subplot()函数的 3 个参数中：第一个表示行，第二个表示列，第三个表示所处位置的索引。

```
import cv2
# 第一张图片
fig = plt.figure()
a = fig.add_subplot(2, 2, 1)
# 设置图题
a.set_title('Result')
# 图片变量 image 是在上面读取到的
# 将 BGR 转为 RGB 模式显示图片
plt.imshow(cv2.cvtColor(image, cv2.COLOR_BGR2RGB))
# 不显示网格
plt.grid(False)
# 显示右边的颜色条
plt.colorbar()
# 第二张图片
bgimg = cv2.cvtColor(image.astype(np.uint8), cv2.COLOR_BGR2RGB)
# 将转换后的图片重置大小，插值参数 interpolation 指重新采样
bgimg = cv2.resize(bgimg,
                   (estimator.heatMat.shape[1], estimator.heatMat.shape[0]),
                   interpolation=cv2.INTER_AREA)
a = fig.add_subplot(2, 2, 2)
# 以半透明状态显示图片，用参数 alpha 来控制，参数值在 0 到 1 之间，1 表示完全显示，0 表示完全透明
plt.imshow(bgimg, alpha=0.5)
# 沿着轴 2 反转 heatmat 的数组，取最大值，就是突出检测到的点
tmp = np.amax(estimator.heatMat[:, :, :-1], axis=2)
# 以灰度的、半透明的方式显示图片
plt.imshow(tmp, cmap=plt.cm.gray, alpha=0.5)
# 设置图题
a.set_title('Dot Network')
plt.grid(False)
plt.colorbar()
# 第三张图片
# 转置 pafMat 的数组
tmp2 = estimator.pafMat.transpose((2, 0, 1))
# 沿着轴 0 取数组的奇数的最大值
tmp2_odd = np.amax(np.absolute(tmp2[::2, :, :]), axis=0)
# 沿着轴 0 取数组的偶数的最大值
tmp2_even = np.amax(np.absolute(tmp2[1::2, :, :]), axis=0)
a = fig.add_subplot(2, 2, 3)
# 设置图题
a.set_title('Vectormap-X')
# 以灰度的、半透明的方式显示奇数图片
plt.imshow(tmp2_odd, cmap=plt.cm.gray, alpha=0.5)
plt.grid(False)
```

```
plt.colorbar()
# 第四张图片
a = fig.add_subplot(2, 2, 4)
# 设置图题
a.set_title('Vectormap-Y')
# 以灰度的、半透明的方式显示偶数图片
plt.imshow(tmp2_even, cmap=plt.cm.gray, alpha=0.5)
plt.colorbar()
plt.grid(False)
plt.show()
```

输出结果如图 13.2 所示。

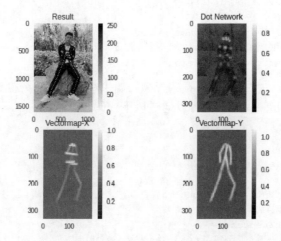

图 13.2　测试推理计算过程中 pafmap 和 heatmap 的 4 张图片

13.1.4　视频中的人体姿态识别

先实例化 TfPoseEstimator 类来加载预训练模型和定义目标的显示窗口大小；然后定义处理视频中每一帧的 process_frame() 函数，原视频文件名是 test_video_1.mp4，识别后保存的目标视频文件名是 test_video_1_out.mp4；最后通过 moviepy 模块的 VideoFileClip 类来处理视频。

```
# 导入 moviepy 模块
from moviepy.editor import VideoFileClip
import cv2
# 导入 tf_pose 模块
from tf_pose.estimator import TfPoseEstimator
from tf_pose.networks import import get_graph_path
# 设置目标窗口的宽和高
width = 375
height = 667
# 实例化 TfPoseEstimator 对象
estimator = TfPoseEstimator(get_graph_path("cmu"), target_size=(width, height))
# 定义处理视频中每一帧的函数
```

```python
def process_frame(frame):
    # 对当前这一帧进行推理计算
    humans = estimator.inference(frame, resize_to_default=False, upsample_size=4.0)
    # 将计算的人体姿态结果数据绘制到原帧(图像)上
    image = TfPoseEstimator.draw_humans(frame, humans, imgcopy=False)
    # 返回该帧,相当于把视频中的当前这一帧修改了
    return image
# 输入视频的文件名
input_filename = 'test_video_1.mp4'
# 输出视频的文件名
output_filename = 'test_video_1_out.mp4'
# 初始化视频剪辑对象,剪切视频在 0 秒到 10 秒之间
clip_playing = VideoFileClip(input_filename).subclip(0, 10)
# 处理视频每一帧(图像)
white_clip = clip_playing.fl_image(process_frame)
# 将处理后的视频写入到文件,%time 表示处理时显示进度,仅在 Jupyter Notebook 中有效
%time white_clip.write_videofile(output_filename)
```

写入视频文件成功后,我们来查看新视频。如果是在本地打开新视频文件,可以直接观看;如果是在 Jupyter Notebook 中,可以通过以下代码查看视频。

```python
import io
import base64
from IPython.display import HTML
# 通过 io 模块加载视频
video = io.open('test_video_1_out.mp4', 'r+b').read()
# 对视频的二进制数据进行 base64 编码
encoded = base64.b64encode(video)
# 以 HTML 的 video 标签方式显示到 Jupyter Notebook 的输出中
HTML(data='''<video alt="test" controls>
                <source src="data:video/mp4;base64,{0}" type="video/mp4" />
             </video>'''.format(encoded.decode('ascii')))
```

随机截取视频中的两张图片,输出如图 13.3 和图 13.4 所示。

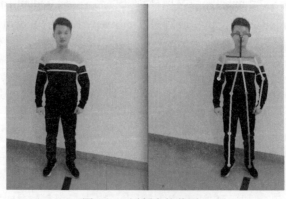

图 13.3 视频中的截图 1

图 13.4 视频中的截图 2

13.1.5 实时摄像识别

我们还是先实例化 TfPoseEstimator 类来加载预训练模型和定义目标的显示窗口大小，然后通过 cv2.VideoCapture(0)函数来开启第 0 个摄像头设备（就是当前计算机的默认摄像头），再通过 read()函数把读取到的每一帧（图片）进行人体的姿态推理计算识别，最后通过 cv2.imshow()函数将处理后的帧（图片）显示到窗口上。

```
import time
import cv2
import numpy as np
# 导入 tf_pose 模块
from tf_pose.estimator import TfPoseEstimator
from tf_pose.networks import get_graph_path, model_wh
fps_time = 0
# 使用的预训练模型
model = "mobilenet_thin"
# 第 0 个摄像头设备，就是计算机默认的摄像头
camera = 0
resize_out_ratio = 4.0
# 设置目标窗口的宽和高
window_width = 600
window_height = 400
# 实例化 TfPoseEstimator 对象
estimator = TfPoseEstimator(get_graph_path(model), target_size=(window_width, window_height))
# 开启第一个摄像头
video_capture = cv2.VideoCapture(0)
# 循环不停止地从摄像头读取图片（帧）
while True:
    # 读取摄像头拍摄到的图片（帧）
    ret_val, image = video_capture.read()
    # 推理计算图片
    humans = estimator.inference(image,
                                 resize_to_default=(window_width > 0 and window_height > 0),
```

```
                                 upsample_size=resize_out_ratio)
        # 将计算出来的人体姿态绘制到原帧(图片)上
        image = TfPoseEstimator.draw_humans(image, humans, imgcopy=False)
        # 显示 FPS 的计时文字到图片上
        cv2.putText(image,
                    "FPS: {}".format(1.0 / (time.time() - fps_time)),
                    (10, 10),  cv2.FONT_HERSHEY_SIMPLEX, 0.5,
                    (0, 255, 0), 2)
        # 将图片显示到窗口上
        cv2.imshow('Pose Estimation in Realtime', image)
        fps_time = time.time()
        # 当焦点聚集在窗口上时,按下 q 键,停止拍摄
        if cv2.waitKey(1) & 0xFF == ord('q'):
            break
# 当不需要拍摄时,释放摄像头对象的内存
video_capture.release()
# 当不需要拍摄时,释放显示窗口的内存
cv2.destroyAllWindows()
```

摄像头随机拍摄到的一张图片如图 13.5 所示。

图 13.5　摄像头随机拍摄到的一张图片

13.2　基于 Keras 实现人体姿态识别

人体姿态识别除了可以通过基于 OpenPose 库的 TensorFlow 实现外,还可以通过 Keras 的深度卷积神经网络(DCNN)实现。该 Keras 的实现在首届 COCO 2016 关键点挑战赛中获胜,还在 2016 ECCV 上获得了最佳演示奖。我们将使用这个库来做单张图片、视频中的人体姿态

识别和摄像头拍摄中的人体姿态实时识别。

13.2.1 环境准备

- numpy = 1.14.6。
- scipy = 1.1.0。
- keras = 2.2.4。
- configobj = 5.0.6。
- cv2 = 3.4.3。
- PIL = 4.0.0。
- matplotlib = 3.0.2。
- IPython = 5.0.5。

13.2.2 下载仓库

我们先通过 git 克隆代码库,该仓库的名称就是 Keras 的多人姿态实时评估,它是在 CVPR 2017 论文的基础上用 Keras 实现的。克隆下载该代码库命令如下。

git clone 请指定本 keras Realtime Multi-Person Pose Estimation.git 代码仓库地址

克隆完后,进入 keras_Realtime_Multi-Person_Pose_Estimation 代码库目录。

```
import os
os.chdir("keras_Realtime_Multi-Person_Pose_Estimation")
```

该仓库主要的文件解释分别如下。
- model/get_keras_model.sh:已训练的 Keras 模型下载。
- demo_image.py:单张图像测试推理计算人体姿态。
- demo_video.py:视频中的人体姿态识别。
- demo_camera.py:实时摄像头拍摄的人体姿态识别。

模型的下载地址可以在本章仓库的 README.md 文件中看到。读者需下载该模型 model.h5 文件,将之放置在本仓库的 keras 目录下。

13.2.3 单张图片识别

我们通过 demo_image.py 文件来识别单张图片中的人体姿态,传入的 3 个参数分别解释如下。
- 参数 image:输入图片,文件名是 Victor_test_1.jpg。
- 参数 output:识别后的输出图片,文件名是 Victor_test_1_out.jpg。
- 参数 model:指定已训练的模型,在 keras 目录下的 model.h5 文件。

```
python demo_image.py \
--image "Victor_test_1.jpg" \
```

```
--output "Victor_test_1_out.jpg" \
--model "./keras/model.h5"
```

输出结果如图 13.6 所示。

图 13.6　左图是原图，右图是识别后的图

13.2.4　视频中的人体姿态识别

通过 demo_video.py 文件可以对视频中的人体姿态进行识别，我们需要指定 4 个参数，它们分别解释如下。

- 参数 video：输入的视频文件，名称是 test_video_1.mp4，和上一节的原视频是一样的。
- 参数 model：已训练的模型，我们用它来对视频里的每一帧进行推理计算识别。
- 参数 frame_ratio：帧率。
- 参数 process_speed：处理速度。

```
python demo_video.py \
--video "test_video_1.mp4" \
--model "./keras/model.h5" \
--frame_ratio 1 \
--process_speed 4
```

随机输出一张视频中的截图，如图 13.7 所示。

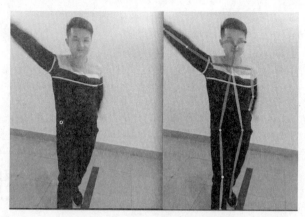

图 13.7 视频中随机的一张截图

13.2.5 实时摄像识别

通过 demo_camera.py 文件,用摄像头实时拍摄人的姿态并识别,需要传入 3 个参数,分别解释如下。

- 参数 model:已训练模型,在 keras 目录下的 model.h5 模型文件。
- 参数 frame_ratio:帧率。
- 参数 process_speed:处理速度。

```
python demo_camera.py \
--model "./keras/model.h5" \
--frame_ratio 1 \
--process_speed 4
```

输出的拍摄结果根据读者实际情况而定,此处拍摄的就是作者的一些动作,与 13.1.5 小节中的输出一样。

13.3 小结

我们所使用的开源库是基于 TensorFlow 和 Keras 实现的。原版是用 Caffe 实现的 OpenPose,现在直接使用下载的预训练模型,进行单张图片、视频中的人体姿态识别和摄像头的实时人体姿态识别。

第 14 章

皮肤癌分类

皮肤癌是最常见的癌症之一，一般不容易被发现，可是一旦被发现，病情就很严重了。目前的皮肤科检查都是从临床筛查开始，然后通过皮肤镜分析、活检和组织病理学检查。国内已经有一些医院正在使用深度学习的技术来分类和识别病理图像中的皮肤癌症状，大大提高了对患者病情的诊断速度。本章将通过卷积神经网络（CNN）来对损伤皮肤的黑素瘤进行检测识别。

14.1 数据准备

数据集中的病理图片包含 3 个类别，分别是 melanoma（黑素瘤）、nevus（痣）和 seborrheic keratosis（脂溢性角化病），共计 600 张图片。名称是 "ISIC 2017: Skin Lesion Analysis Towards Melanoma Detection"。

数据集可以通过本书前言中提到的 Github 地址下载。该数据集包含 3 个类别，每个类别里就是对应病理类别的一张张病理图片，目录结构如下。

```
datasets/
    melanoma/
        ISIC_0012099.jpg
        ISIC_0012151.jpg
        ……
    nevus/
        ISIC_0001769.jpg
        ISIC_0001852.jpg
        ……
    seborrheic_keratosis/
        ISIC_0012143.jpg
        ISIC_0012204.jpg
        ……
```

14.1.1 环境准备

- numpy = 1.14.6。
- sklearn = 0.19.2。
- matplotlib = 2.1.2。
- tensorflow = 1.12.0。
- keras = 2.2.4。
- tqdm = 4.28.1。
- PIL = 4.0.0。

14.1.2 数据下载

在训练前，我们先下载数据集。在 sklearn 中有一个非常易用的加载函数可供使用，就是 load_files()。它可以将目录及子目录的嵌套加载成 NumPy 数组，每个元素都是一个个文件（图片）的相对路径。

```
from sklearn import datasets
filepath = "dataset"
data = datasets.load_files(filepath)
filename_list = data["filenames"]
target_list = data["target"]
target_name_list = data["target_names"]
```

加载完后，如果想看 data 变量中有哪些 key 可以取值，可使用 Python 内置的 dir() 函数，再将 data 作为参数传入，即可看到相应的 key。其中 filename_list 变量就是文件的相对路径，我们先输出，看着有多少张图片。

```
filename_list.shape
```

输出如下。

```
(600,)
```

共计 600 张图片，查看前 20 个元素的值。

```
filename_list[:20]
```

输出如下。

```
array(['dataset/nevus/ISIC_0015990.jpg', 'dataset/nevus/ISIC_0012216.jpg',
       'dataset/nevus/ISIC_0014952.jpg', 'dataset/nevus/ISIC_0016040.jpg',
       'dataset/nevus/ISIC_0014867.jpg',
       'dataset/seborrheic_keratosis/ISIC_0014278.jpg',
       'dataset/nevus/ISIC_0015988.jpg',
       'dataset/seborrheic_keratosis/ISIC_0012330.jpg',
       'dataset/nevus/ISIC_0015995.jpg', 'dataset/nevus/ISIC_0013897.jpg',
       'dataset/nevus/ISIC_0015270.jpg', 'dataset/nevus/ISIC_0015954.jpg',
       'dataset/nevus/ISIC_0014697.jpg', 'dataset/nevus/ISIC_0015160.jpg',
       'dataset/melanoma/ISIC_0014110.jpg',
```

```
        'dataset/nevus/ISIC_0016014.jpg', 'dataset/nevus/ISIC_0016043.jpg',
        'dataset/melanoma/ISIC_0014790.jpg',
        'dataset/nevus/ISIC_0014773.jpg', 'dataset/nevus/ISIC_0015544.jpg'],
       dtype='<U45')
```

查看目标数据,显示 target_list 变量的代码如下。

```
target_list
```

输出如下。

```
array([1, 1, 1, 1, 1, 2, 1, 2, 1, 1, 1, 1, 1, 1, 0, 1, 1, 0, 1, 1, 2, 1,
       1, 1, 1, 1, 1, 1, 1, 2, 1, 1, 1, 0, 1, 2, 0, 1, 1, 0, 1, 1, 1, 1,
       1, 0, 1, 0, 1, 1, 0, 1, 0, 0, 1, 1, 1, 2, 1, 1, 0, 1, 1, 2, 1,
       ......
       0, 1, 1, 1, 0, 1, 1, 2, 1, 0, 1, 1, 1, 1, 2, 0, 2, 1, 0, 1, 1, 2,
       1, 1, 2, 2, 1, 1, 1, 1, 0, 1, 1, 0, 0, 2, 1, 0, 2, 1, 1, 1, 1,
       0, 1, 0, 1, 1, 2])
```

0、1、2 分别代表 3 个病理类别。因为 filename_list 里都是一张张病理图片的相对路径,所以每个路径对应的类别就用 target_list 中的元素来表示。查看 0、1、2 分别对应的病理类别名称列表,显示 target_name_list 变量的代码如下。

```
target_name_list
```

输出如下。

```
['melanoma', 'nevus', 'seborrheic_keratosis']
```

对应的 3 个病理类别:melanoma 是黑素瘤,nevus 是痣,seborrheic keratosis 是脂溢性角化病。

14.1.3 数据可视化

我们随机查看 9 张病理图片。读者在运行此代码后,跟作者运行的结果可能不一致,因为我们使用了随机数,每次产生的结果都不一样。图片以 3 行 3 列的方式显示,每张图片下面都显示出对应的病理类别。

```python
import matplotlib.pyplot as plt
# 设置 matplotlib 在绘图时的默认样式
plt.style.use('default')
from matplotlib import image
import numpy as np
# 创建 9 个绘图对象,3 行 3 列
fig, axes = plt.subplots(nrows=3, ncols=3)
# 设置绘图的总容器大小
fig.set_size_inches(10, 9)
# 随机选择 9 个数,也就是 9 张病理图片(可能重复,且每次都不一样)
random_9_nums = np.random.choice(len(filename_list), 9)
# 从数据集中选出 9 张图片和它的路径
random_9_imgs = filename_list[random_9_nums]
print(random_9_imgs)
```

```
# 根据这随机的 9 张图片路径，截取相应的皮肤癌病理名称
imgname_list = []
for imgpath in random_9_imgs:
    imgname = imgpath[len(filepath) + 1:]
    imgname = imgname[:imgname.find('/')]
    imgname_list.append(imgname)
index = 0
for row_index in range(3): # 行
    for col_index in range(3): # 列
        # 读取图片的数值内容
        img = image.imread(random_9_imgs[index])
        # 获取绘图 Axes 对象，根据[行索引,列索引]
        ax = axes[row_index, col_index]
        # 在 Axes 对象上显示图片
        ax.imshow(img)
        # 设置在绘的图下面显示皮肤癌病理名称
        ax.set_xlabel(imgname_list[index])
        # 索引加 1
        index += 1
```

输出如下。

```
['dataset/melanoma/ISIC_0013814.jpg' 'dataset/nevus/ISIC_0016060.jpg'
 'dataset/melanoma/ISIC_0013678.jpg'
 'dataset/seborrheic_keratosis/ISIC_0012928.jpg'
 'dataset/nevus/ISIC_0015526.jpg' 'dataset/melanoma/ISIC_0013813.jpg'
 'dataset/seborrheic_keratosis/ISIC_0012314.jpg'
 'dataset/nevus/ISIC_0015237.jpg' 'dataset/nevus/ISIC_0015992.jpg']
```

随机显示的 9 张病理图片如图 14.1 所示。

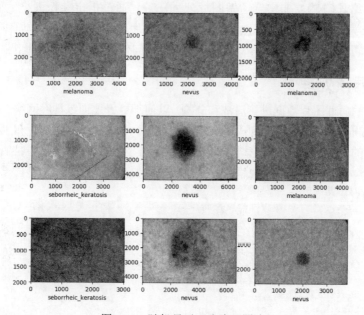

图 14.1　随机显示 9 张病理图片

因为我们定义的神经网络接收的图片大小是(224, 224, 3)，对以上随机显示的 9 张图片，我们需要查看它的 shape。

```python
def print_img_shape(i, filepath):
    shape = image.imread(filepath).shape
    print("第{}张的 shape 是{}".format(i + 1, shape))

print("查看病理图片的大小：\r")
for i, img_path in enumerate(random_9_imgs):
    print_img_shape(i, img_path)
```

输出如下。

```
查看病理图片的大小：
第 1 张的 shape 是(2848, 4288, 3)
第 2 张的 shape 是(2848, 4288, 3)
第 3 张的 shape 是(2000, 3008, 3)
第 4 张的 shape 是(2592, 3872, 3)
第 5 张的 shape 是(4401, 6621, 3)
第 6 张的 shape 是(2848, 4288, 3)
第 7 张的 shape 是(2000, 3008, 3)
第 8 张的 shape 是(4459, 6708, 3)
第 9 张的 shape 是(2592, 3872, 3)
```

从输出的结果可以看出，我们最终需要对图片大小进行处理。

14.2 基于 Keras 的卷积神经网络实现分类

卷积神经网络（CNN）非常合适做图片分类任务。通过指定卷积大小、窗口移动大小，一步步地移动来学习数据特征。每次学习计算卷积层后，计算一次最大池化层，就是将最大值取出来。这样可以防止过拟合，降低维度，经过反向传播训练，直到最优。

14.2.1 数据预处理

我们对所有的病理图片数据进行清洗和分割，以便于更好地训练神经网络。我们通过 sklearn 库的 train_test_split() 把图片数据分割成训练集和测试集，分别有 480 张图片和 120 张图片；然后我们再把测试集的 120 张图片分割一半给验证集，另一半给测试集。

```python
from sklearn import model_selection
# 分割训练集和测试集
X_train, X_test, y_train, y_test = model_selection.train_test_split(filename_list, target_list, test_size=0.2)
# 将测试集数据分割一半给验证集
half_test_count = int(len(X_test) / 2)
X_valid = X_test[:half_test_count]
y_valid = y_test[:half_test_count]
X_test = X_test[half_test_count:]
y_test = y_test[half_test_count:]
print("X_train.shape={}, y_train.shape={}.".format(X_train.shape, y_train.shape))
```

14.2 基于 Keras 的卷积神经网络实现分类

```
print("X_valid.shape={}, y_valid.shape={}.".format(X_valid.shape, y_valid.shape))
print("X_test.shape={}, y_test.shape={}.".format(X_test.shape, y_test.shape))
```

输出如下。

```
X_train.shape=(480,), y_train.shape=(480,).
X_valid.shape=(60,), y_valid.shape=(60,).
X_test.shape=(60,), y_test.shape=(60,).
```

将每张图片加载到内存，以数值表示，并转换成神经网络期待的大小。然后把图片归一化处理，就是将 RGB 对应的具体数值转换成 0 到 1 之间的值，便于神经网络计算。

```python
from keras.preprocessing import image
from tqdm import tqdm
def path_to_tensor(img_path):
    """
    定义一个函数，将每张图片都转换成卷积神经网络期待的大小(1, 224, 224, 3)
    """
    # 使用 PIL 库的 load_img()方法加载图片，它返回一个 PIL 对象
    img = image.load_img(img_path, target_size=(224, 224, 3))
    # 将 PIL 图片对象类型转化成格式为(224, 224, 3)的三维数组
    x = image.img_to_array(img)
    # 将三维数组转化成格式为(1, 224, 224, 3)的四维张量并返回
    return np.expand_dims(x, axis=0)
def paths_to_tensor(img_paths):
    """
    定义一个函数，将数组里的所有路径的图片都转换成图片数值类型并返回
    """
    # tqdm 模块表示使用进度条显示，传入一个所有图片的数组对象
    # 将所有图片的对象一个个都转换成 NumPy 数值对象张量，并返回成数组
    list_of_tensors = [path_to_tensor(img_path) for img_path in tqdm(img_paths)]
    # 将对象垂直堆砌排序摆放
    return np.vstack(list_of_tensors)
import numpy as np
from PIL import ImageFile
# 为了防止 PIL 读取图片对象时出现 IO 错误，设置截断图片为 True
ImageFile.LOAD_TRUNCATED_IMAGES = True
# 将所有图片都转换成标准大小的数值图片对象，然后除以 255
# 进行归一化处理（简单地说，就是将 RGB 值转换成 0 到 1 之间的值，便于神经网络计算和处理）
# RGB 的颜色值，最大为 255，最小为 0
# 对训练集数据进行处理
train_tensors = paths_to_tensor(X_train).astype(np.float32) / 255
# 对验证集数据进行处理
valid_tensors = paths_to_tensor(X_valid).astype(np.float32) / 255
# 对测试集数据进行处理
test_tensors = paths_to_tensor(X_test).astype(np.float32) / 255
```

没有输出信息，但是我们通过 tqdm 对象在循环迭代时可以看到处理的进度。

14.2.2 创建 CNN 模型

我们创建的 CNN 模型主要由卷积层、最大池化层和 Dropout 层组成，最后有一个全连接

层作为输出层。卷积层的深度从 16、32、64 到 128，每个卷积层的内核大小都是 1×1，移动跨步也是 1×1，边缘填充都使用 same；最大池化层大小也都是 1×1；Dropout 层丢弃的比例分别由 0.5、0.3、0.2 和 0.2 组成，减去 Dropout 的比例，剩下的表示在本次网络模型计算时被保留了下来。最后对计算结果做一个全局平均池化计算，为了防止过拟合，再添加一个 Dropout 层，输出 3 个类别的全连接层。

```python
from keras.layers import Conv2D, MaxPooling2D, GlobalAveragePooling2D
from keras.layers import Dropout, Dense
from keras.models import Sequential
# 图片的shape
input_shape = train_tensors[0].shape
# 有多少个类别
num_classes = len(target_name_list)
# 创建 Sequential 模型
model = Sequential()
# 创建输入层，输入层必须传入 input_shape 参数以表示图片大小；输入深度从 16 开始
model.add(Conv2D(filters=16, kernel_size=(1, 1), strides=(1, 1), padding='same',
                 activation='relu', input_shape=input_shape))
# 添加最大池化层，卷积层大小都是 1×1，有效填充范围默认是 valid
model.add(MaxPooling2D(pool_size=(1, 1)))
# 添加 Dropout 层，每次丢弃 50%的网络节点，防止过拟合
model.add(Dropout(0.5))
# 添加卷积层，深度是 32，内核大小是 1×1，跨步是 1×1，使用 ReLU 来调节神经网络
model.add(Conv2D(filters=32, kernel_size=(1, 1), strides=(1, 1), padding='same',
activation='relu'))
model.add(MaxPooling2D(pool_size=(1, 1)))
model.add(Dropout(0.3))
# 添加卷积层，深度是 64
model.add(Conv2D(filters=64, kernel_size=(1, 1), strides=(1, 1), padding='same',
activation='relu'))
model.add(MaxPooling2D(pool_size=(1, 1)))
model.add(Dropout(0.2))
# 添加卷积层，深度是 128
model.add(Conv2D(filters=128, kernel_size=(1, 1), strides=(1, 1), padding='same',
activation='relu'))
model.add(MaxPooling2D(pool_size=(1, 1)))
model.add(Dropout(0.2))
# 添加全局平均池化层，对空间数据进行处理
model.add(GlobalAveragePooling2D())
# 添加 Dropout，每次丢弃 50%
model.add(Dropout(0.5))
# 添加输出层，有 3 个类别输出
model.add(Dense(num_classes, activation="softmax"))
# 打印输出网络模型架构
model.summary()
```

输出神经网络模型架构的信息如图 14.2 所示。可训练参数有 11427 个，最后一个全连接层（dense）是 3，表示网络模型最终有 3 个类别的概率输出。

```
Layer (type)                 Output Shape              Param #
=================================================================
conv2d_1 (Conv2D)            (None, 224, 224, 16)      64
_____
max_pooling2d_1 (MaxPooling2 (None, 224, 224, 16)      0
_____
dropout_1 (Dropout)          (None, 224, 224, 16)      0
_____
conv2d_2 (Conv2D)            (None, 224, 224, 32)      544
_____
max_pooling2d_2 (MaxPooling2 (None, 224, 224, 32)      0
_____
dropout_2 (Dropout)          (None, 224, 224, 32)      0
_____
conv2d_3 (Conv2D)            (None, 224, 224, 64)      2112
_____
max_pooling2d_3 (MaxPooling2 (None, 224, 224, 64)      0
_____
dropout_3 (Dropout)          (None, 224, 224, 64)      0
_____
conv2d_4 (Conv2D)            (None, 224, 224, 128)     8320
_____
max_pooling2d_4 (MaxPooling2 (None, 224, 224, 128)     0
_____
dropout_4 (Dropout)          (None, 224, 224, 128)     0
_____
global_average_pooling2d_1 ( (None, 128)               0
_____
dropout_5 (Dropout)          (None, 128)               0
_____
dense_1 (Dense)              (None, 3)                 387
=================================================================
Total params: 11,427
Trainable params: 11,427
Non-trainable params: 0
```

图 14.2 卷积神经网络模型架构信息

14.2.3 编译模型

我们在编译模型时需要指定 Loss Function（损失函数）。categorical_crossentropy（分类交叉熵）和 sparse_categorical_crossentropy（稀疏分类交叉熵）的使用区别：如果 target 的值经过了独热编码处理，则损失函数使用分类交叉熵；如果 target 是数值，且没有经过独热编码处理，那就使用稀疏分类交叉熵。优化器是 Adam，它是两种优化器 AdaGrad 和 RMSProp 优化的算法，能自动相应地调整学习率，实现简单，计算高效，适用于大规模数据和参数的场景。

```
model.compile(optimizer='adam', loss='sparse_categorical_crossentropy', metrics=['accuracy'])
# 比如我们这里使用 sparse_categorical_crossentropy 作为损失函数，所以以下的 one-hot encoding 就不需要
# 如果读者使用 categorical_crossentropy，则取消注释以下代码来运行
# from keras import utils
# y_train = utils.to_categorical(y_train)
# y_valid = utils.to_categorical(y_valid)
# y_test = utils.to_categorical(y_test)
```

14.2.4 训练模型

设置 epochs 为 10，表示迭代训练 10 次。在训练模型时，我们将最佳的检查点通过模型

权重路径保存为扩展名为.hdf5 的文件，参数 save_best_only 表示仅保存最佳的权重。训练时，每批次（batch）20 张图片，然后验证。检查点对象以回调的形式传入。

```
from keras.callbacks import ModelCheckpoint
# 创建检查点对象
checkpointer = ModelCheckpoint(filepath='saved_models/skin_cancer.best_weights.hdf5',
                               verbose=1,
                               save_best_only=True)
epochs = 10
# 训练模型
model.fit(train_tensors,
          y_train,
          validation_data=(valid_tensors, y_valid),
          epochs=epochs,
          batch_size=20,
          callbacks=[checkpointer],
          verbose=1)
```

训练时输出如下。

```
Train on 480 samples, validate on 60 samples
Epoch 1/10
480/480 [==============================] - 320s 666ms/step - loss: 0.9674 - acc: 0.6229 - val_loss: 0.8786 - val_acc: 0.6667
Epoch 00001: val_loss improved from inf to 0.87865, saving model to saved_models/skin_cancer.best_weights.hdf5
Epoch 2/10
……
Epoch 00009: val_loss did not improve from 0.86875
Epoch 10/10
480/480 [==============================] - 302s 630ms/step - loss: 0.8844 - acc: 0.6521 - val_loss: 0.8689 - val_acc: 0.6667
Epoch 00010: val_loss did not improve from 0.86875
<keras.callbacks.History at 0x7f7215e54828>
```

最终模型的精确度是 0.6667。

14.2.5　评估模型和图像测试

我们先加载刚才训练的权重文件，然后通过 evaluate()函数来对测试集图像数据进行模型评估。

```
# 加载刚才训练的权重到模型中
model.load_weights("saved_models/skin_cancer.best_weights.hdf5")
# 评估模型精确度
score = model.evaluate(test_tensors, y_test, verbose=1)
print("Test {}: {:.2f}. Test {}: {:.2f}.".format(model.metrics_names[0],
                                                  score[0]*100,
                                                  model.metrics_names[1],
                                                  score[1]*100))
```

输出如下。

```
Test loss: 88.00. Test acc: 66.67.
```

另一种模型评估的方式是，只要有模型训练后的权重文件，就可以直接加载权重以进行评估。这里作者事先准备了一张未经过模型训练的图片来测试它的精确度。

```python
from keras.models import load_model
# 加载模型
model = load_model('saved_models/skin_cancer.best_weights.hdf5')

# 加载一张病理图片来测试模型精确度
test_img_path = "nevus_ISIC_0007332.jpg"
# 将图片转换成四维的 NumPy 数值数组，path_to_tensor()函数已在上文定义过
image_tensor = path_to_tensor(test_img_path)
# 归一化，转换成 0 到 1 之间的数值
image_tensor = image_tensor.astype(np.float32) / 255
# 模型预测概率
predicted_result = model.predict(image_tensor)
# 打印输出概率
print(predicted_result)
```

输出如下。

```
[[0.22132273 0.59772515 0.18095216]]
```

我们通过预测概率绘制原图和概率文本。

```python
import matplotlib
def draw_predicted_figure(img_path, X, y):
    """
    绘制测试图片和显示预测概率
    """
    # 创建一个绘图对象
    fig, ax = plt.subplots()
    # 设置绘图的总容器大小
    fig.set_size_inches(5, 5)
    # 拼接病理图片对应的名称和它的概率值的字符串
    fig_title = "\n".join(["{}: {:.2f}%\n".format(n, y[i]) for i, n in enumerate(X)])
    # 设置在图片右上角的注解文字
    ax.text(1.01, 0.7,
            fig_title,
            horizontalalignment='left',
            verticalalignment='bottom',
            transform=ax.transAxes)
    # 读取图片的数值内容
    img = matplotlib.image.imread(img_path)
    # 在 Axes 对象上显示图片
    ax.imshow(img)
draw_predicted_figure(test_img_path, target_name_list, predicted_result[0])
```

输出结果如图 14.3 所示。

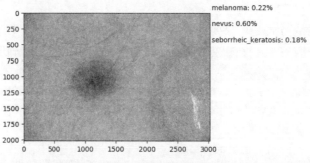

图 14.3 预测该图的病理类别是 nevus

14.3 基于 TensorFlow 的迁移学习实现分类

本节采用 Inception v3 预训练模型训练新类别的图片数据。训练前需要将训练集转换成瓶颈（Bottleneck）文件；Bottleneck 是一个非正式的术语，常用于实际执行分类的最后输出层之前的层。计算 Bottleneck 会让分类器更好地识别所有的图片类别，这些计算的值就是一个小而有意义的摘要，该摘要必须包含足够的识别信息和特征。Bottleneck 的作用就是加快训练速度，减少训练耗时，避免每个图片在模型训练时重复计算。

14.3.1 数据准备

我们先下载代码，该代码已经整理完，请读者自行下载。它包含了重新训练新图片类别和验证模型在新图片上的表现的代码文件。

```
!git clone 请指定本 simple transfer learning.git 代码仓库地址
```

图片数据集就用原生的图片文件即可。

14.3.2 训练模型

在下载的代码文件中，retrain.py 是用来训练模型的，参数解释如下。

- 参数 bottleneck_dir：保存训练图片的 Bottleneck 的目录。
- 参数 how_many_training_steps：所有的图片要训练多少次。
- 参数 learning_rate：学习率，默认是 0.01。
- 参数 model_dir：下载的模型保存目录。
- 参数 summaries_dir：训练模型时保存的训练记录目录。
- 参数 output_graph：训练模型后保存的 pb 文件。
- 参数 output_labels：训练模型后保存的病理类别目录。
- 参数 architecture：指定模型，这里是 Inception v3 模型。
- 参数 dataset_small：数据集目录地址。

```
python -m retrain \
  --bottleneck_dir=tf_files/bottlenecks \
  --how_many_training_steps=1000 \
  --learning_rate=0.05 \
  --model_dir=tf_files/models/ \
  --summaries_dir=tf_files/training_summaries/"inception_v3" \
  --output_graph=tf_files/retrained_graph.pb \
  --output_labels=tf_files/retrained_labels.txt \
  --architecture="inception_v3" \
  --image_dir=dataset_small
```

输出如下。

```
INFO:tensorflow:2018-11-28 00:35:03.956169: Step 0: Train accuracy = 58.0%
INFO:tensorflow:2018-11-28 00:35:03.957484: Step 0: Cross entropy = 0.974238
INFO:tensorflow:2018-11-28 00:35:08.078154: Step 0: Validation accuracy = 55.0% (N=100)
INFO:tensorflow:2018-11-28 00:35:09.188099: Step 10: Train accuracy = 45.0%
INFO:tensorflow:2018-11-28 00:35:09.188387: Step 10: Cross entropy = 1.859571
INFO:tensorflow:2018-11-28 00:35:09.300822: Step 10: Validation accuracy = 36.0% (N=100)
……
INFO:tensorflow:2018-11-28 00:36:57.044767: Step 980: Validation accuracy = 81.0% (N=100)
INFO:tensorflow:2018-11-28 00:36:57.947833: Step 990: Train accuracy = 100.0%
INFO:tensorflow:2018-11-28 00:36:57.947999: Step 990: Cross entropy = 0.012269
INFO:tensorflow:2018-11-28 00:36:58.044394: Step 990: Validation accuracy = 65.0% (N=100)
INFO:tensorflow:2018-11-28 00:36:58.858748: Step 999: Train accuracy = 100.0%
INFO:tensorflow:2018-11-28 00:36:58.858923: Step 999: Cross entropy = 0.013884
INFO:tensorflow:2018-11-28 00:36:58.958107: Step 999: Validation accuracy = 85.0% (N=100)
INFO:tensorflow:Final test accuracy = 45.5% (N=11)
INFO:tensorflow:Froze 2 variables.
```

14.3.3 验证模型

训练完毕后，我们对一张没有参与过训练的图片进行评估和预测，参数解释如下。

- 参数 graph：刚才训练的文件。
- 参数 image：要预测分类的任务。

```
python -m label_image \
  --graph=tf_files/retrained_graph.pb \
  --image=nevus_ISIC_0007332.jpg
```

输出如下。

```
Evaluation time (1-image): 0.923s
nevus (score=0.98778)
seborrheic keratosis (score=0.01222)
melanoma (score=0.00001)
```

14.3.4 Tensorboard 可视化

打开一个终端，切换目录到刚才训练的位置，然后使用 Tensorboard 命令来启动，最终会在浏览器上展示训练的各种记录数据。training_summaries 就是在训练模型时保存的记录文件的目录。

```
tensorboard --logdir tf_files/training_summaries
```

稍等几秒，终端就会自动生成一个链接，然后将它复制到浏览器上，以预览训练模型的记录文件，如图 14.4 和图 14.5 所示。

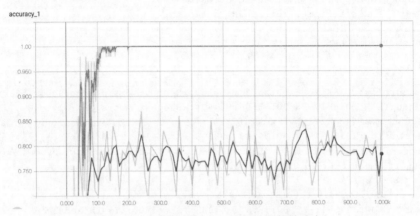

图 14.4　训练模型时，训练和验证的精确度

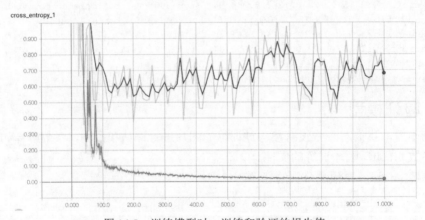

图 14.5　训练模型时，训练和验证的损失值

14.4　小结

我们通过开源的数据集 ISIC 2017 来训练检测黑色素瘤皮肤癌病理图片的模型，使用基于 Keras 构建的卷积神经网络模型和 TensorFlow 的 Inception v3 迁移学习训练出来的模型对 3 个类别的皮肤癌进行了分类与识别。

第 15 章
对象检测

前面的章节已经介绍和使用过简单的图像识别和复杂的图像识别、简单的神经网络模型和卷积神经网络模型,本章仍然讲解卷积神经网络(CNN),但所涉及的是对象检测(Object Detection),对象检测的应用领域十分广泛,例如自动驾驶汽车、无人机、无人超市、实时监控等。对象检测,是指输入一张图片,检测到图片中的人和物后,返回他们的位置。通常,我们在人和物对应的位置标上矩形框,如图 15.1 所示。

图 15.1 对象检测效果预览

15.1 对象检测的应用领域

近几年,对象检测在 3 个领域发展得越来越好,分别是无人机(Drone)、自动驾驶汽车(Auto Driving Car)和无人超市(Unmanned Supermarket)。本节会介绍它们的应用场景。当然现实中还有很多的应用场景和领域等待我们挖掘,也许读者看完后会有更多的灵感。

15.1.1 无人机应用领域

航空成像(Aerial Imagery)。假如您是建筑公司老板,想查看工地的施工情况,您可以派一驾无人机飞往工地上空,然后由无人机自动进行对象检测,实时地将图像传回到您的计算机或者其他设备上。

航拍(Aerial Photo)。例如出去旅游的时候,您想抓拍优美风景的全貌、跟拍人物等,就可以通过对象检测,实时拍照或录制视频,并自动上传到您的智能设备(如手机)上。

农牧业航拍(Aerial Photo)。例如您有一个大型养殖场,现在您想看看牛羊的情况,它们是不是在正常活动;看看西瓜地的情况,西瓜长熟了没、可以不可以采摘,有没有人偷西瓜……这都可以通过实时对象检测技术做到。

15.1.2 自动驾驶汽车应用领域

自动驾驶汽车行业这几年发展得如火如荼,例如百度的阿波罗(Apollo)自动驾驶技术、Google 的 Waymo 自动驾驶技术、Uber 的自动驾驶技术等,都或多或少地应用了实时对象检测技术。但是对于自动驾驶技术,实时检测技术的优化会比一般的实时对象检测更好一些,因为它们具有覆盖的类别更多、层次更深、识别精准度更高等特点。

15.1.3 无人超市应用领域

现在的无人超市省去了收银台的位置,将其改成了摄像头实时检测和识别。从一位顾客过闸机进入超市开始,他就被标记为是否是超市会员、是否使用支付宝等,然后当顾客从货架上取货物后准备出门时,通过摄像头实时检测和货物的 RFID(射频识别)可识别物品的类别和价格,自动从顾客的支付宝或其他账户上扣费。这就是无人超市购物流程,其中实时检测识别是非常重要的。

15.2 原理分析

R-CNN 的全称是 Regional Based CNN,即基于区域的卷积神经网络。到目前为止,它是在快速对象识别和检测方向中发展得最好的一个算法。在这里,我们将分析和使用 R-CNN、Faster R-CNN、Mask R-CNN 和 SSD。

自从 CNN 在 ILSVRC（ImageNet Large Scale Visual Recognition Competition，ImageNet 大规模视觉识别挑战赛）上取得优秀的表现后，由 CNN 演变来的各种图像分类算法和网络模型越来越出色，也越来越实用，尤其是 VGG、Inception、ResNet 50 等神经网络模型。

15.2.1 R-CNN 的介绍与分析

在现实生活中，拍一张图片，上面包含的人或物具有多种形态姿势：独立的或重叠的。我们的目标就是使用分类算法找出图片中人或物的边缘位置，分类出不同的对象。

R-CNN 创建这些对象的边界框时，使用的是选择性搜索（Selective Search）算法。选择性搜索算法通过不同的窗口查看图片，并且对于每个尺寸，尝试通过纹理、颜色和强度将相邻的像素组合到一起以识别对象。

我们来看 R-CNN 的区域特征分类。在创建一组区域提案（Region Proposals）后，R-CNN 将图片传递给 AlexNet 的修改版本，以确定它是否是有效区域，如图 15.2 所示。

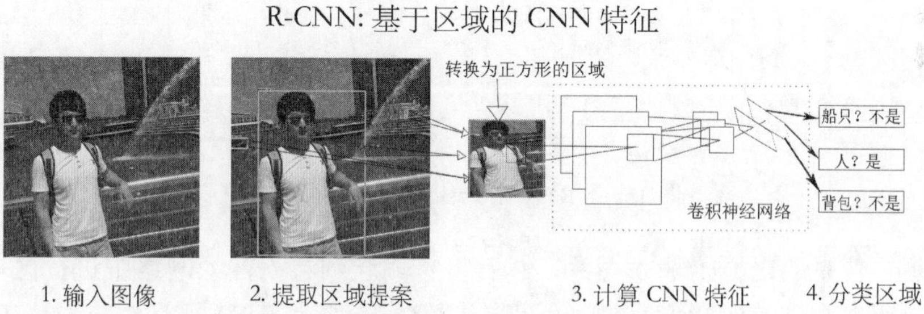

图 15.2　R-CNN 基于大约 2000 个类别分类过程

一旦区域提案创建后，R-CNN 就会将该区域变为标准的正方形大小，并将其传递给 AlexNet 的修改版本（这是 ImageNet 2012 的获奖提交版本，由此启发了 R-CNN）。在 CNN 的最后一层，R-CNN 增加了一个支持向量机（SVM）操作，它简单地分类对象：是否是一个对象、是什么对象。

R-CNN 得益于 CNN 优异的特征提取能力。例如，VOC 2007 mAP（mean Average Precision，平均精度）为 40%左右，现已提升到了 66.0%。选择性搜索在一张图像上提取 2000 个区域，每个区域有不同的作用，这些不同区域的图像大小会被馈送到训练网络中。其中，它为每个区域提取特征向量，然后将向量作为一组线性 SVM 的输入，这些线性 SVM 针对每个类别的训练输出分类，这组向量也被送入边界框以获得最精确的坐标。

15.2.2 Faster R-CNN 的介绍与分析

Fast R-CNN（Fast Region-based Convolutional Neural Networks），中文是快速的基于区域的卷积神经网络。它由 Microsoft 的研究员 Ross Girshick 创建，目前已被弃用，且不再继续

维护，因为升级了一个新版本，叫作 Faster R-CNN。Fast R-CNN 有一个缺点，就是在选择性搜索时比较花费时间，导致大部分时间不是花在计算神经网络分类上。所以，选择性搜索就被替换成了 RPN（Region Proposal Network，区域提案网络）。

Faster R-CNN 原始版本是 MATLAB 的，由任少卿实现；后由 Fast R-CNN 研究员实现了它的 Python 版本。Faster R-CNN 在训练数据 VOC 2007 和 2012 trainval 上，mAP 测试表现为 70.4%。接下来，我们来看 RPN 的关注点，如图 15.3 所示。

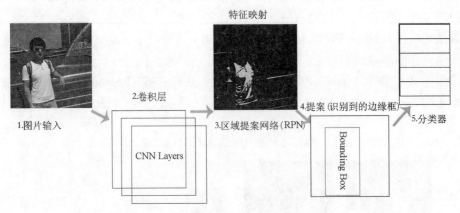

图 15.3　RPN 在 Faster R-CNN 网络中

步骤分解：首先取出原始输入图片，然后将它输送到卷积神经网络中进行区域提案网络，形成边界框。所以在训练时我们需要训练两个网络，一个是 RPN 网络，一个是在得到边界框后的分类网络。常规做法是交替训练，操作就是在一个批次训练中，先训练一次 RPN 网络，再训练一次分类网络，以此类推。

15.2.3　Mask R-CNN 的介绍与分析

Mask R-CNN，中文是掩码 R-CNN，是基于 Faster R-CNN 的像素级别的分割扩展。到目前为止，我们看到了前面的 R-CNN 的选择性搜索和 Faster R-CNN 的 RPN 是如何有效地定位图片中的对象和边界框的。通过 Mask R-CNN 可以更进一步地定位对象的像素位置，而不是边界框。

Mask R-CNN 通过向更快的 R-CNN 添加分支来完成此操作，该分支输出二进制掩码，该掩码表示给定像素是否是对象的一部分。分支是基于 CNN 的特征映射的完全卷积网络，其包含输入是 CNN 特征图。输出是矩阵，在像素所属对象的位置上为 1，在其他位置上为 0（这被称为二进制掩码）。如图 15.4 所示。

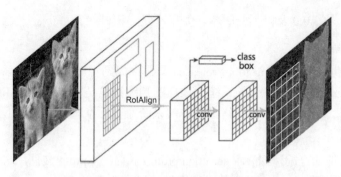

图 15.4　基于 Mask R-CNN 的对象识别

为此，Mask R-CNN 的开发者进行了一次小调整，使这条管道能够顺畅。即当在 Faster R-CNN 的架构上运行对象检测时，Mask R-CNN 的开发者意识到由 RoI Pool 选择的特征图的区域与原始图像的区域略有不对齐。这是因为图像分割需要像素级特性，与边界框不同，自然不准确。他通过使用 RoI Align 巧妙地调整 RoI Pool，以更精确地对齐来解决这个问题。

在 RoI Pool 中，我们将它向下舍入并选择 2 个像素，导致了轻微的错误；但是，在 RoI Align 中，我们避免了这种舍入。相反，我们使用双线性差值（Bilinear Interpolation）来精准地了解像素 2.93 处的内容。这在很大程度上使我们避免了引起 RoI Pool 错位的原因。所以，当生成这些蒙版后，Mask R-CNN 将它们与 Faster R-CNN 中的分类和边界框组合在一起，生成相当精准的分割。

15.3　基于 Mask R-CNN Inception COCO 的图片对象检测

对象检测是基于 TensorFlow Object Detection API 完成的，它从 2017 年 7 月 15 日在 GitHub 上发布后，到现在经历了多次发布更新，最新的一次发布时间是 2018 年 7 月 13 日。本节讲解单张图片的对象检测。

15.3.1　环境准备

- TensorFlow = 1.10.0。
- Protocol Buffer = 3.6.0。
- NumPy = 1.15.0。
- Matplotlib = 2.2.2。
- Pillow = 1.0。

1. 下载 tensorflow/models

git clone 请指定 tensorflow/models.git 代码仓库地址

下载完成后，会有一个 models 目录，而 Object Detection 的相关代码都在目录 models/research/object_detection 里。本次运行这个代码，需要用到 Google Protocol Buffer，所以请确保系统上或虚拟环境中安装了 protoc。如果没有安装，请安装 3.0.0 及其以上的版本；如果已安装，请查看是否是 3.0.0 及以上的版本，如果不是，请升级。

2. protoc 下载

protoc 下载地址在 google/protobuf/releases 页面。找到对应系统的预编译压缩包并下载解压，这里的示例就是 ubuntu 64 位的系统，选择的预编译包是 protoc-3.6.1-linux-x86_64.zip。

在当前目录下，将 protoc 命令拷贝到你的系统命令行默认读取环境下。

```
cp bin/protoc /bin/protoc
```

然后使用 protoc 命令编译 object_detection/protos/ 下所有的 .proto 文件，这就是将 TensorFlow Object Detection API 下的 protobuf 文件进行编译，以供 Python 使用。

```
protoc object_detection/protos/*.proto --python_out=.
```

3. 将 slim 目录模块添加到 PYTHONPATH 环境变量中

```
export PYTHONPATH=$PYTHONPATH:`pwd`:`pwd`/slim
```

执行命令后，打开一个 Python 运行环境，导入 import slim 看看是否无输出，没有任何输出则表示正确。

4. Object Detection 安装完成后测试

在 models/research 目录下，执行如下命令。

```
python object_detection/builders/model_builder_test.py
```

这个命令会自动检查 TensorFlow Object Detection API 是否安装正确，若输出如下所示，则表示安装正确。

```
................
----------------------------------------------------------------------
Ran 18 tests in 0.108s
OK
```

15.3.2　导入 Packages

TensorFlow Object Detection API 默认提供可供使用的 6 个预训练模型，都是基于 COCO 的数据集训练而来的，结构分别是 SSD+MobileNet、SSD+ResNet、SSD+Inception、Faster R-CNN+ResNet、Mask R-CNN+Inception 和 Mask R-CNN+ResNet。每种预训练模型都有一些小区别的扩展模型。读者可以在 models/research/object_detection/g3doc/ 目录下的 detection_model_zoo.md 文件里找到对应的链接。

另外，TensorFlow 官方给出了一个 Jupyter Notebook 示例版本，在 models/research/object_detection/ 目录下的 object_detection_tutorial.ipynb 文件中。在命令行中启动 Jupyter Notebook，可以使用如下命令。

```
jupyter notebook
```

如果读者的计算机上没有安装 Jupyter Notebook，需要自行安装。运行 jupyter notebook

命令后，会在默认浏览器上打开一个页面，默认打开的地址是 http://localhost:8888，紧接着找到 object_detection_tutorial.ipynb 文件并打开，就会看到如图 15.5 所示的内容。

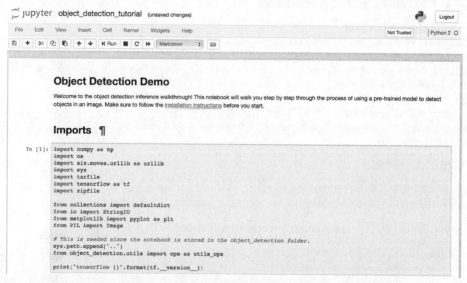

图 15.5　TensorFlow Object Detection Demo Jupyter Notebook 示例

现在，开始导入一些需要用到的包并设置环境。

```
import numpy as np
import os
import six.moves.urllib as urllib
import sys
import tarfile
import tensorflow as tf
import zipfile
from collections import defaultdict
from io import StringIO
from matplotlib import pyplot as plt
from PIL import Image
# 该命令是为了在 Jupyter Notebook 中使用 matplotlib 显示图形
%matplotlib inline
# 查看打印 TensorFlow 版本号
print("tensorflow {}".format(tf.__version__))
```

导入 Object Detection 的 utils 模块。

```
# 将上层目录导入，以便执行下面的模块导入
sys.path.append("..")
from object_detection.utils import ops as utils_ops
# 导入 Object Detection 的 utils 模块
from object_detection.utils import label_map_util
from object_detection.utils import visualization_utils as vis_util
```

15.3.3 下载 Mask R-CNN Inception 2018 预训练模型

指定要下载预训练模型的名称和地址。

```
# 模型的名称和下载地址拼接
MODEL_NAME = 'mask_rcnn_inception_v2_coco_2018_01_28'
MODEL_FILE = MODEL_NAME + '.tar.gz'
DOWNLOAD_BASE = 'http://download.tensorflow.org/models/object_detection/'
# 模型下载解压后的目录里冻结了 graph，此 graph 保存了预训练模型网络的架构
PATH_TO_FROZEN_GRAPH = MODEL_NAME + '/frozen_inference_graph.pb'
# mscoco_label_map.pbtxt 保存了类别和索引的映射关系
PATH_TO_LABELS = os.path.join('data', 'mscoco_label_map.pbtxt')
NUM_CLASSES = 90
```

下载模型的代码如下。

```
opener = urllib.request.URLopener()
opener.retrieve(DOWNLOAD_BASE + MODEL_FILE, MODEL_FILE)
```

下载后，解压模型到 mask_rcnn_inception_v2_coco_2018_01_28 目录。

```
tar_file = tarfile.open(MODEL_FILE)
for file in tar_file.getmembers():
    file_name = os.path.basename(file.name)
    if 'frozen_inference_graph.pb' in file_name:
        tar_file.extract(file, os.getcwd())
```

15.3.4 加载模型到内存中

加载已冻结的预训练模型到内存中。

```
detection_graph = tf.Graph()
with detection_graph.as_default():
    od_graph_def = tf.GraphDef()
    with tf.gfile.GFile(PATH_TO_FROZEN_GRAPH, 'rb') as fid:
        serialized_graph = fid.read()
        od_graph_def.ParseFromString(serialized_graph)
        tf.import_graph_def(od_graph_def, name='')
```

其中，tf.Graph()表示获取 TensorFlow 的默认计算图。tf.GraphDef()表示图的序列化版本，它可以有任何 TensorFlow 前端编写的代码来打印、存储和恢复计算图，一般它存储的文件的扩展名为 pb。

15.3.5 加载类别映射

加载类别和索引的映射关系，此后即可通过对应的索引找到标签类别。比如，索引 5 表示飞机这个类别。这里使用内部的 util 函数来完成映射和对应关系。

```
label_map = label_map_util.load_labelmap(PATH_TO_LABELS)
categories = label_map_util.convert_label_map_to_categories(label_map,
max_num_classes=NUM_CLASSES, use_display_name=True)
category_index = label_map_util.create_category_index(categories)
```

我们输出 category_index 前 10 个类别，以便读者更清晰地知道这块数据显示的是什么。

```
i = 0
for k, v in category_index.items():
    print("索引：{} 对应的类别：{}".format(k, v))
    i += 1
    if i == 10:
        break
```

输出如下。

```
索引：1 对应的类别：{'id': 1, 'name': u'person'}
索引：2 对应的类别：{'id': 2, 'name': u'bicycle'}
索引：3 对应的类别：{'id': 3, 'name': u'car'}
索引：4 对应的类别：{'id': 4, 'name': u'motorcycle'}
索引：5 对应的类别：{'id': 5, 'name': u'airplane'}
索引：6 对应的类别：{'id': 6, 'name': u'bus'}
索引：7 对应的类别：{'id': 7, 'name': u'train'}
索引：8 对应的类别：{'id': 8, 'name': u'truck'}
索引：9 对应的类别：{'id': 9, 'name': u'boat'}
索引：10 对应的类别：{'id': 10, 'name': u'traffic light'}
```

15.3.6　定义函数将图片转为 NumPy 数组

定义一个函数，将输入的图片转换成 NumPy 的三维数组。

```
def load_image_into_numpy_array(image):
    (im_width, im_height) = image.size
    return np.array(image.getdata()).reshape(
        (im_height, im_width, 3)).astype(np.uint8)
```

15.3.7　定义图片对象检测函数

定义图片对象检测函数。

```
# 定义单张图片的检测函数
def run_inference_for_single_image(image, graph):
    # 获取计算图
    with graph.as_default():
        # 开启一个 TensorFlow 会话
        with tf.Session() as sess:
            # 获取输入和输出的张量的句柄
            ops = tf.get_default_graph().get_operations()
            # 获取所有张量的名称
            all_tensor_names = {output.name for op in ops for output in op.outputs}
            tensor_dict = {}
            for key in [ 'num_detections', 'detection_boxes', 'detection_scores',
                        'detection_classes', 'detection_masks' ]:
                tensor_name = key + ':0'
                if tensor_name in all_tensor_names:
                    tensor_dict[key] = tf.get_default_graph().get_tensor_by_name(tensor_name)

            if 'detection_masks' in tensor_dict:
```

```
        # 下面的过程仅仅是针对单张图片的处理
        detection_boxes = tf.squeeze(tensor_dict['detection_boxes'], [0])
        detection_masks = tf.squeeze(tensor_dict['detection_masks'], [0])
        # 需要使用 reframe 将蒙版从框坐标转换为图片坐标并适合图片大小
        real_num_detection = tf.cast(tensor_dict['num_detections'][0], tf.int32)
        detection_boxes = tf.slice(detection_boxes, [0, 0], [real_num_detection, -1])
        detection_masks = tf.slice(detection_masks, [0, 0, 0], [real_num_detection,-1,-1])
        detection_masks_reframed = utils_ops.reframe_box_masks_to_image_masks(
            detection_masks, detection_boxes, image.shape[0], image.shape[1])
        detection_masks_reframed = tf.cast(
            tf.greater(detection_masks_reframed, 0.5), tf.uint8)
        # 通过添加批量维度来遵循惯例
        tensor_dict['detection_masks'] = tf.expand_dims(detection_masks_reframed, 0)

    image_tensor = tf.get_default_graph().get_tensor_by_name('image_tensor:0')
    # 运行对象推理检测,这里就是真正的检测
    output_dict = sess.run(tensor_dict, feed_dict={image_tensor: np.expand_dims(image, 0)})
    # 所有输出都是 float32 NumPy 数组,因此需要适当地转换类型
    # num_detections 表示检测框的个数
    output_dict['num_detections'] = int(output_dict['num_detections'][0])
    # detection_classes 表示每个框对应的检测类别
    output_dict['detection_classes'] = output_dict['detection_classes'][0].astype(np.uint8)
    # detection_boxes 表示检测到检测框
    output_dict['detection_boxes'] = output_dict['detection_boxes'][0]
    # detection_scores 表示检测到的检测结果评分
    output_dict['detection_scores'] = output_dict['detection_scores'][0]
    if 'detection_masks' in output_dict:
        output_dict['detection_masks'] = output_dict['detection_masks'][0]
    return output_dict
```

15.3.8 检测图片中的对象

为了简单地操作,这里我们就用默认提供的两张图片和我新加入的两张图片作为检测目标。

```
# 如果你想测试你自己的图片,就把图片拷贝到 test_images 目录
PATH_TO_TEST_IMAGES_DIR = 'test_images'
TEST_IMAGE_PATHS = [ os.path.join(PATH_TO_TEST_IMAGES_DIR, 'image{}.jpg'.format(i)) for i in range(1, 5) ]
# 可视化显示结果的输出大小,单位:英寸
IMAGE_SIZE = (20, 16)
```

遍历图片,检测每张图片上的对象和边缘框。

```
for image_path in TEST_IMAGE_PATHS:
    image = Image.open(image_path)
    # 将读取到的图片转换成 NumPy 多维数组
    image_np = load_image_into_numpy_array(image)
    # 由于模型需要的维度 shape 是 [1, None, None, 3],所以我们需要扩展维度
    image_np_expanded = np.expand_dims(image_np, axis=0)
    # 运行对象检测
    output_dict = run_inference_for_single_image(image_np, detection_graph)
```

```
# 可视化检测的结果
vis_util.visualize_boxes_and_labels_on_image_array(
    image_np,
    output_dict['detection_boxes'],
    output_dict['detection_classes'],
    output_dict['detection_scores'],
    category_index,
    instance_masks=output_dict.get('detection_masks'),
    use_normalized_coordinates=True,
    line_thickness=8)
plt.figure(figsize=IMAGE_SIZE)
plt.imshow(image_np)
```

15.3.9 效果预览

检测结果预览，如图 15.6 和图 15.7 所示。

图 15.6 效果图 1

图 15.7 效果图 2

15.4 基于 Faster R-CNN Inception COCO 的视频实时对象检测

视频里的实时对象检测技术也是基于 TensorFlow Object Detection API 来完成的，本节我们使用 Faster R-CNN Inception COCO 模型 V2 版本，它是一个预训练模型。本节主要内容是在视频的一帧图像中检测到对象后，先标记出边界框，然后把该帧更新到视频中。

15.4.1 环境准备

环境的准备和上节是一致的，所以 tensorflow/models 库和 protoc 就已经准备好了。在执行以下代码前，请将代码的执行路径定位到 models/research/object_detection。

```
import os
os.chdir("models/research/object_detection")
```

15.4.2 导入 Packages

导入一些需要用到的包并设置环境。

```
import numpy as np
import os
import six.moves.urllib as urllib
import sys
import tarfile
import tensorflow as tf
import zipfile
from collections import defaultdict
from io import StringIO
from matplotlib import pyplot as plt
from PIL import Image
# 这是为了在 Jupyter Notebook 中显示图片
%matplotlib inline
# 查看打印 TensorFlow 版本号
print("tensorflow {}".format(tf.__version__))
```

导入 Object Detection 的 utils 模块。其中，sys.path.append()函数用来将指定的路径导入，然后在这个路径下导入模块。

```
sys.path.append("..")
from object_detection.utils import ops as utils_ops
# 导入 Object Detection 的 utils 模块
from object_detection.utils import label_map_util
from object_detection.utils import visualization_utils as vis_util
```

15.4.3 下载 Faster R-CNN Inception 2018 预训练模型

指定要下载的预训练模型名称和地址。

```
# 模型的名称和下载地址拼接
MODEL_NAME = 'faster_rcnn_inception_v2_coco_2018_01_28'
MODEL_FILE = MODEL_NAME + '.tar.gz'
DOWNLOAD_BASE = 'http://download.tensorflow.org/models/object_detection/'
# 在模型下载解压后的目录里，冻结了graph，此 graph 保存了预训练网络的架构，我们在对象检测时经常这么用
PATH_TO_FROZEN_GRAPH = MODEL_NAME + '/frozen_inference_graph.pb'
# mscoco_label_map.pbtxt 保存了类别和索引的映射关系
PATH_TO_LABELS = os.path.join('data', 'mscoco_label_map.pbtxt')
NUM_CLASSES = 90
```

通过 urllib 网络请求模块进行模型的下载。

```
opener = urllib.request.URLopener()
opener.retrieve(DOWNLOAD_BASE + MODEL_FILE, MODEL_FILE)
```

下载后，解压模型到 faster_rcnn_inception_v2_coco_2018_01_28 目录。

```
tar_file = tarfile.open(MODEL_FILE)
for file in tar_file.getmembers():
    file_name = os.path.basename(file.name)
    if 'frozen_inference_graph.pb' in file_name:
        tar_file.extract(file, os.getcwd())
```

15.4.4 加载模型到内存中

加载已冻结的预训练模型到内存中。

```
detection_graph = tf.Graph()
with detection_graph.as_default():
    od_graph_def = tf.GraphDef()
    with tf.gfile.GFile(PATH_TO_FROZEN_GRAPH, 'rb') as fid:
        serialized_graph = fid.read()
        od_graph_def.ParseFromString(serialized_graph)
        tf.import_graph_def(od_graph_def, name='')
```

其中，tf.Graph()表示获取 TensorFlow 的默认计算图。tf.GraphDef()表示图的序列化版本，它可以用任何 TensorFlow 前端编写的代码来打印、存储和恢复计算图，一般它存储的文件的扩展名为 pb。

15.4.5 加载类别映射

加载类别和索引的映射关系，即此后通过索引，我们可以找到对应的标签类别。比如：索引 7 表示火车这个类别。这里使用内部的 util 函数来完成类别的映射和对应关系。

```
label_map = label_map_util.load_labelmap(PATH_TO_LABELS)
categories = label_map_util.convert_label_map_to_categories(label_map,
max_num_classes=NUM_CLASSES, use_display_name=True)
category_index = label_map_util.create_category_index(categories)
```

15.4.6 定义视频中的图像对象检测函数

定义视频中的图像对象检测函数,将检测到的对象边界框修改到图像中,并返回该图像。

```python
# 定义在视频中的实时对象检测函数
def detect_objects_in_videos(image_np, sess, detection_graph):
    # 获取计算图
    with detection_graph.as_default():
        # 获取输入和输出的张量的句柄
        ops = tf.get_default_graph().get_operations()
        all_tensor_names = {output.name for op in ops for output in op.outputs}
        tensor_dict = {}
        for key in [
                'num_detections', 'detection_boxes', 'detection_scores',
                'detection_classes', 'detection_masks'
          ]:
            tensor_name = key + ':0'
            if tensor_name in all_tensor_names:
                tensor_dict[key] = tf.get_default_graph().get_tensor_by_name(tensor_name)
        if 'detection_masks' in tensor_dict:
            # 下面的过程仅仅是针对单个图片的处理
            detection_boxes = tf.squeeze(tensor_dict['detection_boxes'], [0])
            detection_masks = tf.squeeze(tensor_dict['detection_masks'], [0])

            # 需要使用reframe将蒙版从框坐标转换为图像坐标并适合图像大小
            real_num_detection = tf.cast(tensor_dict['num_detections'][0], tf.int32)
            detection_boxes = tf.slice(detection_boxes, [0, 0], [real_num_detection, -1])
            detection_masks = tf.slice(detection_masks, [0, 0, 0], [real_num_detection, -1, -1])
            detection_masks_reframed = utils_ops.reframe_box_masks_to_image_masks(
                detection_masks, detection_boxes, image_np.shape[0], image_np.shape[1])
            detection_masks_reframed = tf.cast(
                tf.greater(detection_masks_reframed, 0.5), tf.uint8)
            # 通过添加批量维度来遵循惯例
            tensor_dict['detection_masks'] = tf.expand_dims(
                detection_masks_reframed, 0)
        image_tensor = tf.get_default_graph().get_tensor_by_name('image_tensor:0')
        # 运行对象推理检测,这里就是真正的检测
        output_dict = sess.run(tensor_dict, feed_dict={image_tensor: np.expand_dims(image_np, 0)})
        # 所有输出都是float32 NumPy数组,因此需要适当地转换类型
        # num_detections 表示检测框的个数
        output_dict['num_detections'] = int(output_dict['num_detections'][0])
        # detection_classes 表示每个框对应的检测类别
        output_dict['detection_classes'] = output_dict['detection_classes'][0].astype(np.uint8)
        # detection_boxes 表示检测到边界框
        output_dict['detection_boxes'] = output_dict['detection_boxes'][0]
        # detection_scores 表示检测到的检测结果评分
        output_dict['detection_scores'] = output_dict['detection_scores'][0]
```

```
        if 'detection_masks' in output_dict:
            output_dict['detection_masks'] = output_dict['detection_masks'][0]
        # 将检测到的对象边界框修改到图像中
        vis_util.visualize_boxes_and_labels_on_image_array(
          image_np,
          output_dict['detection_boxes'],
          output_dict['detection_classes'],
          output_dict['detection_scores'],
          category_index,
          instance_masks=output_dict.get('detection_masks'),
          use_normalized_coordinates=True,
          line_thickness=1)

    return image_np
```

15.4.7 定义视频中的图像处理函数

定义视频的图像处理函数,并启动一个 TensorFlow 会话,然后进行图像对象检测。

```
def process_image(image):
    global counter
    if counter%1 ==0:
        with detection_graph.as_default():
            with tf.Session(graph=detection_graph) as sess:
                image_np = detect_objects_in_videos(image, sess, detection_graph)
    counter +=1
    return image
```

15.4.8 视频中的图像对象检测

我们需要用到 moviepy 模块,所以先安装它。通过 moviepy 可以对视频进行读取、编辑、保存等操作。

```
pip install moviepy
```

给定一个输入视频名称地址和一个输出视频名称地址,通过 VideoFileClip 类初始化该视频,然后通过 subclip 函数将视频的某一段时间间隔作为检测目标。起始时间和终止时间在此例子中是(0,10),表示 0 秒到 10 秒的视频剪辑片段。

```
# 输入的视频文件
input_filename = 'cars.mp4'
# 输出的视频文件
output_filename = 'cars_on_the_road.mp4'
counter = 0
# 指定视频地址,通过函数 suclip(t_start, t_end),指定剪切的起始时间和终止时间
clip_playing = VideoFileClip(input_filename).subclip(0,10)
# 通过 process_image 函数修改每一帧的图片,将修改后的图片再更新到视频帧中
white_clip = clip_playing.fl_image(process_image)
# 将视频导出到指定的文件中,并且带上百分比进度
%time white_clip.write_videofile(output_filename)
```

15.4.9 效果预览

视频中的对象检测结果预览如图 15.8 所示。由于是纸质书籍，所以随机截取了视频中的两张效果图。读者运行代码后，可以观看到更佳的视频效果。

图 15.8 视频中的实时对象检测效果图

15.5 基于 SSD MobileNet COCO 的实时对象检测

15.3 节和 15.4 节分别讲解了对单张图片和视频里的对象检测技术的使用。本节我们更深入地讲解如何对摄像头提取到的图像进行实时对象检测，并返回到拍摄的视频中显示出来。本节的预训练模型和技术还是基于 TensorFlow Object Detection API，使用的是 SSD MobileNet COCO 模型。

> 注意　本次的实时对象检测需要使用摄像头。如果读者的运行环境是服务器端的 Linux，那么需要在本地计算机上搭建一个虚拟环境。本地计算机不管是 Windows 系统，还是 Mac OS X 系统，或者是 Ubuntu 等 Linux 系列的桌面系统都可以，只要能读取到摄像头的信息即可。

15.5.1 环境准备

环境的准备和 15.3.1 节中是一致的，所以 tensorflow/models 库和 protoc 已经准备好了。在执行以下代码前，请将代码的执行路径定位到 models/research/object_detection 下。

```
import os
os.chdir("models/research/object_detection")
```

15.5.2 导入 Packages

导入一些需要用到的包并设置环境。

```python
# 导出相关的模块
import numpy as np
import os
import six.moves.urllib as urllib
import sys
import tarfile
import tensorflow as tf
# 查看打印 TensorFlow 版本号
print("tensorflow {}".format(tf.__version__))
```

导入 Object Detection 的 utils 模块,其中 sys.path.append()函数是用来将指定的路径导入进来,然后在这个路径下导入模块的。

```python
sys.path.append("..")
from object_detection.utils import ops as utils_ops
# 导入 Object Detection 的 utils 模块
from object_detection.utils import label_map_util
from object_detection.utils import visualization_utils as vis_util
```

15.5.3　下载 SSD MobileNet 2018 预训练模型

指定要下载的预训练模型名称和地址。SSD 全称是 Single Shot Detector,中文是单帧检测器。

```python
# 模型的名称和下载地址拼接
MODEL_NAME = 'ssd_mobilenet_v1_coco_2018_01_28'
MODEL_FILE = MODEL_NAME + '.tar.gz'
DOWNLOAD_BASE = 'http://download.tensorflow.org/models/object_detection/'
# 在模型下载解压后的目录里,冻结了 graph,此 graph 保存了预训练网络的架构,这在对象检测里是很常见的
PATH_TO_FROZEN_GRAPH = MODEL_NAME + '/frozen_inference_graph.pb'
# mscoco_label_map.pbtxt 保存了类别和索引的映射关系
PATH_TO_LABELS = os.path.join('data', 'mscoco_label_map.pbtxt')
NUM_CLASSES = 90
```

通过 urllib 网络请求模块进行模型的下载。

```python
opener = urllib.request.URLopener()
opener.retrieve(DOWNLOAD_BASE + MODEL_FILE, MODEL_FILE)
```

下载后,解压模型到 ssd_mobilenet_v1_coco_2018_01_28 目录。

```python
tar_file = tarfile.open(MODEL_FILE)
for file in tar_file.getmembers():
    file_name = os.path.basename(file.name)
    if 'frozen_inference_graph.pb' in file_name:
        tar_file.extract(file, os.getcwd())
```

15.5.4　加载模型

加载已冻结的预训练模型。

```
detection_graph = tf.Graph()
with detection_graph.as_default():
    od_graph_def = tf.GraphDef()
    with tf.gfile.GFile(PATH_TO_FROZEN_GRAPH, 'rb') as fid:
        serialized_graph = fid.read()
        od_graph_def.ParseFromString(serialized_graph)
        tf.import_graph_def(od_graph_def, name='')
```

其中，tf.Graph()表示获取 TensorFlow 的默认计算图。tf.GraphDef()表示图的序列化版本，它可以用任何 TensorFlow 前端编写的代码来打印、存储和恢复计算图，一般它存储的文件的扩展名为 pb。

15.5.5 加载类别映射

加载类别和索引的映射关系，即此后我们通过索引可以找到对应的标签类别。比如：索引 7 表示火车这个类别。这里使用内部的 util 函数来完成类别的映射和对应关系。

```
label_map=label_map_util.load_labelmap(PATH_TO_LABELS)
categories = label_map_util.convert_label_map_to_categories(label_map,
max_num_classes=NUM_CLASSES, use_display_name=True)
category_index = label_map_util.create_category_index(categories)
```

15.5.6 开启实时对象检测

由于需要用到 OpenCV 来开启摄像头并读取摄像头拍摄到的画面，所以我们先下载它。

```
pip install cv2
```

整个步骤就是：先通过摄像头拍摄图像，并把图像用 TensorFlow 来进行模型运算和推理；然后把检测到的对象（包括类别、置信水平值、个数和边界框）再修改更新到图像上；最后把图像添加到窗口上显示。这里有一个 while 循环，所以检测程序一直不停地检测和显示，按下字母键 Q 才会退出检测程序。

```
# 导入 OpenCV2
import cv2
# 初始化 web camera
cap = cv2.VideoCapture(0)
# 获取 TensorFlow 默认计算图
with detection_graph.as_default():
    # 获取 TensorFlow 会话
    with tf.Session(graph=detection_graph) as sess:
        # 从 web camera 上开启无限循环
        ret = True
        while (ret):
            # 从 web cameras 读取图片
            ret,image_np = cap.read()
            # 由于模型期望的图片维度是四维shape = [1, None, None, 3]，所以这里将图片扩展为四维
            image_np_expanded = np.expand_dims(image_np, axis=0)
            # 获取图片的张量
            image_tensor = detection_graph.get_tensor_by_name('image_tensor:0')
```

```python
# 获取图片中检测到的对象边界框的张量
boxes = detection_graph.get_tensor_by_name('detection_boxes:0')
# 获取图片中检测到的对象的置信水平值的张量，该值会与该对象的类别标签一起显示
scores = detection_graph.get_tensor_by_name('detection_scores:0')
# 获取图片中检测到的对象的类别的张量
classes = detection_graph.get_tensor_by_name('detection_classes:0')
# 获取图片中检测到的对象个数的张量
num_detections = detection_graph.get_tensor_by_name('num_detections:0')
# 开始检测对象
(boxes, scores, classes, num_detections) = sess.run(
    [boxes, scores, classes, num_detections],
    feed_dict={image_tensor: image_np_expanded})
# 将检测结果可视化到图片上
vis_util.visualize_boxes_and_labels_on_image_array(
    image_np,
    np.squeeze(boxes),
    np.squeeze(classes).astype(np.int32),
    np.squeeze(scores),
    category_index,
    use_normalized_coordinates=True,
    line_thickness=8)
# 将结果的图片显示到窗口上
cv2.imshow('image', cv2.resize(image_np, (960,700)))

# 当按下字母键Q的时候，退出检测程序
if cv2.waitKey(25) & 0xFF == ord('q'):
    cv2.destroyAllWindows()
    cap.release()
    break
```

15.5.7 效果预览

拍摄的视频画面中的对象检测效果预览，如图 15.9 和图 15.10 所示。由于是纸质图书，所以以下两张效果图是视频截图，读者运行代码后，可以观看到更佳的拍摄检测效果。

图 15.9　实时对象检测效果图 1

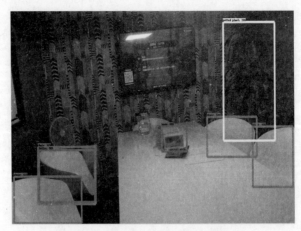

图 15.10　实时对象检测效果图 2

15.6　小结

我们首先介绍了 R-CNN、Faster R-CNN 和 Mask R-CNN。不仅检测速度越来越快，而且检测的精确度也在不断地提升。它们都是基于 VOC 2007 和 VOC 2012 的数据集来训练的，得出的最高平均精准度从一开始的大约 40%，到现在的大约 85%。这 3 种对象检测算法框架是一脉相承的，而且还在不断地研究和改进。之后，我们用 Mask R-CNN、Faster R-CNN 和 SSD MobileNet 的预训练模型进行了单张图片、视频中的对象检测和 Web 摄像头的实时对象检测。

第三篇

生成类项目实战

- 第 16 章　看图写话
- 第 17 章　生成电视剧剧本
- 第 18 章　风格迁移
- 第 19 章　生成人脸
- 第 20 章　图像超分辨率
- 第 21 章　移花接木

第 16 章

看图写话

各位读者肯定都做过看图写话一类的题。在小学，老师总是布置这样的作业，笔者却总是只能答对一半。这涉及将图片中的主要行为和物体进行语言的组织和描述，最后串联出一句话来清晰地表达。现在笔者长大了，要回答一张图片上有什么物体、有什么行为，这很容易。倘若计算机要执行这样的操作，它是如何根据图片中的物体和行为来表达图片上有什么，以及他在做什么的呢？

看图写话在深度学习中可以被专业地称作 Image to Text。著名的神经图像字幕生成器模型 Show and Tell，就是一个由 Google Brain 研究团队实现的看图写话模型。它首先通过计算机视觉模型来理解图片上有什么内容，然后通过自然语言处理将理解的内容转成正确并且通顺的语句。

16.1 数据准备

图片数据集来自 MSCOCO，它是一个大型图片数据集，对训练集、验证集和测试集数据、标注、字幕等都有相应的预处理。本章数据集经过处理后的下载地址在前言中可以找到。

此图片数据集中是来自 2017 年的图片，训练集、验证集和测试集里分别包含 11.8 万张、5000 张和 4.1 万张图片，以及相应的训练、验证和测试的标注（Annotations）文件。这些图片数据集可以通过以下的代码进行训练。但是如果要完整地训练结束，在单个 Tesla K20m GPU 上训练模型需耗时 2 周左右；如果微调模型再重新训练的话，需耗时几周，这取决于读者的计算机配置。

对我们学习者来说，这种稍大规模的图片数据集模型训练耗费的成本高、时间多。因此，笔者将数据集的图片总数量调整了一下，我们只训练 4500 张图片，并只用 500 张图片来验证。这样的模型训练既可以在双核 CPU 上训练，也可以在单个 GPU 上训练，且只会耗费 3~5 小时。减少训练模型的样本图片，虽然会相应地降低精确度，但是这样我们可以拿自己的计算机或者普通的 GPU 云服务器来训练模型。

16.1.1 环境准备

- numpy = 1.14.6。
- matplotlib = 2.1.2。
- nltk = 3.2.5。
- tensorflow = 1.12.0。

通过 TensorFlow 我们可以查看 GPU 设备信息。

```
import tensorflow as tf
tf.test.gpu_device_name()
```

输出如下。

```
'/device:GPU:0'
```

查看 NVIDIA-SMI 信息。

```
!/opt/bin/nvidia-smi
```

输出的是一个 Tesla K80 显卡的相关信息。

16.1.2 数据下载

本章所需要的代码库都在 image2text 的项目里,读者通过 git clone 可以将其克隆下来。

`git clone 请指定本 image2text.git 代码仓库地址`

image2text 库的代码是从 Google 的 models 项目中迁移出来的,然后笔者对之进行了一些与本章内容相关的调整。

数据集完整地取自 val2017,然后被拆分为训练集和验证集。我们通过 curl -O 命令加上文件地址,就可以下载数据集。

`curl -O 请指定本 val2017.zip 数据集下载地址`

通过 unzip 命令解压 val2017.zip 文件。

`unzip val2017.zip`

我们同样通过 curl -O 加上文件地址下载标注文件(Annotations)。

`curl -O 请指定本 annotation trainval2017.zip 压缩包下载地址`

解压标注文件。

`unzip annotations_trainval2017.zip`

此时目录下有名为 annotations 的目录和 annotations_trainval2017.zip 压缩文件,名为 val2017 的目录和 val2017.zip 压缩文件,还有一个文件是 image2text 代码库。

16.1.3 数据预处理

我们在 annotations 目录下看到有多个 json 文件,其中 captions_val2017.json 文件就是此次的标注文件,先看看它包含了什么样的数据内容。

```python
import json
json_path = "annotations/captions_val2017.json"
with open(json_path, encoding='utf-8') as f:
    json_dict = json.load(f)
print(json_dict.keys())
```

输出如下。

```
dict_keys(['info', 'licenses', 'images', 'annotations'])
```

图片的相关属性值在 images 下,标注在 annotations 下。

```python
images = json_dict["images"]
annotations = json_dict["annotations"]
print("images.length={}, annotations.length={}".format(len(images), len(annotations)))
```

输出如下。

```
images.length=5000, annotations.length=25014
```

查看变量 images 的前 3 个值。

```python
json_dict["images"][:3]
```

输出如下。

```
[{'coco_url': 'http://images.cocodataset.org/val2017/000000397133.jpg',
  'date_captured': '2013-11-14 17:02:52',
  'file_name': '000000397133.jpg',
  'flickr_url': 'http://farm7.staticflickr.com/6116/6255196340_da26cf2c9e_z.jpg',
  'height': 427,
  'id': 397133,
  'license': 4,
  'width': 640},
 {'coco_url': 'http://images.cocodataset.org/val2017/000000037777.jpg',
  'date_captured': '2013-11-14 20:55:31',
  'file_name': '000000037777.jpg',
  'flickr_url': 'http://farm9.staticflickr.com/8429/7839199426_f6d48aa585_z.jpg',
  'height': 230,
  'id': 37777,
  'license': 1,
  'width': 352},
 {'coco_url': 'http://images.cocodataset.org/val2017/000000252219.jpg',
  'date_captured': '2013-11-14 22:32:02',
  'file_name': '000000252219.jpg',
  'flickr_url': 'http://farm4.staticflickr.com/3446/3232237447_13d84bd0a1_z.jpg',
  'height': 428,
  'id': 252219,
```

```
 'license': 4,
 'width': 640}]
```

数组里的每个元素都是一个字典，对里面的值的解释如下。

- coco_url：表示该图片在 COCO 网站的托管地址。
- date_captured：表示获取图片的日期。
- file_name：文件名，可以在 val2017 目录下找到。
- flickr_url：图片的原始托管网站 flickr 的地址。
- height：图片的高度。
- id：图片的 id。
- license：图片的证书类别号。
- width：图片的宽度。

查看变量 annotations 的前 3 个值。

```
json_dict["annotations"][:3]
```

输出如下。

```
[{'caption': 'A black Honda motorcycle parked in front of a garage.',
  'id': 38,
  'image_id': 179765},
 {'caption': 'A Honda motorcycle parked in a grass driveway',
  'id': 182,
  'image_id': 179765},
 {'caption': 'An office cubicle with four different types of computers.',
  'id': 401,
  'image_id': 190236}]
```

数组里每个元素的值的解释如下。

- caption：表示 image_id 对应的图片的字幕/描述。
- id：当前文件排列顺序中的 id。
- image_id：图片的 id。

查看变量 json_dict 里的 info 和 license 的值。

```
json_dict["info"]
```

输出如下。

```
{'contributor': 'COCO Consortium',
 'date_created': '2017/09/01',
 'description': 'COCO 2017 Dataset',
 'url': '',
 'version': '1.0',
 'year': 2017}
```

这就是数据集的由来，它是由 COCO 联盟贡献的。查看图片的证书信息。

```
json_dict["licenses"]
```

输出如下。

```
[{'id': 1,
  'name': 'Attribution-NonCommercial-ShareAlike License',
  'url': 'http://×××/licenses/by-nc-sa/2.0/'},
……
 {'id': 7,
  'name': 'No known copyright restrictions',
  'url': 'http://×××/commons/usage/'},
……
]
```

这个证书的类型可以在上面的 images 下看到，每张图片都对应着不同的证书类型。之所以提到这些是因为后面需要将这个 json 文件进行拆分和合并，以便于模型训练和测试。

我们输出一张图片以及它的字幕来查看他们的关系。假设我们想看文件名为 000000037777.jpg 的图片以及它的相关属性值，那么可通过如下代码进行查看。

```
import matplotlib.pyplot as plt
import matplotlib.image as pltimg
def plot_preview_image(filename):
    """
    定义一个根据文件名来预览图片的函数
    """
    # 对标注数组进行遍历
    for img_anno in annotations:
        # 每个标注的元素都包含一个 image_id 的 key，用它的值和我们指定的文件名来判断是否是同一个
        # 如果是同一个 image_id，就等于找到了图片标注属性
        if img_anno["image_id"] == filename:
            # 对图片数组进行遍历
            for img_prop in images:
                # 将文件名中的扩展名去掉
                file_name = img_prop["file_name"][:-4]
                # 如果图片文件名和我们指定的文件名匹配上了，则视为找到了该图片
                if int(file_name) == filename:
                    # 读取找到的图片
                    img = pltimg.imread("val2017/" + img_prop["file_name"])
                    # 不显示网格
                    plt.grid(False)
                    # 显示图片
                    plt.imshow(img)
                    print("图片的相应属性有：")
                    print(img_prop)
                    break
            print("图片的字幕属性有：")
            print(img_anno)
            break
# 调用函数，输出图片和相关值
plot_preview_image(int("000000037777"))
```

输出如下。

图片的相应属性有：
{'license': 1, 'file_name': '000000037777.jpg',
'coco_url': 'http://×××/val2017/000000037777.jpg',
'height': 230, 'width': 352, 'date_captured': '2013-11-14 20:55:31',
'flickr_url': 'http://×××/8429/7839199426_f6d48aa585_z.jpg',
'id': 37777}
图片的字幕属性有：
{'image_id': 37777, 'id': 597185,
'caption': 'The dining table near the kitchen has a bowl of fruit on it.'}

输出如图 16.1 所示。图 16.1 对应的字幕是 "The dining table near the kitchen has a bowl of fruit on it."，中文翻译是 "厨房旁边的餐桌上有一碗水果"。

图 16.1　37777 图片预览

我们再来预览一张文件名为 000000005037.jpg 的图片。

```
plot_preview_image(int("000000005037"))
```

输出如下。

图片的相应属性有：
{'license': 6, 'file_name': '000000005037.jpg',
'coco_url': 'http://×××/val2017/000000005037.jpg',
'height': 425, 'width': 640, 'date_captured': '2013-11-16 20:17:28',
'flickr_url': 'http://×××/7379/9599671465_8a2f486da1_z.jpg',
'id': 5037}
图片的字幕属性有：
{'image_id': 5037, 'id': 450868,
'caption': 'A passenger bus pulling up to the side of a street.'}

输出如图 16.2 所示。图 16.2 对应的字幕是 "A passenger bus pulling up to the side of a street."，中文翻译是 "一辆客车停靠在一条街的一侧"。

图 16.2 5037 图片预览

现在开始将数据分割成训练集和验证集。我们首先读取所有的图片路径，然后将其移动到训练集文件夹 train 和验证集文件夹 valid。

```
from glob import glob
# 读取 val2017 目录下所有图片的路径，星号表示通配符所有
image_path = "val2017/*"
image_paths = glob(image_path)
# 输出前 10 个图片路径的地址
image_paths[:10]
```

输出如下。

```
['val2017/000000190236.jpg',
 'val2017/000000032038.jpg',
 'val2017/000000388056.jpg',
 'val2017/000000150417.jpg',
 'val2017/000000415238.jpg',
 'val2017/000000192716.jpg',
 'val2017/000000262895.jpg',
 'val2017/000000569059.jpg',
 'val2017/000000075393.jpg',
 'val2017/000000416256.jpg']
```

我们把 val2017 目录下的 5000 张图片分给训练集 90%、验证集 10%，即训练集和验证集分别有 4500 张图片和 500 张图片。我们先创建 3 个目录，分别是 dataset 和在 dataset 目录下的 train_images（用来存放训练集图片）、valid_images（用来存放验证集图片），创建目录的代码如下。

```
mkdir dataset
mkdir dataset/train_images
mkdir dataset/valid_images
```

然后开始分割图片，其实就是调用图片移动的函数。这就需要用到 shutil 模块，它是一个高级文件操作模块，可进行拷贝、删除、移动等文件操作。

```python
import os
import shutil
# 设置训练集和验证集的图片路径
train_path = os.getcwd() + "/dataset/train_images/"
valid_path = os.getcwd() + "/dataset/valid_images/"
# 分割 4500 张图片作为训练集
for img_path in image_paths[:4500]:
    filename = img_path[len("val2017/"):]
    shutil.move(img_path, train_path + filename)

# 分割 500 张图片作为验证集
for img_path in image_paths[4500:]:
    filename = img_path[len("val2017/"):]
    shutil.move(img_path, valid_path + filename)
```

数据集分割完后,我们需要对数据集相应的标注的 json 文件进行处理。首先读取所有的训练集和验证集图片路径。

```python
train_img_paths = glob(train_path + "*")
valid_img_paths = glob(valid_path + "*")
train_img_paths[:10]  # 输出前 10 个图片路径
```

输出如下。

```
['/content/dataset/train_images/000000190236.jpg',
 '/content/dataset/train_images/000000032038.jpg',
 '/content/dataset/train_images/000000388056.jpg',
 '/content/dataset/train_images/000000150417.jpg',
 '/content/dataset/train_images/000000415238.jpg',
 '/content/dataset/train_images/000000192716.jpg',
 '/content/dataset/train_images/000000262895.jpg',
 '/content/dataset/train_images/000000569059.jpg',
 '/content/dataset/train_images/000000075393.jpg',
 '/content/dataset/train_images/000000416256.jpg']
```

现在,我们确认图片的标注,保证全部信息都一致。所以需要对训练集和验证集进行处理,保证数据一致性,这样在训练时才能保证精确度。

```python
def generate_images_and_annotations(img_paths, is_train=False):
    # 图片路径
    img_path_prefix = len(os.getcwd() + "/dataset/" + ("train_images/" if is_train else "valid_images/"))
    # 图片文件名(不包含扩展名)
    img_prefix_len = len("000000570664")
    # 定义数组
    train_anntate_json = []
    train_image_json = []
    for img_path in img_paths:
        # 将训练集的图片相关属性都找出来,并添加到数组中
        img_filename = img_path[img_path_prefix:img_path_prefix+img_prefix_len]
        for img_annotate in annotations:
            if img_annotate["image_id"] == int(img_filename):
```

```
                    train_anntate_json.append(img_annotate)
                    break
            # 将训练集的图片字幕都找出来,并添加到数组中
            img_filename = img_path[img_path_prefix:]
            for img_prop in images:
                if img_prop["file_name"] == img_filename:
                    train_image_json.append(img_prop)
                    break
    return train_image_json, train_anntate_json
# 处理训练集的标注信息
train_image_json, train_anntate_json = generate_images_and_annotations(train_img_paths)
# 处理验证集的标注信息
valid_image_json, valid_anntate_json = generate_images_and_annotations(valid_img_paths, is_train=True)
```

训练集和验证集的标注信息已分割好,现在将它们合并起来,保存到 dataset 目录下。

```
def merge_dataset(json_dict, image_json, annotate_json, save_json_path):
    # 读取 info 信息
    info_dict = json_dict["info"]
    # 读取 license 信息
    licenses_dict = json_dict["licenses"]
    # 构建 json 结构
    json_ds = {
        "info" : info_dict,
        "licenses" : licenses_dict,
        "images" : image_json,
        "annotations" : annotate_json
    }
    # 保存 json 文件
    with open(save_json_path, 'w') as f:
        f.write(json.dumps(json_ds))

# 合并训练集的 json 文件
merge_dataset(json_dict, train_image_json, train_anntate_json, "dataset/train_images.json")
# 合并验证集的 json 文件
merge_dataset(json_dict, valid_image_json, valid_anntate_json, "dataset/valid_images.json")
```

现在我们查看 dataset 目录,有 4 个文件,其中训练集 2 个,验证集 2 个。每个数据集中有一个文件包含图片,另一个是图片的标注信息 json 文件。然后我们将 dataset 目录移动到 image2txt/im2txt 目录下,方便后续的训练和验证。

```
mv dataset image2text/im2txt/
```

16.2 基于 TensorFlow 的 Show and Tell 实现看图写话

看图写话使用的是 Show and Tell 模型,它是一个编码器—解码器的神经网络。其工作原理是先将图片编码为固定长度的矢量,然后将其解码为自然语言描述。该模型是一个深度卷积神经网络(DCNN),常用于图片分类识别的任务。在本次训练模型中,我们采用的是预训

练模型 Inception v3，它作为一个编码器。而解码器是我们构建的 LSTM 网络，LSTM 常用在语言建模上，两者友好地配合就是 Show and Tell 模型，如图 16.3 所示。

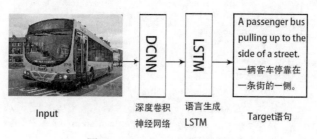

图 16.3 Show and Tell 模型

16.2.1 介绍

Show and Tell 模型在 Inception v3 模型训练的参数的第一阶段保持固定，因为那时它只是一个静态图片编码器。然后在 Inception v3 模型的顶部添加单个可训练层，将图像嵌入转换为单词嵌入的向量空间。该模型是根据单词嵌入的参数，Inception v3 顶层的参数和 LSTM 的参数进行训练的。在训练的第二阶段，训练所有的参数（包括 Inception v3 的参数）以共同微调图像编码器和 LSTM。

假如给定一个已训练的模型和图像，我们使用 BeamSearch 来生成图片的标题。标题是逐字生成的，而其中每个步骤 t 是使用已经生成的长度为 $t-1$ 的句子集来生成长度为 t 的新句子集。我们在每个步骤中仅保留前 k 个候选者，其中超参数 k 被称为 Beam Size，最后我们会发现，当 $k=3$ 时性能最佳。

16.2.2 数据统计

在训练模型前，我们先来确认图片和标注的数据，确保他们的数量是一致的、ID 是对应的，训练集 4500 个和验证集 500 个。

```
import os
import json
from glob import glob
# 切换目录到代码仓库下
os.chdir("image2text/im2txt/")
# 设置训练集和验证集的路径
cwd_path = os.getcwd()
train_img_path = cwd_path + "/dataset/train_images/"
train_json_path = cwd_path + "/dataset/train_images.json"
valid_img_path = cwd_path + "/dataset/valid_images/"
valid_json_path = cwd_path + "/dataset/valid_images.json"
# 获取所有训练图片路径
train_img_paths = glob(train_img_path + "*")
# 获取所有的验证图片路径
valid_img_paths = glob(valid_img_path + "*")
```

```python
# 加载训练标注数据
with open(train_json_path, encoding='utf-8') as f:
    train_json_dict = json.load(f)
# 加载验证标注数据
with open(valid_json_path, encoding='utf-8') as f:
    valid_json_dict = json.load(f)
# 打印基本的 key
print(train_json_dict.keys())
print(valid_json_dict.keys())
```

输出如下。

```
dict_keys(['info', 'licenses', 'images', 'annotations'])
dict_keys(['info', 'licenses', 'images', 'annotations'])
```

统计训练集和验证集的数量。

```python
train_json_img_paths = train_json_dict["images"]
valid_json_img_paths = valid_json_dict["images"]
train_json_annotations_paths = train_json_dict["annotations"]
valid_json_annotations_paths = valid_json_dict["annotations"]
print("train:\r")
print("图片有{}张,图片属性有{}条,图片标注有{}个。".format(
                                        len(train_img_paths),
                                        len(train_json_img_paths),
                                        len(train_json_annotations_paths)))
print("valid:\r")
print("图片有{}张,图片属性有{}条,图片标注有{}个。".format(
                                        len(valid_img_paths),
                                        len(valid_json_img_paths),
                                        len(valid_json_annotations_paths)))
```

输出如下。

```
train:
图片有 4500 张,图片属性有 4500 条,图片标注有 4500 个。
valid:
图片有 500 张,图片属性有 500 条,图片标注有 500 个。
```

16.2.3　构建 TFRecords 格式数据

要训练模型，我们需要提供原生的 TFRecords 格式的训练数据。TFRecords 格式由一系列序列化 tf.SequenceExample 协议缓冲区的分片文件组成，每个 tf.SequenceExample proto 包含一张图片（JPEG 格式）、一个标题和元数据，如图片 ID。将图片构建成 TFRecords 格式数据的代码文件是 build_mscoco_data.py。

```python
import nltk
# 先把 nltk 的 punkt 的组件安装上
nltk.download('punkt')
```

然后调用 build_mscoco_data.py 文件以构建 TFRecords 格式文件，已生成的 TFRecords 文件存储在 output 目录下。构建 TFRecords 格式文件的代码如下。

```
python data/build_mscoco_data.py \
  --train_image_dir="dataset/train_images/" \
  --val_image_dir="dataset/valid_images/" \
  --train_captions_file="dataset/train_images.json" \
  --val_captions_file="dataset/valid_images.json" \
  --output_dir="output" \
  --word_counts_output_file="output/word_counts.txt" \
```

这个过程需要花几分钟才能完成。当然图片数量越多，所花时间就越长。它会生成 256 个训练集文件、4 个验证集文件和 8 个测试集文件。

调用 build_mscoco_data.py 文件的参数解释如下。

- train_image_dir：训练集图片数据的路径。
- val_image_dir：验证集图片数据的路径。
- train_captions_file：训练集的标注文件。
- val_captions_file：验证集的标注文件。
- output_dir：TFRecords 格式文件最终要生成的路径。
- word_counts_output_file：统计的单词的输出路径。

统计 TFRecords 格式的文件数量。在 output 目录下，我们还是一样通过 glob 模块读取图片路径，再对每个图片路径的前缀进行匹配。前缀为 train 的是训练集文件，前缀为 val 的是验证集文件，前缀为 test 的是测试集文件。

```
from glob import glob
tfrecords_output_files = glob("output/*")
tfrecords_output_files[:10]
train_files = []
valid_files = []
test_files = []
for output_file in tfrecords_output_files:
    if output_file.startswith("output/train"):
        train_files.append(output_file)
    elif output_file.startswith("output/val"):
        valid_files.append(output_file)
    elif output_file.startswith("output/test"):
        test_files.append(output_file)
print("TFRecords 的文件个数分别有：\r")
print("训练集有{}个；".format(len(train_files)))
print("验证集有{}个；".format(len(valid_files)))
print("测试集有{}个。".format(len(test_files)))
```

输出如下。

TFRecords 的文件个数分别有：
训练集有 256 个；
验证集有 4 个；
测试集有 8 个。

16.2.4 训练模型

我们使用的是预训练模型 Inception v3，所以需要先下载该预训练模型，然后再通过 train.py 文件进行模型训练。

```
curl -O 请指定本预训练模型地址 models/inception_v3_2016_08_28.tar.gz
# 解压到当前目录
tar -xvf inception_v3_2016_08_28.tar.gz
# 将预训练模型移动到 data 目录下
mv inception_v3.ckpt data/
```

通过 train.py 文件训练模型，脚本参数解释如下。

- input_file_pattern：输入文件，就是 TFRecords 格式的训练文件。有 256 个文件，它们的序号是从"00000"开始的（在文件名中间的数字），然后以 1 为步长递增。
- inception_checkpoint_file：预训练模型 Inception v3 的检查点文件。
- train_dir：要训练的训练集图片目录地址。
- train_inception：是否训练 Inception 子模型变量，这里不训练。
- number_of_steps：要训练多少步，我们给定 1 万步。

```
python train.py \
  --input_file_pattern="output/train-00000-of-00256" \
  --inception_checkpoint_file="data/inception_v3.ckpt" \
  --train_dir="output_model/train" \
  --train_inception=false \
  --number_of_steps=10000
```

训练模型的耗时取决于计算机性能。这里测试时用的是单机 GPU 云服务器，2 小时训练完毕，如果是 CPU 会更慢些。输出部分信息如下。

```
……
INFO:tensorflow:global_step/sec: 0
INFO:tensorflow:global step 1: loss = 9.3996 (18.145 sec/step)
INFO:tensorflow:Recording summary at step 1.
INFO:tensorflow:global step 2: loss = 8.9013 (0.615 sec/step)
INFO:tensorflow:global step 3: loss = 7.2473 (0.573 sec/step)
INFO:tensorflow:global step 4: loss = 11.0709 (0.533 sec/step)
……
INFO:tensorflow:global step 9999: loss = 0.0001 (0.482 sec/step)
INFO:tensorflow:global step 10000: loss = 0.0002 (0.497 sec/step)
INFO:tensorflow:Stopping Training.
INFO:tensorflow:Finished training! Saving model to disk.
```

基本上每 0.4～0.5 秒训练完毕一个训练步，损失值从最初的 9.3996 降到了最后的 0.0002。

16.2.5 评估模型

使用 evaluate.py 文件可以进行模型评估，我们通过 Python 执行该脚本，相应的参数解释如下。

- input_file_pattern：输入验证文件的起始文件，也就是 TFRecords 格式的文件。有 4 个这样的文件，它们的序号是从"00000"开始的（在文件名中间的数字），然后以 1 为步长递增。
- checkpoint_dir：训练模型后的检查点文件目录。
- eval_dir：评估后的事件日志文件保存在 output_model/valid 目录下。
- min_global_step：评估模型的最小全局步数，默认是 5000 步，我们也可以自定义评估步数。

```
python evaluate.py \
  --input_file_pattern="output/val-00000-of-00004" \
  --checkpoint_dir="output_model/train" \
  --eval_dir="output_model/valid" \
  --min_global_step=500
```

验证模型的耗时还是取决于计算机性能。这里测试时用的是单机 GPU 云服务器，4～5 小时验证完毕。读者也可以降低 min_global_step 步数，以达到验证更快的目的。但是这样做的话，验证效果会变差。输出部分信息如下。

```
INFO:tensorflow:Loading model from checkpoint: output_model/train/model.ckpt-10000
INFO:tensorflow:Restoring parameters from output_model/train/model.ckpt-10000
INFO:tensorflow:Successfully loaded model.ckpt-10000 at global step = 10000.
INFO:tensorflow:Computed losses for 1 of 317 batches.
INFO:tensorflow:Computed losses for 101 of 317 batches.
INFO:tensorflow:Computed losses for 201 of 317 batches.
INFO:tensorflow:Computed losses for 301 of 317 batches.
INFO:tensorflow:Perplexity = 110967.077435 (1.3e+02 sec)
INFO:tensorflow:Finished processing evaluation at global step 10000.
```

16.2.6 测试模型

run_inference.py 文件可以测试模型在新图片上的表现，我们通过 Python 执行该脚本，相应的参数解释如下。

- checkpoint_path：训练的模型检查点目录。
- vocab_file：在标注文件中所有图片标注字幕的句子单词统计的 word_counts.txt 文件。
- input_files：需要测试的图片路径，使用逗号将多个图片路径分隔开。测试图片是作者拍摄的新加坡滨海湾金融中心。

```
python run_inference.py \
  --checkpoint_path="output_model/train" \
  --vocab_file="output/word_counts.txt" \
  --input_files="test_singapore_financial_center.jpeg"
```

输出部分信息如下。

```
Captions for test_singapore_financial_center.jpeg:
  0) a <UNK> filled with boats floating in <UNK> blue water . (p=0.377559)
```

1) a <UNK> filled with boats floating on the blue water . (p=0.092945)
2) a <UNK> standing with boats floating in <UNK> blue water . (p=0.058053)

显示该图片的原貌。

```
import matplotlib.pyplot as plt
import matplotlib.image as pltimg
img = pltimg.imread("test_singapore_financial_center.jpeg")
plt.grid(False)
plt.imshow(img)
```

输出结果如图 16.4 所示。

图 16.4　预测的图片

模型生成了三个语句，每个语句后面都有一个概率 P，表示对该图片最佳描述的概率值。其中<UNK>这个 token 表示 Unknown，说明这个模型被训练得不是很理想，因为图片数量和训练步数都不高。

给出的最佳训练图片数量是 2017 年的 MSCOCO 图片数据集，训练集有 11.8 万张图片，验证集有 5000 张图片，测试集有 4.1 万张图片。在训练模型时的步数是 1000000 步，验证模型时是 3000000 步。

16.3　小结

我们使用公开的 MSCOCO 图片数据集，对其做预处理，将用到的图片都转换成 TFRecords 格式，然后训练、评估和测试模型。我们下载的 MSCOCO 图片数据集并不是 cocodataset.org 网站上（训练集、验证集和测试集）的一套图片数据集，而是取自 val2017，原因是希望任何一位读者都可以低成本地学会此技术。最后，如果想要训练完整的 MSCOCO 图片数据集，可以根据本章的代码，适当修改读取数据的代码即可。

第 17 章

生成电视剧剧本

生成电视剧剧本，听起来很棒，对吗？实际上，我们完全可以通过神经网络模型来生成剧本，当然文章也行。因为它是基于文本生成（Text Generation）的技术。运用文本生成技术来拟合，拟合得好的话，以后小说、诗词和文章等是不是都可以拿此技术来生成了？

我们将使用《辛普森一家》(*The Simpsons*) 的剧本文本，通过基于字符的循环神经网络来生成有趣的剧本文本。我们不是让神经网络学习所有的剧本文本，而是取出其中的第 18 个片段来作为学习数据，然后生成剧本。

17.1 数据准备

The Simpsons 有近 600 个片段（Episode），可以回溯到 1989 年。完整剧本的文本数据集已经下载，其中第 18 个片段的文本也处理好了，读者可以直接读取和使用。如果读者想自己取出完整剧本文本数据集中的任意片段，可以使用 17.1.2 小节中的代码来查询和保存剧本文本数据。

17.1.1 环境准备

- numpy = 1.14.6。
- pandas = 0.22.0。
- matplotlib = 2.1.2。
- seaborn = 0.7.1。
- wordcloud = 1.5.0。
- tensorflow = 1.12.0。

为了训练时速度更快、效果更好，我们将使用 GPU 来训练模型。先来检查读者的计算机是否有 GPU。

```python
import tensorflow as tf
from tensorflow.python.client import device_lib
# 查看当前计算机的 GPU 是否可用
if tf.test.is_gpu_available():
    # 查看 GPU 设备名字
    gpu_device_name = tf.test.gpu_device_name()
    print("gpu_device_name={}.".format(gpu_device_name))

    # 列出本地计算机里所有的 GPU/CPU 设备
    local_device_protos = device_lib.list_local_devices()
    # 只打印 GPU 设备
    [print(x.physical_device_desc) for x in local_device_protos if x.device_type == 'GPU' if x.physical_device_desc is not None]
```

输出日志如下，检查结果为本地计算机里有一个 Tesla K80 的 GPU。

```
gpu_device_name=/device:GPU:0.
device: 0, name: Tesla K80, pci bus id: 0000:00:04.0, compute capability: 3.7
```

如果没有 GPU，使用 CPU 来运行也可以。

17.1.2 数据预处理

由于本章我们使用《辛普森一家》完整剧本数据集中第 18 个片段的剧本文本来训练模型，所以，读者想要训练其他片段的模型，也可以使用如下代码，只需要将查询条件"等于 18"改成"等于××"（读者想要的片段值）。

```python
import pandas as pd
# 读取剧本所有的文本
df = pd.read_csv("simpsons_script_lines.csv", error_bad_lines=False)
# 取出所有的剧本片段 id 和正文
df_1 = df[["episode_id", "raw_text"]]
# 我们拿第 18 个片段作为教程所使用的对象
df_2 = df_1.query("episode_id == 18")
# 取出第 18 个片段的所有对话
raw_text = df_2["raw_text"]
# 遍历第 18 个片段的所有对话
my_list = []
for i, t in enumerate(raw_text):
    # 不要前 5 行数据
    if i < 5:
        continue
    # 分号分割，第一个元素是说话者，第二个元素是对话文本
    t_a = t.split(":")
    # 替换说话者的名字中的空格为下划线，并去掉首尾引号
    a = t_a[0].strip("\"").replace(" ", "_") + ":" + t_a[1].strip("\"")
    # 一个新的场景前使用两个换行分割
    if a.startswith("(") and i > 5:
        a = "\n\n" + a
    my_list.append(a)
```

```
# 将查询到的第18个片段的剧本文本数据写入文件
with open('Simpsons_Episode_18.txt', 'w') as f:
    for item in my_list:
        f.write("{}\n".format(item))
```

加载第18个片段的剧本文本数据,然后打印基本统计信息。

```
import os
def load_data(filepath):
    """
    加载文件
    """
    input_file = os.path.join(filepath)
    with open(input_file, "r") as f:
        data = f.read()
    return data

# 加载剧本的第18个片段文本文件
data_filepath = 'Simpsons_Episode_18.txt'
text = load_data(data_filepath)
import numpy as np
print('数据集基本统计')
print('粗略统计单词的数量: {}。'.format(len({word: None for word in text.split()})))
print("\r")
scenes = text.split('\n\n')
print('有{}个场景。'.format(len(scenes)))
sentence_count_scene = [scene.count('\n') for scene in scenes]
print('每个场景平均句子数量: {}。'.format(np.average(sentence_count_scene)))
print("\r")
sentences = [sentence for scene in scenes for sentence in scene.split('\n')]
print('总计{}行。'.format(len(sentences)))
word_count_sentence = [len(sentence.split()) for sentence in sentences]
print('每一行单词的平均数量: {}。'.format(np.average(word_count_sentence)))
print("\r")
print("前5行句子: ")
print('\n'.join(text.split('\n')[:5]))
```

输出如下。

数据集基本统计
粗略统计单词的数量: 1444。
有27个场景。
每个场景平均句子数量: 9.88888888888889。
总计294行。
每一行单词的平均数量: 10.785714285714286。
前5行句子:
(Moe's_Tavern: int. Moe's tavern - night)
Barney_Gumble: So, Homer. What happened in Capitol City?
Homer_Simpson: Aw, Barney.
Moe_Szyslak: Come on, Homer. We're dyin' of curiosity.

Homer_Simpson: Look, there's only one thing worse than being a loser. It's being one of those guys who sits in a bar telling the story of how he became a loser. And I never want that to happen to me.

17.1.3　数据可视化分析

我们对数据进行词云和直方图分析，主要是对说话者和说话内容的统计，使用开源项目 WordCloud 做词云分析。

```python
from wordcloud import WordCloud
import matplotlib.pyplot as plt
def draw_wordcloud_image(bodytext=None,
                         filepath=None,
                         background_color="white",
                         min_font_size=5,
                         max_font_size=100,
                         width=700,
                         height=500,
                         colormap="Blues"
                         ):
    """
    定义绘制词云图函数
    参数 bodytext: 文本
    参数 filepath: 文本文件路径
    参数 background_color: 绘制的背景颜色
    参数 min_font_size: 最小字体
    参数 max_font_size: 最大字体
    参数 width: 宽度
    参数 height: 高度
    参数 colormap: 颜色表，从单词中随机选择绘制颜色
    """
    data_text = bodytext
    if filepath is not None:
        # 读取文本文件
        with open(filepath) as f:
            data_text = f.read()

    if len(data_text) > 0:
        # 生成词云图 WordCloud 对象
        wordcloud = WordCloud(background_color=background_color,
                              min_font_size=min_font_size,
                              max_font_size=max_font_size,
                                width=width,
                                height=height,
                              colormap=colormap).generate(data_text)
        # 绘制图像
        plt.figure()
        plt.imshow(wordcloud, interpolation='bilinear')
        # 绘图时的网格是否显示
        plt.grid()
```

```
# 绘图时不显示 X 轴和 Y 轴的尺度
# plt.axis("off")
# 显示绘图
plt.show()
```

直接绘制并查看词云图。

```
draw_wordcloud_image(filepath=data_filepath, background_color="black")
```

输出结果如图 17.1 所示。

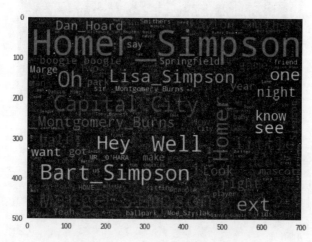

图 17.1　第 18 个片段的词云图

获取剧本文本数据集中说话者的姓名。

```
# 将剧本中所有句子的说话者和说话内容分开
speakers_list = []
bodytext_list = []
for i, line in enumerate(sentences):
    # 分号分割
    line_arr = line.split(":")
    # 分割时,如果元素的长度小于 2,就过滤掉
    if len(line_arr) == 2:
        # 第一个元素是说话者名字
        speaker_name = line_arr[0]
        # 第二个元素是说话内容
        bodytext = line_arr[1]
        if len(speaker_name) > 0 and not speaker_name.startswith("("):
            speakers_list.append(speaker_name)
        if len(bodytext) > 0 and not bodytext.endswith(")"):
            bodytext_list.append(bodytext)
```

查看说话者的词云图。

```
speakers_list_str = " ".join(speakers_list)
draw_wordcloud_image(bodytext=speakers_list_str, background_color="black")
```

输出结果如图 17.2 所示。

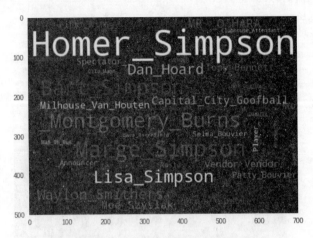

图 17.2 所有说话者的词云图

查看所有说话者说话内容的词云图。

```
bodytext_list_str = " ".join(bodytext_list)
draw_wordcloud_image(bodytext=bodytext_list_str, background_color="black", colormap="BuPu")
```

输出结果如图 17.3 所示。

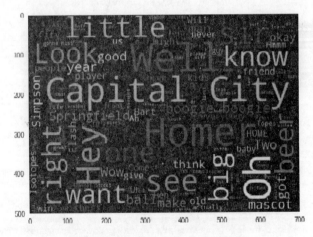

图 17.3 说话者说话内容的词云图

为了以更清晰的方式查看，我们绘制说话者的直方图，使用内置的 collections 模块下的 Counter 对象来统计每个单词出现的次数。

```
from collections import Counter
# 通过 Counter 来统计每个说话者重复的次数
speaker_counter = Counter(speakers_list)
```

```
# 打印最常见的单词，简单来说，就是从大到小排序，默认是全部说话者的信息
speakers_name_most_common = speaker_counter.most_common()
speakers_name_most_common
```

输出如下。

```
[('Homer_Simpson', 84),
 ('Marge_Simpson', 23),
 ('Bart_Simpson', 19),
 ('C._Montgomery_Burns', 19),
 ('Lisa_Simpson', 13),
 ('Waylon_Smithers', 9),
 ('Dan_Hoard', 9),
 ("MR._O'HARA", 6),
 ('Moe_Szyslak', 5),
 ……
 ('Clubhouse_Attendant', 1),
 ('Dave_Rosenfield', 1)]
```

绘制直方图，只显示前 15 个说话者的信息。

```
# 取前 15 个元素
speaker_X = [name[0] for name in speakers_name_most_common][:15]
speaker_y = [name[1] for name in speakers_name_most_common][:15]
import seaborn as sns
# 绘制直方图
# 参数 orient：h 表示水平（horizontal），v 表示垂直（vertical）
sns.barplot(x=speaker_y, y=speaker_X, orient='h')
```

输出结果如图 17.4 所示。

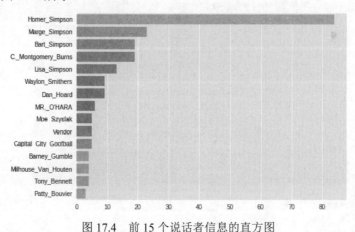

图 17.4　前 15 个说话者信息的直方图

17.2　基于 TensorFlow 的循环神经网络实现电视剧剧本生成

使用循环神经网络来实现文本生成是非常合适的。我们先将检查表、数据预处理、创建 LSTM 的 Cell、嵌入层、交叉熵、优化器和构建神经网络模型的函数定义好，然后在训练前

组成起来，最后我们会根据生成的模型来生成一段剧本文本。

17.2.1 创建检查表

使用检查表（Lookup Table）需要做单词嵌入。首先我们创建两个字典，第一个字典是让每个 ID 对应一个单词，第二个字典是让每个单词都对应一个 ID。

```
import numpy as np
from collections import Counter
def create_lookup_tables(text):
    # Counter 对象是一个便利、快速的计数器
    word_counts = Counter(text)
    # 每个 id 对应一个单词
    int_to_vocab = {counts: word for counts, word in enumerate(word_counts)}
    # 每个单词对应一个 ID
    vocab_to_int = {word: counts for counts, word in int_to_vocab.items()}
    return vocab_to_int, int_to_vocab
```

17.2.2 数据 token 化预处理

为什么要将标点符号 token 化呢？因为标点符号，比如问号、感叹号、逗号等，对于神经网络来说是难以训练的，也就是很难分辨 thanks!和 thanks 的区别。

```
def token_lookup():
    """
    生成一个字典，将标点符号映射成 token
    标记化（Tokenization）字典，key 是标点符号，value 是 token
    """
    token_dict = { "." : "||Period||",
                   "," : "||Comma||",
                   "\"" : "||Quotation_Mark||",
                   ";" : "||Semicolon||",
                   "!" : "||Exclamation_Mark||",
                   "?" : "||QuestionMark||",
                   "(" : "||Left_Parentheses||",
                   ")" : "||Right_Parentheses||",
                   "--" : "||Dash||",
                   "\n" : "||Return||" }
    return token_dict
import pickle
def save_preprocessed_data(int_text, vocab_to_int, int_to_vocab, token_dict):
    """
    保存预处理数据
    """
    pickle.dump((int_text, vocab_to_int, int_to_vocab, token_dict), open('preprocessed_data.p', 'wb'))

def load_preprocessed_data():
    """
    加载预处理数据
    """
    return pickle.load(open('preprocessed_data.p', mode='rb'))
```

```python
def preprocess_scripts_data(text):
    """
    预处理剧本文本数据
    """
    # 将剧本文本里的标点符号都转换成 token
    token_dict = token_lookup()
    for key, token in token_dict.items():
        text = text.replace(key, ' {} '.format(token))
    # 全部转换成小写
    text = text.lower()
    # 全部分割,默认分隔符是空格
    text = text.split()
    # 将剧本文本数据转换成检查表
    vocab_to_int, int_to_vocab = create_lookup_tables(text)
    # 将单词对应的 id 取出
    int_text = [vocab_to_int[word] for word in text]
    # 保存预处理数据
    save_preprocessed_data(int_text, vocab_to_int, int_to_vocab, token_dict)
    # 以元组形式返回
    return int_text, vocab_to_int, int_to_vocab, token_dict

int_text, vocab_to_int, int_to_vocab, token_dict = preprocess_scripts_data(text)
```

17.2.3 创建 Tensor 占位符和学习率

创建 TensorFlow 训练模型时所需的 input 占位符、targets 占位符和 learning rate 占位符。

```python
import tensorflow as tf
def get_inputs():
    # tf.placeholder(dtype, shape=None, name=None)
    # 参数说明
    # dtype:在张量中的元素的数据类型
    # shape:张量的大小。如果不指定大小,就是任意大小
    # name:操作的名称
    input = tf.placeholder(tf.int32, [None, None], name="input")
    targets = tf.placeholder(tf.int32, [None, None], name="targets")
    learning_rate = tf.placeholder(tf.float32, name="learning_rate")
    return (input, targets, learning_rate)
```

17.2.4 初始化 RNN Cell

创建 BasicLSTMCell,并初始化 RNN Cell,堆砌一个或者多个 Cell。

```python
def get_init_cell(batch_size, rnn_size, n_layers=1):
    # 创建 BasicLSTMCell
    def make_lstm(rnn_size):
        return tf.contrib.rnn.BasicLSTMCell(rnn_size)

    # 根据 n_layers 来创建多个 RNN Cell
    cell = tf.contrib.rnn.MultiRNNCell([make_lstm(rnn_size) for _ in range(n_layers)])
    # 使用 zero 来初始化 RNN Cell 的状态
    initial_state = cell.zero_state(batch_size, tf.float32)
    # 返回一个和所输入的大小一样的张量
```

```
            initial_state = tf.identity(input=initial_state, name="initial_state")
            return cell, initial_state
```

17.2.5　创建 Embedding

对输入的数据创建 Embedding，返回嵌入层。

```
        def get_embed(input_data, vocab_size, embed_dim):
            # 创建一个随机均匀分布的 TensorFlow 变量
            embedding = tf.Variable(tf.random_uniform((vocab_size, embed_dim), -1, 1))
            # 在嵌入的张量列表中查找 id，返回嵌入矩阵
            embed = tf.nn.embedding_lookup(embedding, input_data)
            return embed
```

17.2.6　创建神经网络

先创建循环神经网络，然后构建此次的 TensorFlow 的神经网络。

```
        def build_rnn(cell, inputs):
            # 通过指定的 Cell 来构建 RNN
            outputs, final_state = tf.nn.dynamic_rnn(cell, inputs, dtype=tf.float32)
            # 应用一个最终状态，名称为 final_state
            final_state = tf.identity(input=final_state, name="final_state")
            return outputs, final_state
        def build_nn(cell, rnn_size, input_data, vocab_size, embed_dim):
            """
            通过以上定义的函数，我们来构建神经网络
            参数 cell：RNN cell
            参数 rnn_size：RNN 的大小
            参数 input_data：输入数据
            参数 vocab_size：单词大小
            参数 embed_dim：嵌入的维度的数量
            返回：返回一个 Tuple 类型的(Predictions, FinalState)
            """
            # 获取嵌入层
            embedded_layer = get_embed(input_data, vocab_size, embed_dim)
            # 构建 RNN
            outputs, final_state = build_rnn(cell, embedded_layer)
            # 添加一个全连接层，输出层大小是 vocab_size
            predictions = tf.contrib.layers.fully_connected(outputs, vocab_size, activation
_fn=None)
            # 返回预测和最后状态
            return predictions, final_state
```

17.2.7　创建超参数和优化器

先定义超参数，经过调试训练 200 次，我们发现该参数的性能还可以，读者也可以调整训练次数以获得更好的训练效果。然后创建 Adam 优化器，并进行梯度裁剪。

```
        # Epochs 的数量
        num_epochs = 200
        # 批次大小
        batch_size = 128
```

```python
# RNN 大小
rnn_size = 256
# 嵌入维度大小
embed_dim = 500
# 序列长度
seq_length = 10
# 学习率
learning_rate = 0.01
# 每 50 个批次打印一次统计日志信息
show_every_n_batches = 50
from tensorflow.contrib import seq2seq
# 获取 TensorFlow 计算图
train_graph = tf.Graph()
with train_graph.as_default():
    vocab_size = len(int_to_vocab)
    # 获取输入文本、预测目标和学习率的占位符
    input_text, targets, lr = get_inputs()
    input_data_shape = tf.shape(input_text)
    # 初始化 RNN Cell
    cell, initial_state = get_init_cell(input_data_shape[0], rnn_size)
    # 构建神经网络
    logits, final_state = build_nn(cell, rnn_size, input_text, vocab_size, embed_dim)
    # 生成单词的概率,因为是 N 类分类问题,使用 softmax 刚好
    probs = tf.nn.softmax(logits, name='probs')
    # 计算损失值,就是计算 sequence 的加权交叉熵的损失值
    cost = seq2seq.sequence_loss(
        logits,
        targets,
        tf.ones([input_data_shape[0], input_data_shape[1]]))
    # 创建优化器
    optimizer = tf.train.AdamOptimizer(lr)
    # 梯度裁剪
    gradients = optimizer.compute_gradients(cost)
    capped_gradients = [(tf.clip_by_value(grad, -1., 1.), var) for grad, var in gradients if grad is not None]
    train_op = optimizer.apply_gradients(capped_gradients)
```

17.2.8 训练神经网络模型

我们先定义函数,可以将输入数据进行分割和分批,然后启动 TensorFlow 计算图。训练时每 50 个批次打印一次日志信息。保存以 my_tvscripts_generator 开头的模型检查点。

```python
def get_batches(int_text, batch_size, seq_length):
    """
    获取所有批次
    参数 int_text: 所有单词的 id 列表
    参数 batch_size: 每个 batch 的大小
    参数 seq_length: 序列的长度
    """
    # 计算出需要多个批次
    n_batches = int(len(int_text) / (batch_size * seq_length))
```

```python
        # 为了训练完整的批次，我们去掉最后面的几个字符
        # X 的数据从 0 开始, y 数据从 1 开始, 这是因为训练 X, 得出 y
        xdata = np.array(int_text[:n_batches * batch_size * seq_length])
        ydata = np.array(int_text[1:n_batches * batch_size * seq_length + 1])
        # np.split() 函数是将一个数组分割成多个子数组
        x_batches = np.split(xdata.reshape(batch_size, -1), n_batches, 1)
        y_batches = np.split(ydata.reshape(batch_size, -1), n_batches, 1)
        # 最后将 X 和 y 组成元组, 以 NumPy 数组的形式返回
        return np.array(list(zip(x_batches, y_batches)))
# 获得所有的剧本文本数据的批次
batches = get_batches(int_text, batch_size, seq_length)
# 定义 ckpt 保存路径
load_ckpt_path = "./my_tvscripts_generator"
# 启动 TensorFlow 计算图
with tf.Session(graph=train_graph) as sess:
        # 初始化 TensorFlow 计算图的全局变量
        sess.run(tf.global_variables_initializer())
        # 遍历训练
        for epoch_i in range(num_epochs):
                # 初始化状态
                state = sess.run(initial_state, {input_text: batches[0][0]})
                # 分批训练所有的数据
                for batch_i, (x, y) in enumerate(batches):
                        feed = {
                            input_text: x,
                            targets: y,
                            initial_state: state,
                            lr: learning_rate}
                        # 训练模型
                        train_loss, state, _ = sess.run([cost, final_state, train_op], feed)
                        # 每训练 50 个批次, 打印一次统计日志信息
                        if (epoch_i * len(batches) + batch_i) % show_every_n_batches == 0:
                            print('Epoch {:>3} Batch {:>4}/{}   train_loss = {:.3f}'.format(
                                epoch_i,
                                batch_i,
                                len(batches),
                                train_loss))
        # 保存模型
        saver = tf.train.Saver()
        saver.save(sess, load_ckpt_path)
        print('Model Trained and Saved')
```

输出如下。

```
Epoch   0 Batch    0/3   train_loss = 6.933
Epoch  16 Batch    2/3   train_loss = 0.239
Epoch  33 Batch    1/3   train_loss = 0.136
Epoch  50 Batch    0/3   train_loss = 0.145
Epoch  66 Batch    2/3   train_loss = 0.140
Epoch  83 Batch    1/3   train_loss = 0.127
Epoch 100 Batch    0/3   train_loss = 0.142
Epoch 116 Batch    2/3   train_loss = 0.137
Epoch 133 Batch    1/3   train_loss = 0.125
```

```
Epoch 150 Batch    0/3    train_loss = 0.141
Epoch 166 Batch    2/3    train_loss = 0.136
Epoch 183 Batch    1/3    train_loss = 0.125
Model Trained and Saved
```

17.2.9　生成电视剧剧本

我们先恢复保存的预处理数据，然后获得 RNN 模型的输入、初始状态、最终状态和概率的 Tensor。在选词时，选择概率最大的那个。生成剧本时，我们定义的剧本文本最大长度是 200，并从 Moe_Szyslak 开始说话。启动 TensorFlow 计算图，开始预测剧本文本。

```
# 恢复之前保存的预处理数据
int_text, vocab_to_int, int_to_vocab, token_dict = load_preprocessed_data()
def get_tensors(loaded_graph):
    """
    在 TensorFlow 计算图中，获取 input、initial_state、final_state 和 probabilities 张量
    参数 loaded_graph: 从文件中加载的 TensorFlow 的计算图
    返回: (InputTensor, InitialStateTensor, FinalStateTensor, ProbsTensor)
    """
    InputTensor = loaded_graph.get_tensor_by_name("input:0")
    InitialStateTensor = loaded_graph.get_tensor_by_name("initial_state:0")
    FinalStateTensor = loaded_graph.get_tensor_by_name("final_state:0")
    ProbsTensor = loaded_graph.get_tensor_by_name("probs:0")
    return InputTensor, InitialStateTensor, FinalStateTensor, ProbsTensor

def pick_word(probabilities, int_to_vocab):
    """
    在生成的文本中，选择下一个单词
    参数 probabilities: 下一个单词的概率
    参数 int_to_vocab: id 对应着单词的元素的字典
    返回预测单词的字符串
    """
    return int_to_vocab[np.argmax(probabilities)]
# 最大生成长度,我们定义为 200,读者也可以自定义
gen_length = 200
# 3 位说话者 homer_simpson, moe_szyslak 和 barney_gumble
# 我们从 moe_szyslak 开始
prime_word = 'moe_szyslak'
# 获取 TensorFlow 计算图
loaded_graph = tf.Graph()
with tf.Session(graph=loaded_graph) as sess:
    # 重新创建保存在 MetaGraphDef 中的图表
    loader = tf.train.import_meta_graph(load_ckpt_path + '.meta')
    # 从文件中恢复模型
    loader.restore(sess, load_ckpt_path)
    # 通过加载的模型,获取神经网络的张量
    input_text, initial_state, final_state, probs = get_tensors(loaded_graph)
    # 生成序列设置
    gen_sentences = [prime_word + ':']
    # 计算初始状态
    prev_state = sess.run(initial_state, {input_text: np.array([[1]])})
```

```python
# 生成句子
for n in range(gen_length):
    # 动态输入
    dyn_input = [[vocab_to_int[word] for word in gen_sentences[-seq_length:]]]
    dyn_seq_length = len(dyn_input[0])
    # 获取预测
    probabilities, prev_state = sess.run(
        [probs, final_state],
        {input_text: dyn_input, initial_state: prev_state})
    # 从变量probabilities的维度中删除一个维度
    probabilities = np.squeeze(probabilities)
    # 选择预测的单词
    pred_word = pick_word(probabilities[dyn_seq_length-1], int_to_vocab)
    # 将单词添加起来
    gen_sentences.append(pred_word)

# 将变量gen_sentences数组拼接成字符串
tv_script = ' '.join(gen_sentences)
# 遍历所有的token字典
for key, token in token_dict.items():
    ending = ' ' if key in ['\n', '(', '"'] else ''
    # 将拼接的剧本字符串中的token替换成原本的标点符号
    tv_script = tv_script.replace(' ' + token.lower(), key)
# 去掉字符串文本中的所有"\n"和"("后面的空格
tv_script = tv_script.replace('\n ', '\n')
tv_script = tv_script.replace('( ', '(')
# 输出打印生成的剧本
print(tv_script)
```

输出如下。

```
INFO:tensorflow:Restoring parameters from ./my_tvscripts_generator
moe_szyslak:(moe's_tavern: int. moe's tavern - night)
barney_gumble:... some may, to capital city.
homer_simpson:(reverential) the duff brewery, homer!
bart_simpson: wow!
lisa_simpson: the pennyloafer!
homer_simpson: as i got up in front of this thing, if the players," expecting".
waylon_smithers:(sotto) the card needs needs, marge.
homer_simpson: you mean--?
homer_simpson:(sotto) oh, no.
homer_simpson: well, i'm ready to take"...
 (ballpark: ext. ballpark - evening)
dan_hoard: bases loaded. 'topes have one out, too.
mr. _o'hara: excuse me, homer.
 (ballpark: ext. ballpark - evening)
dan_hoard: bases loaded. 'topes have one out, too.
mr. _o'hara: excuse me, homer.
 (ballpark: ext. ballpark - evening)
dan_hoard: bases loaded. 'topes have one out, too.
mr. _o'hara: excuse me, homer.
```

生成的剧本文本并不是最优的，但这给读者展示了一种最简单的文本生成实现方式。读者可以尝试用更多的剧本文本来训练模型并生成剧本文本，以获得更佳的剧本。

17.3 基于 textgenrnn 来实现电视剧剧本生成

textgenrnn 是一个流行的文本生成（Text Generation）开源库，它是基于 char-rnns，由 Keras/TensorFlow 实现的。textgenrnn 使用起来很简单，只需要传递给它一个任意大小的文本文件，然后调用 generate() 方法就能训练模型。并且它的性能不错，因为它可以基于预训练模型来训练我们自己的模型。

17.3.1 介绍

textgenrnn 利用注意加权和跳过嵌入来加速训练并提高模型质量，在字符级别或单词级别上训练并生成文本，配置 RNN 大小、RNN 层数以及是否使用双向 RNN，训练任何通用的输入文本文件，包括大文件。在 GPU 上训练时，它利用强大的 CuDNN RNN 来实现，与典型的 LSTM 实现方式相比，大大缩短了训练时间。

在默认的模型下，最多接收 40 个字符的输入，将每个字符转换为 100-D 字符嵌入向量，然后将其馈送到有 128 个 cell 的 LSTM 循环层，这些输出接着又会把它们馈送到另一个有 128 个 cell 的长短期记忆网络层；最后将这 3 层都送入 Attention 层，以对最重要的时间特征加权并将它们加在一起求均值。该输出被映射到最多 394 个不同字符的概率，它们是序列中的下一个字符，包括大写字母、小写字母、标点符号和表情符号。如果我们在新的数据集上训练新模型，那么上面所有的数字参数都是可以配置的。

17.3.2 训练模型

使用前，我们首先要安装 textgenrnn，通过 pip install textgenrnn 就可以完成安装。然后将《辛普森一家》第 18 个片段的剧本文本文件传递给 textgenrnn 对象，训练 30 次。

```
from textgenrnn import textgenrnn
textgen = textgenrnn()
# 从文件中训练模型
textgen.train_from_file('Simpsons_Episode_18.txt', num_epochs=30)
# 根据 Temperature 的值来生成剧本文本
textgen.generate()
```

训练时输出日志信息如下。

```
Epoch 1/30
155/155 [==============================] - 8s 49ms/step - loss: 1.7872
####################
Temperature: 0.2
####################
 (CAPITC) The Capital Capital Capital Capital City.
 (CAPITS: Capital missimation - City - Capital Capital Capital City (CAPITT)
 Homer_Simpson: Homer - Simpson: (CAPITT) The ball of the studios into the bigger of a big ball of the bigger.
```

```
……
Epoch 30/30
155/155 [==============================] - 5s 30ms/step - loss: 0.1413
……
####################
Temperature: 1.0
####################
Homer_Simpson: Sorry, honey.
Marge_Simpson: (TO SOUGHS) Simp this year, "Racin's danch!
Homer_Simpson: You know, boy, some of he losest art, butto for Capital City.
Homer_Simpson: (CHANTING) LITTLE BABY BATTER / CAN'T CONTROL HIS BLADDER! (LAUGHS)
```

17.3.3　生成剧本文本

生成 10 行剧本文本，并写入文件，文件名为 my_generated_texts.txt。

```
textgen.generate_to_file('my_generated_texts.txt', n=10)
```

用 Linux 的 cat 命令查看生成的文本。

```
cat my_generated_texts.txt
```

输出如下。

```
Marge_Simpson: Look! The Crosstown Bridge!
C._Montgomery_Burns: Oh, shut up.
Homer_Simpson: You know, boy, some of the players you see tonight may make it to the
big leagues, one day.
Homer_Simpson: Oooh, red hots.
Marge_Simpson: (SOTTO) Oh, no. Marge, sir.
Homer_Simpson: (INTERRUPTING) Oh, oh, oh, the late gentlemen.
Homer_Simpson: I can't help but feel that if we had gotten to know each other better,
my leaving would actually have meant something.
Homer_Simpson: Bart was seen the ball spit. but too complex.
Homer_Simpson: (REVERENTIAL) The Duff Brewery!
```

17.4　小结

我们采用开放的 *The Simpsons* 剧本文本中的第 18 个片段的剧本文本作为训练集。一开始我们构建了自己的循环神经网络和 LSTM 模型，进行训练、预测以及新剧本生成；后来我们采用了开源库 textgenrnn 构建 Text Generation 模型来生成新剧本。两个模型生成的剧本文本很不一样，如果剧本的文本数据更多，或者调整超参数，模型的效果可能会更好。

第 18 章

风格迁移

风格迁移技术能让无绘画基础的人也能像巴勒罗·毕加索或文森特·梵高这样的顶级画家一样作画。如果你想让自拍照或者风景照犹如画家画出来的一样,那本章就非常适合你。通过本章的风格迁移技术,我们将几种画作风格迁移到指定的风景或街景图像上,让深度学习的算法模型帮我们作画。我们将通过 TensorFlow 和 Keras 的迁移学习技术来实现图像的风格迁移。

18.1 基于 TensorFlow 实现神经风格迁移

神经风格迁移的原则是定义两个距离函数,一个用来描述两个图像的内容有多么不同,另一个用来描述两个图像之间的风格差异;然后给定 3 个图像,分别是风格图像、内容图像和输入图像,输入图像是内容图像初始化后得到的。我们将采用反向传播给输入图像转换最小化内容图像和风格样式的距离。

18.1.1 环境准备

- numpy = 1.14.6。
- matplotlib = 2.1.2。
- PIL = 4.0.0。
- tensorflow = 1.12.0。

18.1.2 图像预览

我们先预览风格图像和内容图像,定义两个函数,它们分别是加载图片函数 load_img() 和图像显示函数 imshow()。

```
import matplotlib.pyplot as plt
import matplotlib as mpl
```

```python
mpl.rcParams['figure.figsize'] = (12,15)
mpl.rcParams['axes.grid'] = False
import numpy as np
import time
from PIL import Image
from tensorflow.python.keras import preprocessing
def load_img(path_to_img):
    # 加载图片
    img = Image.open(path_to_img)
    # 获取图片的最大边长（要么是宽，要么是高）
    long = max(img.size)
    # 修改图片大小，最大边长为512
    max_dim = 512
    scale = max_dim/long
    img = img.resize((round(img.size[0]*scale), round(img.size[1]*scale)), Image.ANTIALIAS)
    # 预处理图片，转换成三维数组
    img = preprocessing.image.img_to_array(img)
    # 将图片数组维度扩展成batch dimension，也就是CNN期待输入的shape
    img = np.expand_dims(img, axis=0)
    return img
def imshow(img, title):
    # 先移除batch这个维度
    out_img = np.squeeze(img, axis=0)
    plt.title(title)
    plt.imshow(out_img)
# 定义图片路径
style_path = "udnie.jpg"
content_path = "wangjing_selfie.jpg"
# 绘制内容图像
plt.subplot(1, 2, 1)
p_img = load_img(content_path).astype('uint8')
imshow(p_img, "Content Image")
# 绘制风格图像
plt.subplot(1, 2, 2)
p_img = load_img(style_path).astype('uint8')
imshow(p_img, "Style Image")
```

输出结果如图18.1所示。

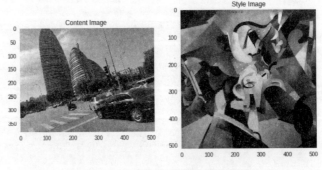

图18.1　内容图像和风格图像预览

18.1.3 处理图像

根据 VGG 的训练过程，我们对输入图像进行同样的预处理。VGG 网络在图像上进行训练，每个信道通过均值和信道 BGR 进行归一化处理，均值通常是[103.939, 116.779, 123.68]。

```
import tensorflow as tf
import tensorflow.contrib.eager as tfe
from tensorflow.python.keras import models
from tensorflow.python.keras import losses
from tensorflow.python.keras import layers
from tensorflow.python.keras import backend as K
# 启用 eager execution 的目的是使我们能够用更清晰和易读的方式完成训练
tf.enable_eager_execution()
print("Eager execution: {}".format(tf.executing_eagerly()))
def load_and_process_img(path_to_img):
    # 加载图像
    img = load_img(path_to_img)
    # 使用 VGG19 来预处理图像
    img = tf.keras.applications.vgg19.preprocess_input(img)
    return img
```

为了查看优化后的输出结果，这就要求我们进行反转处理。在反转处理时，可能会遇到处理的值是在-∞和∞（负无穷大到正无穷大）之间，所以必须要对这个值进行有效范围控制，我们将其控制在 0~255。

```
def deprocess_img(processed_img):
    # 拷贝一个副本
    x = processed_img.copy()
    # 如果 x 的维度是 4，那就移除 batch 那一维度
    if len(x.shape) == 4:
        x = np.squeeze(x, 0)
    assert len(x.shape) == 3, ("输入的反转处理图像维度必须是[1, height, width, channel]"
                                "或者[height,width,channel]")
    if len(x.shape) != 3:
        raise ValueError("无效的图像！")
    # 执行图像预处理反转的步骤，为了通过和信道的归一化处理，这 3 个信道均值分别如下
    x[:, :, 0] += 103.939
    x[:, :, 1] += 116.779
    x[:, :, 2] += 123.68
    # 反转
    x = x[:, :, ::-1]
    # 裁切数组上的值在 0~255
    x = np.clip(x, 0, 255).astype('uint8')
    return x
```

18.1.4 模型获取

我们通过 VGG19 模型的中间层来创建模型，这些中间层代表越来越高阶的特征，这对我们在图像中定义内容和风格是必要的。因此，我们将尝试对这些中间层匹配相应的样式和内容。

```python
# 特征映射层的内容层名字
content_layers = ['block5_conv2']
# 这是我们需要的风格化的层的名字
style_layers = ['block1_conv1',
                'block2_conv1',
                'block3_conv1',
                'block4_conv1',
                'block5_conv1'
               ]
num_content_layers = len(content_layers)
num_style_layers = len(style_layers)
```

我们定义 get_model()函数，加载 VGG19 模型并访问中间层，通过这些中间层来创建新模型，然后返回一个 Keras 模型，接受图像输入，并输出样式和内容中间层；也就是我们使用 Keras 的 Functional API，通过期待的输出来激活和定义我们的模型。

```python
def get_model():
    # 加载预训练VGG19 模型，它是在 imagenet 数据上训练而来的
    # include_top 等于 False 表示不包含 VGG19 神经网络模型顶层的全连接层
    vgg = tf.keras.applications.vgg19.VGG19(include_top=False, weights='imagenet')
    # 设置这些层不参与训练
    vgg.trainable = False
    # 获取与样式对应的输出层
    style_outputs = [vgg.get_layer(name).output for name in style_layers]
    # 获取与内容对应的输出层
    content_outputs = [vgg.get_layer(name).output for name in content_layers]
    model_outputs = style_outputs + content_outputs
    # 构建模型
    return models.Model(vgg.input, model_outputs)
```

18.1.5 损失函数计算

向新模型传递基础图像和内容图像，返回中间层输出。然后取这两个图像之间的欧几里得距离（Euclidean Distance），按照常规的方式进行反向传播，以便最小化这种 content 损失值。我们会将每一个层都添加上 content 损失值，这样所有穿过 model 的 content 损失值都会被正确地计算。

```python
def get_content_loss(base_content, target):
    # 计算 content 损失值
    return tf.reduce_mean(tf.square(base_content - target))
```

计算风格损失值（style loss）会复杂些。我们给神经网络提供基本的输入图像和风格图像，但是，我们的目的不是比较基本输入图像和风格图像的中间输出，而是比较两个图像输出的 gram 矩阵。

```python
def gram_matrix(input_tensor):
    # image channels first
    channels = int(input_tensor.shape[-1])
    a = tf.reshape(input_tensor, [-1, channels])
```

```
        n = tf.shape(a)[0]
        # 特征映射的点积运算
        gram = tf.matmul(a, a, transpose_a=True)
        return gram / tf.cast(n, tf.float32)
    def get_style_loss(base_style, gram_target):
        # 每一层都有 height, width, num_filters
        # 我们通过特征映射的大小和过滤层的数量在给定的层上计算损失值
        height, width, channels = base_style.get_shape().as_list()
        gram_style = gram_matrix(base_style)
        # 计算风格损失值
        return tf.reduce_mean(tf.square(gram_style - gram_target))
```

定义计算当前 style 和 content 的特征，它会简单地从路径中获取 style 和 content 图像，然后传递到网络模型以获得中间层的输出。

```
    def get_feature_representations(model, content_path, style_path):
        # 加载图片
        content_image = load_and_process_img(content_path)
        style_image = load_and_process_img(style_path)
        # 批量计算 style 和 content 特征
        style_outputs = model(style_image)
        content_outputs = model(content_image)
        # 从我们的模型上获取 style 和 content 特征
        style_features = [style_layer[0] for style_layer in style_outputs[:num_style_layers]]
        content_features = [content_layer[0] for content_layer in content_outputs[num_style_layers:]]
        return style_features, content_features
```

计算损失和梯度，通过跟踪操作并利用可用的自动区分，方便地计算梯度。它记录前向传递期间的操作，然后能够计算我们的损失函数相对于向后传递的输入图像的梯度。compute_loss()函数参数解释如下。

- 参数 1：model，表示模型是可以访问中间层的。
- 参数 2：loss_weights，表示每个损失函数的每个贡献的权重，它的 shape 用(style weights, content weights, total variation weights)来表示。
- 参数 3：init_image，表示初始化的基础图片，用于优化更新处理，加以计算梯度损失。
- 参数 4：gram_style_features，预计算对应的已定义的 style 层的 gram 矩阵。
- 参数 5：content_features，从已定义的 content 层预计算输出。

```
    def compute_loss(model, loss_weights, init_image, gram_style_features, content_features):
        style_weight, content_weight = loss_weights
        # 通过我们的模型获得初始图像，为我们提供所需的内容和样式。
        # 由于我们正在使用 eager execution，所以我们的模型可以像其他功能一样调用！
        model_outputs = model(init_image)
        # 获取 style 和 content 的输出层
        style_output_features = model_outputs[:num_style_layers]
        content_output_features = model_outputs[num_style_layers:]
        style_score = 0
        content_score = 0
```

```python
    # 从所有层上累计 style 的损失
    # 我们同样地权衡每个损失层的每个贡献
    weight_per_style_layer = 1.0 / float(num_style_layers)
    for target_style, comb_style in zip(gram_style_features, style_output_features):
        style_score += weight_per_style_layer * get_style_loss(comb_style[0], target_style)
    # 从所有层上累计 content 损失
    weight_per_content_layer = 1.0 / float(num_content_layers)
    for target_content, comb_content in zip(content_features, content_output_features):
        content_score += weight_per_content_layer* get_content_loss(comb_content[0], target_content)
    style_score *= style_weight
    content_score *= content_weight
    # 获取全部的损失
    loss = style_score + content_score
    return loss, style_score, content_score
# 计算梯度
def compute_grads(cfg):
    with tf.GradientTape() as tape:
        all_loss = compute_loss(**cfg)
    # 计算输入图像的梯度
    total_loss = all_loss[0]
    return tape.gradient(total_loss, cfg['init_image']), all_loss
```

18.1.6　训练模型与图像生成

现在，用以上定义的各个函数来配置训练前所需的参数。通过 style 和 content 的图片路径来获取它们的特征。添加 Adam 优化器，将各参数用字典对象变量 cfg 存放起来，以便于计算梯度。best_loss 和 best_img 变量用来存储最佳的损失值和风格化后的图像。

```python
# 我们不希望训练完整的 VGG19 神经网络，所以直接将模型的 trainable 属性设置为 False
model = get_model()
for layer in model.layers:
    layer.trainable = False
# 从我们指定的层中获取 style 和 content 的特征
style_features, content_features = get_feature_representations(model, content_path, style_path)
gram_style_features = [gram_matrix(style_feature) for style_feature in style_features]
# 设置初始化图片
init_image = load_and_process_img(content_path)
init_image = tfe.Variable(init_image, dtype=tf.float32)
# 创建 Adam 优化器
opt = tf.train.AdamOptimizer(learning_rate=5, beta1=0.99, epsilon=1e-1)
# 存储我们的结果
best_loss, best_img = float('inf'), None
# 创建基本配置参数
style_weight = 1e-2
content_weight = 1e3
loss_weights = (style_weight, content_weight)
cfg = {
    'model': model,
    'loss_weights': loss_weights,
```

```
        'init_image': init_image,
        'gram_style_features': gram_style_features,
        'content_features': content_features
}
```

开始迭代计算 1000 次，其中每迭代 100 次，记录一次当前状态图像的模样。当全部迭代计算完后，输出 10 个训练时的图像表现，且将最佳的损失值记录和最佳的风格化图像都保存在 best_loss 和 best_img 变量中以便于查看。

```
# 如果使用 Jupyter Notebook 则需要导入该模块
import IPython.display
# 定义迭代 1000 次
num_iterations = 1000
# 定义 2 行 5 列用来指定训练时间隔的图片风格化的显示
num_rows = 2
num_cols = 5
# 定义每多少个训练，保存一次训练的值
display_interval = num_iterations/(num_rows*num_cols)
start_time = time.time()
global_start = time.time()
# 正常的均值
norm_means = np.array([103.939, 116.779, 123.68])
min_vals = -norm_means
max_vals = 255 - norm_means
imgs = []
# 开始训练
for i in range(num_iterations):
    # 计算梯度
    grads, all_loss = compute_grads(cfg)
    loss, style_score, content_score = all_loss
    opt.apply_gradients([(grads, init_image)])
    # 裁切值在范围内
    clipped = tf.clip_by_value(init_image, min_vals, max_vals)
    init_image.assign(clipped)
    end_time = time.time()
    if loss < best_loss:
        # 从全部的损失里更新最佳的 loss 和最佳的图片
        best_loss = loss
        best_img = deprocess_img(init_image.numpy())
    # 每迭代 100 次，保存一次训练的值
    if i % display_interval== 0:
        start_time = time.time()
        plot_img = init_image.numpy()
        plot_img = deprocess_img(plot_img)
        imgs.append(plot_img)
        # 清空当前的 Jupyter 的 cell 的输出内容
        IPython.display.clear_output(wait=True)
        IPython.display.display_png(Image.fromarray(plot_img))
        print('Iteration: {}'.format(i))
        print('Total loss: {:.4e}, '
              'style loss: {:.4e}, '
```

```
                            'content loss: {:.4e}, '
                            'time: {:.4f}s'.format(loss, style_score, content_score, time.time()
- start_time))
    print('Total time: {:.4f}s'.format(time.time() - global_start))
    # 清空当前的 Jupyter 的 cell 的输出内容
    IPython.display.clear_output(wait=True)
    # 最后将记录的 10 个图像打印输出
    plt.figure(figsize=(14,4))
    for i,img in enumerate(imgs):
        plt.subplot(num_rows,num_cols,i+1)
        plt.imshow(img)
        plt.xticks([])
        plt.yticks([])
```

输出结果如图 18.2 所示。

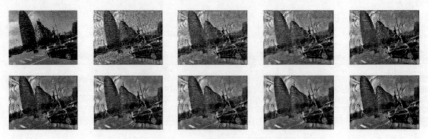

图 18.2　风格化后的图像

查看最佳的风格化后的大图。

```
Image.fromarray(best_img)
```

输出结果如图 18.3 所示。

图 18.3　风格化后的大图

我们再来风格化一张神奈川的大浪的图。把这张图的风格迁移到望京街道的图片上，大浪的图片是 rough_sea.jpeg。先预览图片，如图 18.4 所示。

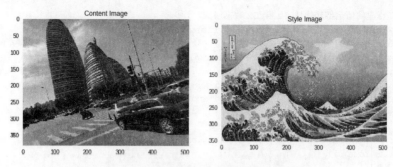

图 18.4　望京街道图和神奈川的大浪图

经过以上的代码计算和风格化后，最终输出效果如图 18.5 所示。

图 18.5　望京街道图经过神奈川大浪风格化后的图

18.2　基于 Keras 实现神经风格迁移

18.1 节中的神经风格迁移是通过 TensorFlow 的 keras 模块实现的，本节就把 Keras 和 SciPy 结合起来使用以实现风格迁移。最重要的是 fmin_l_bfgs_b()函数，它使用 L_BFGS_B 算法来最小化功能函数，然后返回评估的最小值等信息。

18.2.1　图像预览

风格图像的文件名是 the_shipwreck_of_the_minotaur.jpg，内容图像的文件名是 xiangshan.jpeg。我们通过 PIL 模块来加载图像，并获取图像数组对象。

```
from PIL import Image
import numpy as np
# 图像的大小
width = 512
height = 512
# 加载图像函数
def load_img(filepath):
    # 根据图像地址读取图像
    img = Image.open(filepath)
    # 重新设置图像的大小
    img = img.resize((width, height))
```

```python
    # 将图像里的值全部转换成 float32 数据格式
    arr = np.asarray(img, dtype=np.float32)
    # 将数组数据扩展一个维度
    img_arr = np.expand_dims(arr, axis=0)
    return img_arr
content_path = "xiangshan.jpeg"
style_path = "the_shipwreck_of_the_minotaur.jpg"
style_image = load_img(style_path)
content_image = load_img(content_path)
print("content_image.shape={}, style_image.shape={}.".format(content_image.shape,
style_image.shape))
```

输出信息如下。

`content_image.shape=(1, 512, 512, 3), style_image.shape=(1, 512, 512, 3).`

定义 show_img() 函数来加载图像。加载图像时，因为 RGB 图像的数据是三维的，所以我们通过 sequeeze() 函数来去掉其中的第四维度。

```python
import matplotlib.pyplot as plt
def show_img(img_arr, title):
    out_img = np.squeeze(img_arr.astype('uint8'), axis=0)
    plt.title(title)
    plt.grid(False)
    plt.imshow(out_img)
# 显示内容图像
plt.figure(figsize=(18, 9))
plt.subplot(1, 2, 1)
show_img(content_image, "Content Image")
# 显示风格图像
plt.subplot(1, 2, 2)
show_img(style_image, "Style Image")
```

输出结果如图 18.6 所示。

图 18.6　香山风景图和海浪击碎船舶的图

18.2.2　图像处理

对内容图像和风格图像进行同样的预处理，每个信道通过均值和信道 BGR 进行归一化处理,均值通常是[103.939, 116.779, 123.68]。

```
def deprocess_img(processed_img):
    processed_img[:, :, :, 0] -= 103.939
    processed_img[:, :, :, 1] -= 116.779
    processed_img[:, :, :, 2] -= 123.68
    processed_img_arr = processed_img[:, :, :, ::-1]
    return processed_img_arr
content_array = deprocess_img(content_image)
style_array = deprocess_img(style_image)
print("content_array.shape={}, style_array.shape={}.".format(content_array.shape,
style_array.shape))
```

输出如下。

```
content_array.shape=(1, 512, 512, 3), style_array.shape=(1, 512, 512, 3).
```

18.2.3　获取模型

我们直接使用 Keras 的默认后端 TensorFlow,然后使用 VGG16 模型来创建我们的模型。VGG16 模型同样也不包含顶层的全连接层。接下来,我们对每一层计算损失值。

```
from keras import backend
from keras.applications.vgg16 import VGG16
# 创建 TensorFlow 的 content 变量和 style 变量
content_image = backend.variable(content_array)
style_image = backend.variable(style_array)
# 定义输入 tensor
combination_image = backend.placeholder((1, height, width, 3))
input_tensor = backend.concatenate([content_image, style_image, combination_image],
axis=0)
# 定义全局 loss 变量
loss = backend.variable(0.)
# 定义 content 权重
content_weight = 0.05
# 定义 style 权重
style_weight = 5.0
# 定义全局变量权重
total_variation_weight = 1.0
# 获取 VGG16 模型,如果本地不存在 VGG16,就自动下载
model = VGG16(input_tensor=input_tensor, weights='imagenet', include_top=False)
# 获取该模型下所有的层
layers = dict([(layer.name, layer.output) for layer in model.layers])
# 模型中我们需要的卷积层的名称
feature_layers = ['block1_conv2', 'block2_conv2',
                  'block3_conv3', 'block4_conv3',
                  'block5_conv3']
# 对层进行损失值计算
for layer_name in feature_layers:
    layer_features = layers[layer_name]
    style_features = layer_features[1,:,:,:]
    combination_features = layer_features[2,:,:,:]
```

```
        sl = style_loss(style_features, combination_features)
        loss += (style_weight / len(feature_layers)) * sl
```

18.2.4 损失函数计算

计算 content loss、gram 矩阵、style loss 和全局 loss，以及梯度。

```
def content_loss(content, combination):
    return backend.sum(backend.square(content - combination))
layer_features = layers['block2_conv2']
content_image_features = layer_features[0,:,:,:]
combination_features = layer_features[2,:,:,:]
loss += content_weight * content_loss(content_image_features, combination_features)
def gram_matrix(x):
    # 将图像通过维度转置，然后扁平化处理
    features = backend.batch_flatten(backend.permute_dimensions(x, (2,0,1)))
    # 进行点积运算
    gram = backend.dot(features, backend.transpose(features))
    return gram
# 计算 style loss
def style_loss(style,combination):
    # 计算 style 的 gram 矩阵
    S = gram_matrix(style)
    # 计算组合图的 gram 矩阵
    C = gram_matrix(combination)
    channels = 3
    size = height * width
    st = backend.sum(backend.square(S - C)) / (4. * (channels ** 2) * (size ** 2))
    return st
# 计算全部的可变 loss
def total_variation_loss(x):
    a = backend.square(x[:,:height-1,:width-1,:]-x[:,1:,:width-1,:])
    b = backend.square(x[:, :height-1, :width-1, :] - x[:, :height-1, 1:, :])
    return backend.sum(backend.pow(a + b, 1.25))
loss += total_variation_weight * total_variation_loss(combination_image)
# 梯度计算
grads = backend.gradients(loss, combination_image)
outputs=[loss]
if isinstance(grads, (list, tuple)):
    outputs += grads
else:
    outputs.append(grads)
# 初始化 Keras 的 function
f_outputs = backend.function([combination_image], outputs)
```

18.2.5 迭代与生成风格图像

为了评估损失值和梯度，我们定义一个 Evaluator 类，该类可以计算损失值和梯度。然后我们定义 generate()函数来迭代和生成风格图像，设置 10 次迭代就可以达到很不错的效果。其中通过 fmin_l_bfgs_b()函数来最小化目标函数也是一种优化算法。

```python
def eval_loss_and_grads(x):
    x = x.reshape((1, height, width, 3))
    outs = f_outputs([x])
    loss_value = outs[0]
    grad_values = outs[1].flatten().astype('float64')
    return loss_value, grad_values
class Evaluator(object):
    def __init__(self):
        self.loss_value=None
        self.grads_values=None
    # 计算 loss
    def loss(self, x):
        assert self.loss_value is None
        loss_value, grad_values = eval_loss_and_grads(x)
        self.loss_value = loss_value
        self.grad_values = grad_values
        return self.loss_value
    # 计算梯度
    def grads(self, x):
        assert self.loss_value is not None
        grad_values = np.copy(self.grad_values)
        self.loss_value = None
        self.grad_values = None
        return grad_values
import time
from scipy.optimize import fmin_l_bfgs_b
def generate():
    x = np.random.uniform(0, 255, (1, height, width, 3)) - 128.0
    evaluator = Evaluator()
    iterations = 10
    for i in range(iterations):
        print('Iteration {}'.format(i + 1))
        start_time = time.time()  # 开始计时
        # 参数 1：我们期望的最小化的 loss 函数
        # 参数 2：一个初始化的 NumPy 数组
        # 参数 3：计算梯度的函数
        # 参数 4：最大功能评估数
        x, min_val, info = fmin_l_bfgs_b(evaluator.loss,
                                    x.flatten(),
                                    fprime=evaluator.grads,
                                    maxfun=20)
        print(min_val)
        end_time = time.time()  # 一次结束计时
        print('Iteration {} 在{:.2f}秒完成'.format(i, end_time - start_time))
    return x
img_x = generate()
```

输出如下。

```
Iteration 1
47021440000.0
Iteration 0 在 15.87 秒完成
Iteration 2
```

```
28353214000.0
Iteration 1 在14.39秒完成
……
18183066000.0
Iteration 8 在14.87秒完成
Iteration 10
18011770000.0
Iteration 9 在14.77秒完成
```

然后我们对最终生成的 img_x 图像变量进行输出查看。因为前面使用 deprocess_img()时将图像反转了,所有的值都减去了 BGR 的均值,所以最后显示的时候需要再反转回来,把均值加回去。

```
def output_process_img(img_arr):
    img_arr = img_arr.reshape((height, width, 3))
    img_arr = img_arr[:, :, ::-1]
    img_arr[:, :, 0] += 103.939
    img_arr[:, :, 1] += 116.779
    img_arr[:, :, 2] += 123.68
    img_arr = np.clip(img_arr, 0, 255).astype('uint8')
    return img_arr

x = output_process_img(img_x)
Image.fromarray(x)
```

输出结果如图 18.7 所示。

图 18.7　生成风格化后的效果图

18.3　小结

我们主要讲的是风格迁移技术,也就是说将画家的画作风格迁移到您指定的图画上。我们构建了两种实现方式,分别是 TensorFlow 的 VGG19 的迁移学习和 Keras 的 VGG16 的迁移学习,并将生成的风格化图片与原图进行了对比。

第 19 章

生成人脸

生成对抗网络（Generative Adversarial Networks）一般称作 GAN。本章我们使用 GAN 来生成人脸图像，在生成人脸图像前，先拿 MNIST 手写数字图像数据集来练习，再生成 LFW 数据集的人脸图像。

GAN 是由 Ian GoodFellow（现任谷歌大脑研究科学家）发明的。GAN 由两个模型组成，分别是生成器（Generator）模型和鉴别器（Discriminator）模型。生成器负责生成假图像，以欺骗鉴别器；鉴别器负责区分输入的真实图像和生成的假图像。在多次迭代中，生成器不停地学习生成新图像，以达到让鉴别器无法区分输入的图像是真还是假的目的。

19.1 基于 TensorFlow 的 GAN 实现 MNIST 数字图像生成

鉴别器模型本质上是一个分类器，用来确定给定的图像是来自数据集的真实图像还是生成的假图像。该模型是一个二进制分类器（其分类结果要么是来自数据集的真实图像，要么是生成的假图像），它通过构建卷积神经网络（CNN）模型实现。

生成器模型采用随机输入值，并通过卷积神经网络将它们转换为图像。在多次训练迭代过程中，鉴别器和生成器的权重和偏差通过反向传播进行训练，然后生成器通过鉴别器的反馈来学习如何产生逼真度较高的图像，以达到让鉴别器不能把假图像与真实图像区分开的目的。现在，我们看看生成对抗网络的架构，如图 19.1 所示。

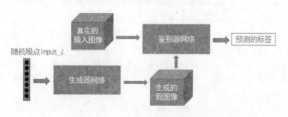

图 19.1　生成对抗网络的架构

19.1.1 环境准备

- numpy = 1.14.6。
- matplotlib = 3.0.2。
- PIL = 4.0.0。
- tensorflow = 1.12.0。
- tqdm=4.28.1。

19.1.2 MNIST 数字图像数据准备

从 MNIST 官方网站下载训练集的 60000 个数字图像数据,下载地址在本代码仓库。下载后,通过 decompress_gzfile()函数将 MNIST 数据转换成图片,并存储到本地目录 mnist_dataset 下,然后使用 gzip 模块加载 mnist 文件中的数据,再通过_read32()函数把数据字节流读取到 NumPy 数组中,最后通过遍历来把每一张 MNIST 数字图片保存到 mnist_dataset 目录下。

```python
import os
import gzip
import numpy as np
from tqdm import tqdm
from PIL import Image
def _read32(bytestream):
    """
    从字节流读取 32 位的整型数据
    参数 bytestream:字节流数据
    返回值:32-bit integer
    """
    dt = np.dtype(np.uint32).newbyteorder('>')
    return np.frombuffer(bytestream.read(4), dtype=dt)[0]

def decompress_gzfile(extract_path, save_path):
    """
    将 MNIST 的压缩包文件解压,然后把其数组数据转换成图片
    参数 extract_path:MNIST 压缩包文件路径
    参数 save_path:压缩包文件解压的路径
    """
    # 直接读取 MNIST 数据集文件
    with open(extract_path, 'rb') as f:
        # gzip 是一个单文件/无损数据压缩实用程序,其中生成的压缩文件扩展名通常为 gz
        # gzip 还指实用程序使用的关联压缩数据格式
        # 在这里,我们将数据集数据读取到字节流中
        with gzip.GzipFile(fileobj=f) as bytestream:
            # 然后对字节流进行处理
            magic = _read32(bytestream)
            # 2051 表示 magic number,在 MNIST 数据集里有说明,用来区分图片数据和标签数据
            # 2051 表示在数据集中的图片文件,而 2049 就是表示数据集中的图片标签
            if magic != 2051:
                raise ValueError('Invalid magic number {} in file: {}'.format(magic, f.name))
            num_images = _read32(bytestream)
```

```
                    rows = _read32(bytestream)
                    cols = _read32(bytestream)
                    buf = bytestream.read(rows * cols * num_images)
                    data = np.frombuffer(buf, dtype=np.uint8)
                    data = data.reshape(num_images, rows, cols)
            # 遍历并把图片数据保存到mnist_datasets目录下，形成一张张的灰度图片
            for image_i, image in enumerate(tqdm(data,
                                                unit='File',
                                                unit_scale=True,
                                                miniters=1,
                                                desc='提取MNIST数据集图像')):
                save_img_path = os.path.join(save_path, 'image_{}.jpg'.format(image_i))
                # 保存图片到指定路径
                Image.fromarray(image, 'L').save(save_img_path)

# MNIST 数据集文件路径
extract_path = "train-images-idx3-ubyte.gz"
# 要保存到的MNIST目录
save_path = "mnist_dataset"
decompress_gzfile(extract_path, save_path)
```

19.1.3 随机查看 25 张图片

通过 glob 模块可以加载所有的 MNIST 图像路径到数组中，然后打印输出前 10 张图片路径。

```
from glob import glob
# 加载所有的jpg图片路径
mnist_all_paths = glob(save_path + '/*.jpg')
mnist_all_paths[:10]
```

前 10 张图片路径输出如下。

```
['mnist_dataset/image_26205.jpg',
 'mnist_dataset/image_53592.jpg',
 'mnist_dataset/image_6381.jpg',
 'mnist_dataset/image_7547.jpg',
 'mnist_dataset/image_22247.jpg',
 'mnist_dataset/image_2355.jpg',
 'mnist_dataset/image_40473.jpg',
 'mnist_dataset/image_32139.jpg',
 'mnist_dataset/image_29817.jpg',
 'mnist_dataset/image_41088.jpg']
```

查看有多少张 MNIST 图片。

```
len(mnist_all_paths)
```

输出如下。

```
60000
```

想要根据图片路径来随机查看 25 张 MNIST 图片，我们可以使用 random 模块的 sample() 随机函数，该函数需传入的参数是"（整个数组，要随机的数量）"。

```python
import matplotlib.pyplot as plt
from matplotlib import image
import random
def plot_random_25_img(filepaths):
    # 返回从总体序列中选择的长度为 k 的唯一元素列表。 用于随机抽样而无需更换。
    # 从指定的数组文件路径中随机获取 25 张图片的路径
    random_25_imgs = random.sample(filepaths, 25)
    # 创建 25 个绘图对象，5 行 5 列
    fig, axes = plt.subplots(nrows=5, ncols=5)
    # 设置绘图的总容器大小
    fig.set_size_inches(8, 8)
    index = 0
    for row_index in range(5): # 行
        for col_index in range(5): # 列
            # 读取图片的数值内容
            img = image.imread(random_25_imgs[index])
            # 根据[行索引，列索引]获取绘图 Axes 对象
            ax = axes[row_index, col_index]
            # 在 Axes 对象上显示图片
            ax.imshow(img)
            # 不显示网格
            ax.grid(False)
            # 索引加 1
            index += 1

plot_random_25_img(mnist_all_paths)
```

输出结果如图 19.2 所示。

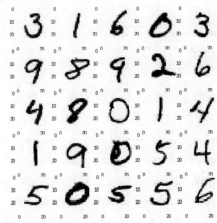

图 19.2　随机查看 25 张图片，每次随机出来的图片都会不一样

19.1.4　构建模型输入

构建模型输入函数，传入图片的宽度、图片的高度、图片的通道数，以及 z 的维度，最后返回它们的 TensorFlow 的张量。

```python
import tensorflow as tf
def model_inputs(image_width, image_height, image_channels, z_dim):
    """
    创建模型的输入占位符
    参数 image_width: 输入的图片宽度
    参数 image_height: 输入的图片高度
    参数 image_channels: 输入的图片通道数
    参数 z_dim: z 的维度
    返回值: 由(tensor of real input images, tensor of z data, learning rate)组成的元组
    """
    inputs_real = tf.placeholder(tf.float32,
                                 [None, image_width, image_height, image_channels],
                                 name="input_real")
    inputs_z = tf.placeholder(tf.float32, [None, z_dim], name="input_z")
    learning_rate = tf.placeholder(tf.float32, name="learning_rate")
    return inputs_real, inputs_z, learning_rate
```

19.1.5 构建鉴别器

ReLU 表示修正线性单元。我们常见的激活函数,只有输入超出阈值时才激活神经元。ReLU 对所有负的输入返回 0,对大于等于 0 的输入则返回自己。缺点就是在输入为负值时,输出始终为 0,这就导致了不能更新参数,也就是神经元不学习了,这种现象叫作死神经元(Dead Neuron)。为了解决这个问题,我们在 ReLU 函数的负半区间引入了泄露(Leaky),这就是 Leaky ReLU。

```python
# 创建 Leaky ReLU 激活函数
def leaky_relu(x, alpha):
    return tf.maximum(x * alpha, x)
```

为了能够重新使用神经网络中的变量,我们构建鉴别器模型时需要使用变量范围,即使用 tf.variable_scope()来初始化。因为模型输出时是二分类,所以就用 sigmoid 激活函数。

```python
def discriminator(images, reuse=False):
    """
    创建鉴别器网络
    参数 image: 输入图像的 Tensor
    参数 reuse: 权重是否要被重新使用
    返回值: (鉴别器的输出 Tensor,鉴别器的 logits 的 Tensor)
    """
    # 用 tf.contrib.layers.xavier_initializer()作为 kernel_initializerparameter,这样可
    以加快训练的收敛。
    alpha = 0.2
    with tf.variable_scope('discriminator', reuse=reuse):
        x1 = tf.layers.conv2d(images, 64, 5, strides=2, padding='same',
                              kernel_initializer=tf.contrib.layers.xavier_initializer())
        x1 = tf.maximum(alpha * x1, x1) # Leaky ReLU
        x1 = tf.nn.dropout(x1, 0.9)

        x2 = tf.layers.conv2d(x1, 128, 5, strides=2, padding='same',
                              kernel_initializer=tf.contrib.layers.xavier_initializer())
        x2 = tf.layers.batch_normalization(x2, training=True)
```

```python
        x2 = tf.maximum(alpha * x2, x2) # Leaky ReLU
        x2 = tf.nn.dropout(x2, 0.9)

        x3 = tf.layers.conv2d(x2, 256, 5, strides=2, padding='same',
                        kernel_initializer=tf.contrib.layers.xavier_initializer())
        x3 = tf.layers.batch_normalization(x3, training=True)
        x3 = tf.maximum(alpha * x3, x3) # Leaky ReLU
        x3 = tf.nn.dropout(x3, 0.9)
        flat = tf.reshape(x3, (-1, 4*4*256))
        logits = tf.layers.dense(flat, 1)
        out = tf.sigmoid(logits)
        return out, logits
```

19.1.6 构建生成器

使用 z 来构建生成器并生成图像。构建生成器模型，需要使用变量范围来初始化模型层级名称的前缀，根据传入的 reuse 来决定是否要重新使用。该函数返回生成的图像，它的维度是 28×28×out_channel_dim。

```python
def generator(z, out_channel_dim, is_train=True, reuse=True):
    """
    创建生成器网络
    参数 z：输入 z
    参数 out_channel_dim：输出图像的通道数
    参数 is_train：生成器是否要被训练
    参数 reuse：是否重新使用
    返回值：生成器的输出 Tensor
    """
    alpha = 0.2
    with tf.variable_scope('generator', reuse=not is_train):
        x1 = tf.layers.dense(z, 7*7*512)
        x1 = tf.reshape(x1, (-1, 7, 7, 512))
        x1 = tf.layers.batch_normalization(x1, training=is_train)
        x1 = tf.maximum(alpha * x1, x1)
        x1 = tf.nn.dropout(x1, 0.5)

        x2 = tf.layers.conv2d_transpose(x1, 256, 5, strides=2, padding='same')
        x2 = tf.layers.batch_normalization(x2, training=is_train)
        x2 = tf.maximum(alpha * x2, x2)
        x2 = tf.nn.dropout(x2, 0.5)

        x3 = tf.layers.conv2d_transpose(x2, 128, 5, strides=2, padding='same')
        x3 = tf.layers.batch_normalization(x3, training=is_train)
        x3 = tf.maximum(alpha * x3, x3)
        x3 = tf.nn.dropout(x3, 0.5)

        logits = tf.layers.conv2d_transpose(x3, out_channel_dim, 5, strides=1, padding='same')
        out = tf.tanh(logits)
    return out
```

19.1.7 计算模型损失

通过输入真实图像和生成的图像来计算 GAN 模型的损失，最后返回鉴别器损失和生成器损失。

```python
def model_loss(input_real, input_z, out_channel_dim):
    """
    获取鉴别器和生成器的损失
    参数 input_real：从真实图像获取的
    参数 input_z：Z 输入
    参数 out_channel_dim：输出图像的通道数
    返回值：(鉴别器 loss, 生成器 loss) 元组
    """
    smooth = 0.1
    # 构建生成器模型
    g_model = generator(input_z, out_channel_dim)
    # 构建鉴别器模型
    d_model_real, d_logits_real = discriminator(input_real)
    d_model_fake, d_logits_fake = discriminator(g_model, reuse=True)
    # 计算鉴别器对真实图像的 loss 和假图像的 loss
    d_loss_real = \
    tf.reduce_mean(
        tf.nn.sigmoid_cross_entropy_with_logits(logits=d_logits_real,
                                    labels=tf.ones_like(d_model_real) * (1 - smooth)))
    d_loss_fake = \
    tf.reduce_mean(
        tf.nn.sigmoid_cross_entropy_with_logits(logits=d_logits_fake,
                                    labels=tf.zeros_like(d_model_fake)))
    d_loss = d_loss_real + d_loss_fake
    # 计算生成器的 loss
    g_loss = tf.reduce_mean(tf.nn.sigmoid_cross_entropy_with_logits(logits=d_logits_fake,
    labels=tf.ones_like(d_model_fake)))
    return d_loss, g_loss
```

19.1.8 构建优化器

我们通过 tf.trainable_variables() 来获取鉴别器和生成器所有的可训练变量，然后使用这些变量来创建 Adam 优化器。

```python
def model_opt(d_loss, g_loss, learning_rate, beta1):
    """
    获取优化器
    参数 d_loss：鉴别器 loss 的张量
    参数 g_loss：生成器 loss 的张量
    参数 learning_rate：学习率占位符
    参数 beta1：优化器中第一个时刻的指数衰减率
    返回值：(鉴别器训练操作，生成器训练操作)
    """
    # 获取权重和偏差以更新操作
```

```
t_vars = tf.trainable_variables()
d_vars = [var for var in t_vars if var.name.startswith('discriminator')]
g_vars = [var for var in t_vars if var.name.startswith('generator')]
# 创建优化器
with tf.control_dependencies(tf.get_collection(tf.GraphKeys.UPDATE_OPS)):
    d_train_opt = tf.train.AdamOptimizer(learning_rate, beta1=beta1).minimize(d_loss, var_list=d_vars)
    g_train_opt = tf.train.AdamOptimizer(learning_rate, beta1=beta1).minimize(g_loss, var_list=g_vars)
return d_train_opt, g_train_opt
```

19.1.9 构建训练模型时的图像输出

训练模型时，我们希望每隔一个小阶段就输出图像以观察训练的结果状态。这里我们定义 TensorFlow 训练模型后的值，并将该值转换成图像输出。

```
from PIL import Image
from PIL import ImageOps
def images_square_grid(images, mode):
    """
    以正方形的网格形式保存图片
    参数 images：要保存网格形状的图片
    参数 mode：该图片的模式，灰度或者 RGB
    返回值：正方形网格图像 Image 类型的对象
    """
    # 获取正方形图片的网格的最大 size
    save_size = int(math.floor(np.sqrt(images.shape[0])))
    # 修改图片的数值，使其在 0 到 255 之间
    images = (((images - images.min()) * 255) / (images.max() - images.min())).astype(np.uint8)
    # 在正方形的排列中放置图片
    images_in_square = np.reshape(images[:save_size*save_size], (save_size, save_size, images.shape[1], images.shape[2], images.shape[3]))
    # 如果模式是 L，就表示灰度图片
    if mode == 'L':
        images_in_square = np.squeeze(images_in_square, 4)
    # 将图片组合构建成新的网格图片
    new_im = Image.new(mode, (images.shape[1] * save_size, images.shape[2] * save_size))
    for col_i, col_images in enumerate(images_in_square):
        for image_i, image in enumerate(col_images):
            im = Image.fromarray(image, mode)
            if mode == "L":
                im = ImageOps.invert(im)
            new_im.paste(im, (col_i * images.shape[1], image_i * images.shape[2]))
    return new_im
```

然后处理生成图像的函数，该图像数值是模型训练时生成器生成的图像的数值。

```
import matplotlib.pyplot as plt
def show_generator_output(sess, n_images, input_z, out_channel_dim, image_mode):
    """
    显示生成器的输出样本图像
    参数 sess：启动 TensorFlow 会话
    参数 n_images：要显示的图片数量
    参数 input_z：输入 z 张量
```

```
参数 out_channel_dim：输出图像的通道数量
参数 image_mode：图像使用的模式，L 或者 RGB
"""
cmap = None if image_mode == 'RGB' else 'gray'
z_dim = input_z.get_shape().as_list()[-1]
example_z = np.random.uniform(-1, 1, size=[n_images, z_dim])
# 生成器生成后数值样本
samples = sess.run(
    generator(input_z, out_channel_dim, False),
    feed_dict={input_z: example_z})
# 转换图像网格数值
images_grid = images_square_grid(samples, image_mode)
plt.imshow(images_grid, cmap=cmap)
plt.grid(False)
plt.show()
```

19.1.10 构建训练模型函数

构建训练模型函数，每迭代训练 10 次，输出一次日志信息；每迭代训练 50 次，输出一次图像。

```
def train(epoch_count, batch_size, z_dim, learning_rate, beta1, get_batches, data_shape, data_image_mode):
    """
    训练生成对抗网络 GAN 模型
    参数 epoch_count: epoch 的数量
    参数 batch_size: 批次大小
    参数 z_dim: z 维度
    参数 learning_rate: 学习率
    参数 beta1: 优化器中第一个时刻的指数衰减率
    参数 get_batches: 获取 batches 的函数
    参数 data_shape: 数据的形状 (shape)
    参数 data_image_mode: 图像使用的模式，L 或者 RGB
    """
    # 获得模型输入的 Tensor
    inputs_real, inputs_z, lr = model_inputs(data_shape[1], data_shape[2], data_shape[3], z_dim)
    # 获得模型的 loss
    d_loss, g_loss = model_loss(inputs_real, inputs_z, data_shape[3])
    # 获得模型的优化器
    d_optimizer, g_optimizer = model_opt(d_loss, g_loss, learning_rate, beta1)
    # 启动 TensorFlow 会话
    with tf.Session() as sess:
        sess.run(tf.global_variables_initializer())
        # 开始迭代训练
        iteration = 0
        for epoch_i in range(epoch_count):
            batches_generator = get_batches(batch_size)
            for batch_images in batches_generator:
                # 开始训练模型
                # 随机生成一个图像数据矩阵
                z_ = np.random.uniform(-1, 1, (batch_size, z_dim))
```

```
        # 更新鉴别器
        _ = sess.run(d_optimizer, feed_dict={inputs_real:batch_images*2, inputs_z:z_})
        # 更新生成器
        _ = sess.run(g_optimizer, feed_dict={inputs_z:z_, inputs_real:batch_images,})
        iteration += 1
        # 每迭代 10 次输出日志信息
        if iteration % 10 == 0:
            d_loss_ = d_loss.eval({inputs_z:z_, inputs_real:batch_images})
            g_loss_ = g_loss.eval({inputs_z:z_})
            print("Iteration: {}, d_loss_={:.5f}, g_loss_={:.5f}".format
                (iteration, d_loss_, g_loss_))
        # 每迭代 50 次输出图像以预览模型训练的效果
        if iteration % 50 == 0:
            show_generator_output(sess, 25, inputs_z, data_shape[3], data_image_mode)
```

19.1.11 训练 MNIST 数据集的 GAN 模型

我们创建一个 Python 的类来方便进行训练和调用，初始化类时传入所有图片的路径数组，然后加载所有的图片到 NumPy 数组。如果是 MNIST 数据集，就是使用灰度模式加载，只有一个通道数；如果是 LFW 数据集，就是使用 RGB 模式加载，有 3 个通道数。其中定义的 get_batches()函数是在训练模型时所需要获取的训练批次数据。

```
import math
class Dataset(object):
    def __init__(self, dataset_name, data_files):
        """
        参数 dataset_name：表示数据集类型名称
        参数 data_files：表示数据集的图片文件数组
        """
        DATASET_LFW_NAME = 'lfw'
        DATASET_MNIST_NAME = 'mnist'
        IMAGE_WIDTH = 28
        IMAGE_HEIGHT = 28
        # 如果是 LFW 数据集
        if dataset_name == DATASET_LFW_NAME:
            self.image_mode = 'RGB'
            image_channels = 3
        # 如果是 MNIST 数据集
        elif dataset_name == DATASET_MNIST_NAME:
            self.image_mode = 'L'
            image_channels = 1
        # 所有的图片路径数组
        self.data_files = data_files
        self.shape = len(data_files), IMAGE_WIDTH, IMAGE_HEIGHT, image_channels

    def get_batches(self, batch_size):
        """
        生成训练时所需的批次数据
        参数 batch_size：批次大小
```

```
        返回值：批次的数据
        """
        IMAGE_MAX_VALUE = 255
        current_index = 0
        while current_index + batch_size <= self.shape[0]:
            data_batch = self.get_batch(
                self.data_files[current_index:current_index + batch_size],
                *self.shape[1:3],
                mode=self.image_mode)
            current_index += batch_size
            yield data_batch / IMAGE_MAX_VALUE - 0.5

    def get_batch(self, image_files, width, height, mode):
        """
        参数 image_files：该批次的图片的文件数
        参数 width：图片的宽度
        参数 height：图片的高度
        参数 mode：表示图片的通道数，L 或者 RGB
        """
        data_batch = np.array(
            [self.get_image(sample_file, width, height, mode) for sample_file in
            image_files]).astype(np.float32)
        # 如果图片不是四维的，就转换成四维的
        if len(data_batch.shape) < 4:
            data_batch = data_batch.reshape(data_batch.shape + (1,))
        return data_batch

    def get_image(self, image_path, width, height, mode):
        """
        从图片路径中读取图片
        参数 image_path：图片的路径
        参数 width：图片的宽度
        参数 height：图片的高度
        参数 mode：图片的模式
        返回值：图片数值数据
        """
        image = Image.open(image_path)
        if image.size != (width, height):
            face_width = face_height = 108
            j = (image.size[0] - face_width) // 2
            i = (image.size[1] - face_height) // 2
            image = image.crop([j, i, j + face_width, i + face_height])
            image = image.resize([width, height], Image.BILINEAR)
        return np.array(image.convert(mode))
```

定义超参数训练 MNIST 的 GAN 模型。我们传入所有的 MNIST 图片的路径，共有 60000 个，每个批次大小是 64，每次迭代训练 937 个，一共要迭代训练 20 次，所以最后一共要训练运行 18740 次。

```
# 每个批次的大小
batch_size = 64
```

```python
# 输入 z 的维度
z_dim = 100
# 学习率
learning_rate = 0.001
# 指数衰减率
beta1 = 0.5
# 一共迭代训练 20 次
epochs = 20
# 通过所有的 MNIST 图片的路径数组初始化 MNIST 数据集
mnist_dataset = Dataset('mnist', mnist_all_paths)
with tf.Graph().as_default():
    train(epochs, batch_size, z_dim, learning_rate,
          beta1, mnist_dataset.get_batches,
          mnist_dataset.shape, mnist_dataset.image_mode)
```

训练到 300 次的时候，认不出来图片是数字几，我们来看输出的信息。

```
Iteration: 10, d_loss_=6.06957, g_loss_=0.14938
Iteration: 20, d_loss_=0.62488, g_loss_=4.46301
Iteration: 30, d_loss_=0.96122, g_loss_=2.96270
……
Iteration: 290, d_loss_=1.54455, g_loss_=1.98669
Iteration: 300, d_loss_=1.13601, g_loss_=1.32676
```

输出结果如图 19.3 所示。

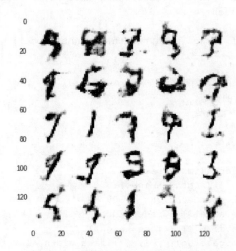

图 19.3　训练到 300 次时的图片显示

训练到 10000 次的时候，基本可以认出来图片是数字几，来看输出的信息。

```
……
Iteration: 9980, d_loss_=9.08727, g_loss_=3.25862
Iteration: 9990, d_loss_=9.34113, g_loss_=3.85601
Iteration: 10000, d_loss_=9.82517, g_loss_=4.62705
```

输出结果如图 19.4 所示。

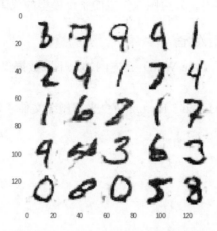

图 19.4 训练到 10000 次时的图像显示

训练到 18740 次的时候（训练完毕），图像完全可以被认出来，来看输出的信息。

……
```
Iteration: 18730, d_loss_=8.35124, g_loss_=3.89784
Iteration: 18740, d_loss_=8.53583, g_loss_=3.50421
```

输出结果如图 19.5 所示。

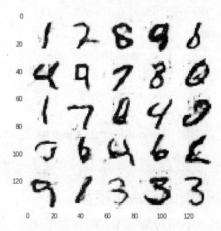

图 19.5 训练完毕的图像显示

19.2 基于 TensorFlow 的 GAN 实现 LFW 人脸图像生成

本节将通过以上的生成对抗网络模型的训练和代码来实现 LFW 数据集的人脸图像生成。MNIST 数据集是单通道的，灰度的图像；LFW 数据集是三通道的，RGB 的图像。它们在图像数据上是有区别的，所以在上一节的代码中，在一些必要的代码处加了判断是 L 模式的图像还是 RGB 模式的图像的代码。人脸图像生成的整个模型的代码仍然使用上一节的。

19.2.1 人脸图像数据准备

从 LFW（Labeled Faces in the Wild）官方网站上下载数据集，共有 13233 个人脸图像数据，下载地址在本代码仓库。

下载后，我们可以通过 tar -xvf lfw.tgz 命令来解压缩文件，然后通过 glob 模块把图片文件路径读取到数组中。

```
from glob import glob
# 加载所有的 jpg 图片路径
lfw_all_paths = glob('lfw/*/*.jpg')
lfw_all_paths[:10]
```

前 10 个图片路径输出如下。

```
['lfw/Lachlan_Murdoch/Lachlan_Murdoch_0001.jpg',
 'lfw/Donald_Hays/Donald_Hays_0001.jpg',
 'lfw/Nelson_Acosta/Nelson_Acosta_0001.jpg',
 'lfw/Sidney_Poitier/Sidney_Poitier_0001.jpg',
 'lfw/Charlie_Deane/Charlie_Deane_0001.jpg',
 'lfw/Joan_Laporta/Joan_Laporta_0002.jpg',
 'lfw/Joan_Laporta/Joan_Laporta_0001.jpg',
 'lfw/Joan_Laporta/Joan_Laporta_0006.jpg',
 'lfw/Joan_Laporta/Joan_Laporta_0005.jpg',
 'lfw/Joan_Laporta/Joan_Laporta_0008.jpg']
```

查看数组里包含的图片张数。

```
len(lfw_all_paths)
```

输出如下。

```
13233
```

随机查看 25 张人脸图片，调用 plot_random_25_img() 函数，该函数在上一节中已定义过。

```
plot_random_25_img(lfw_all_paths)
```

19.2.2 训练 LFW 数据集的 GAN 模型

参数值和上一节是一样的，不同的是迭代训练次数是 100。由于图片有 13233 张，那么每次要迭代 206 个，最终要迭代训练 20600 次。

```
batch_size = 64
z_dim = 100
learning_rate = 0.001
beta1 = 0.5
epochs = 100
lfw_dataset = Dataset(lfw_all_paths)
```

```
with tf.Graph().as_default():
    train(epochs, batch_size, z_dim, learning_rate,
          beta1, lfw_dataset.get_batches,
          lfw_dataset.shape, lfw_dataset.image_mode)
```

训练 600 次、10000 次和 20600 次时的图像显示，如图 19.6 所示。

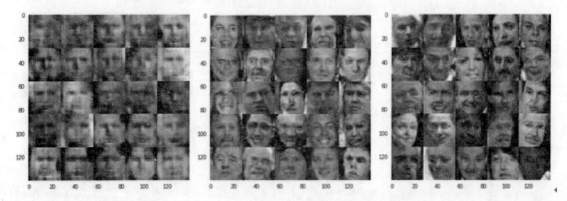

图 19.6　训练 600 次、10000 次和 20600 次时的图像显示

19.3　小结

本章主要讲的是如何生成逼真的图像的模型。一开始我们拿 MNIST 数据集来训练和生成手写数字图像，然后用同一套代码对 LFW 的人脸图像数据集训练模型，生成人脸图像，以达到让生成器模型能够生成出很逼真的图像的目的。

第 20 章

图像超分辨率

图像超分辨率（Image Super-Resolution）有多种实现方案，以生成对抗网络为基础的论文和实现非常多，各有各的优点和缺点。通过图像超分辨率的生成对抗网络来对模糊图像进行清晰化处理，以及用生成对抗神经网络模型对图像进行超分辨率的操作，放大 4 倍仍然清晰，而超分辨率（Super-Resolution，SR）主要的研究就是基于插值和重建。

20.1 效果预览与数据准备

不管是采用 TensorFlow 或者 Keras 实现图像超分辨率模型，大部分都是以深度生成对抗网络（DCGAN）模型为基础的。目前，该技术在更快、更深层的卷积神经网络对图像超分辨率的准确性和速度的研究上取得了一定进展。基于优化的超分辨率方法的行为主要由目标函数的选择来驱动，因此我们最主要的工作是最小化均方重建误差，这是因为生成器网络模型的主要工作是重建图像。

20.1.1 效果预览

在开始之前，我们先来看看在线版的图像超分辨率 Demo 的效果，预览效果链接在本代码仓库。

然后上传一张小图（宽：487px，高：470px），图片名称是 blur_img_test0.jpg。它是由 LFW 数据集随机读取到的 16 张人脸图片，LFW（Labeled Faces in the Wild）数据集我们在之前的章节中经常用到，这次是用的 4×4 排版的数据集。

在第 19 章中，我们生成人脸图片也是用 LFW 数据集，那么通过本章介绍的方法我们可以将第 19 章中生成的有点模糊的人脸图片进行清晰化计算和处理，并且还可以无损放大图片。

经过模型计算处理后，生成的新图像（宽：974px，高：940px），图片名称如下。

```
blur_img_test0_waifu2x_photo_noise3_scale_tta_1.png
```

生成的两倍大小的图片清晰度明显高。除了通过上面的链接可以查看最终的效果，我们还可以试试查看其他类似的在线演示效果，都是用 SRGAN 来实现的，页面地址在本代码仓库。

20.1.2 环境准备

- numpy = 1.14.6。
- matplotlib = 2.1.2。
- tensorflow = 1.12.0。
- scipy = 1.1.0。
- moviepy = 0.2.3.5。
- ffmpeg = 3.4.4。

20.1.3 数据准备

我们是用 LFW 数据集来进行模型训练和图像生成的，所以在训练模型前，需要事先准备好图片数据集。我们可以通过 Python 的内置 glob 模块加载所有图片的路径，之后使用图片的文件路径来加载图片以查看人脸图片数据。这里加载的目录名称是 dataset，我们要把 LFW 数据集压缩包解压后的目录名称修改成 dataset，后面训练时会直接加载该目录路径。

```
from glob import glob
all_filenames = glob("dataset/*/*")
# 查看前 10 个文件名
all_filenames[:10]
```

输出如下。

```
['dataset/Uzi_Landau/Uzi_Landau_0001.jpg',
 'dataset/Selma_Phoenix/Selma_Phoenix_0001.jpg',
 'dataset/Kamel_Morjane/Kamel_Morjane_0001.jpg',
 'dataset/Sargis_Sargsian/Sargis_Sargsian_0001.jpg',
 'dataset/Maria_Shriver/Maria_Shriver_0004.jpg',
 'dataset/Vincent_Gallo/Vincent_Gallo_0003.jpg',
 'dataset/Vincent_Gallo/Vincent_Gallo_0002.jpg',
 'dataset/Vincent_Gallo/Vincent_Gallo_0001.jpg']
```

变量 all_filenames 数组里有全部图片的路径。随机查看 25 张人脸图片，通过 Python 内置的 random 模块的 sample()方法，可以从指定的数组里随机获取图片路径。

```
import matplotlib.pyplot as plt
from matplotlib import image
import random
def plot_random_25_img(filepaths):
    # 随机获取 25 张人脸图片的路径
    random_25_imgs = random.sample(filepaths, 25)
```

```
    fig, axes = plt.subplots(nrows=5, ncols=5)
    fig.set_size_inches(8, 8)
    index = 0
    # 遍历 5 行
    for row_index in range(5):
        # 遍历 5 列
        for col_index in range(5):
            # 读取图片
            img = image.imread(random_25_imgs[index])
            ax = axes[row_index, col_index]
            # 显示图片
            ax.imshow(img)
            ax.grid(False)
            index += 1
plot_random_25_img(all_filenames)
```

20.2 基于 TensorFlow 的 DCGAN 实现超分辨率

ResNet 就是深度残差网络（Deep Residual Networks）模型，它已经成为一种深层的架构，显示出了非常不错的准确性和良好的收敛。通过分析残差构建块背后的传播公式，得出的是：当使用标识映射来跳过连接和加上激活后，前向或者后向的信号可以直接从一个块传播到其他任何块，这使训练更容易且泛化。

20.2.1 下载 srez 代码库

代码文件在 GitHub 上，可以通过 git 命令将其克隆下来，也可以通过 Web 页面的下载按钮直接下载。如果是在 Jupyter Notebook 中，需要在命令前面加一个半角的感叹号才可以执行 git 命令。

```
git clone 请指定本srez.git 代码仓库地址
```

该代码仓库中包含 5 个代码文件，分别解释如下。

- srez_main.py：主要的入口文件。
- srez_train.py：训练模型文件。
- srez_input.py：图片的输入转换文件。
- srez_model.py：生成器、鉴别器和损失函数的模型文件。
- srez_demo.py：根据阶段性训练的图片生成视频的代码文件。

20.2.2 训练模型根据模糊图像生成清晰图像

通过 srez_main.py 代码文件来训练模型，既可以在终端命令行里执行，也可以在 Jupyter Notebook 中执行，命令执行位置取决于读者下载该代码仓库的目录。当我们要执行 Python 的脚本文件时，如果希望从命令行上传入参数，那么可以通过以下方式上传，打开 srez_main.py 文件里的 tf.app.flags.DEFINE_ 后缀的代码。

```
python3 srez_main.py \
--run "train" \
--batch_size 16 \
--checkpoint_dir "checkpoint" \
--checkpoint_period 100 \
--dataset "dataset" \
--epsilon 1e-8 \
--gene_l1_factor 0.90 \
--learning_beta1 0.5 \
--learning_rate_start 0.00020 \
--learning_rate_half_life 5000 \
--sample_size 168 \
--summary_period 10 \
--random_seed 0 \
--test_vectors 16 \
--train_dir "train" \
--train_time 20
```

以上带--前缀的参数解释如下。

- run：我们要执行的操作，train 表示训练，demo 表示生成视频。
- batch_size：每个批次的样本大小。
- checkpoint_dir：存储检查点的文件目录。
- checkpoint_period：每训练多少个批次保存一次检查点。
- dataset：数据集目录名称。
- epsilon：模糊项以避免数值不稳定。
- gene_l1_factor：专为乘数的生成器 L1 的损失项。
- learning_beta1：Adam 优化器的 Beta1 参数。
- learning_rate_start：Adam 优化器一开始使用的学习率。
- learning_rate_half_life：学习率的终止设置。
- sample_size：图片样本的大小，像素单位。
- summary_period：每训练多少个批次保存一次训练效果。
- random_seed：随机种子。
- test_vectors：测试集数量。
- train_dir：训练后的文件存储目录。
- train_time：训练时长，单位为"分钟"。

经过训练后，输出如下。

```
进度[ 1%], ETA[ 19m], Batch [  10], G_Loss[0.212], D_Real_Loss[1.112], D_Fake_Loss[0.464]
进度[ 2%], ETA[ 19m], Batch [  20], G_Loss[0.171], D_Real_Loss[1.141], D_Fake_Loss[0.499]
进度[ 3%], ETA[ 19m], Batch [  30], G_Loss[0.177], D_Real_Loss[0.865], D_Fake_Loss[0.626]
……
进度[98%], ETA[  0m], Batch [1070], G_Loss[0.096], D_Real_Loss[0.601], D_Fake_Loss[0.795]
进度[99%], ETA[  0m], Batch [1080], G_Loss[0.112], D_Real_Loss[0.696], D_Fake_Loss[0.619]
```

进度[100%], ETA[0m], Batch [1090], G_Loss[0.097], D_Real_Loss[0.622], D_Fake_Loss[0.752]
检查点 Checkpoint 文件保存成功! 文件名: checkpoint/checkpoint_new.txt。
训练完毕!

20.2.3 输出效果预览

经过训练后,我们在 train 目录下会发现一些图片,这些图片的张数是由我们在训练时传入的参数 summary_period 决定的。这里定义的是 10 张,所以每训练 10 次就会保存一张图片。现在我们来看第一次生成的图片。

```
import matplotlib.pyplot as plt
from matplotlib import image
def plot_img(filepath):
    fig, ax = plt.subplots()
    fig.set_size_inches(8, 8)
    img = image.imread(filepath)
    ax.imshow(img)
    ax.grid(False)
plot_img("train/No_10.000000_trained_out.png")
```

在训练模型时,每训练 10 次保存一张 8 个人的人脸图片,第 1 列是输入的小图像,所以能看到像素点;第 2 列是标准双三次插值;第 3 列是由神经网络生成的输出图像;第 4 列是原图。输出结果如图 20.1 所示。

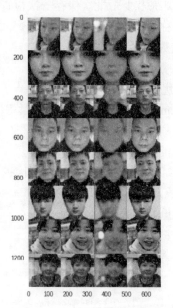

图 20.1 第 10 次训练时生成的训练预览图

再来看第 1090 次输出图像的效果。

```
plot_img("train/No_1090.000000_trained_out.png")
```

输出结果如图 20.2 所示。

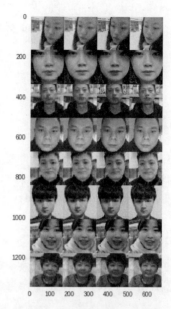

图 20.2　第 1090 次训练时生成的训练预览图

20.2.4　生成效果图视频

依然以 srez_main.py 为主文件入口，这次我们在 run 参数中传入 demo，表示将所有生成的图像生成视频。

```
python3 srez_main.py --run demo
```

20.2.5　图片放大高清化

我们通过 git 命令下载该代码库，该代码库由论文《使用深度卷积神经网络的图片超分辨率》而来，调用起来非常简单。

```
git clone 请指定本 Image-Super-Resolution.git 代码仓库地址
```

然后进入 Image-Super-Resolution 仓库的目录下。生成转换前，我们需要先来看该图片，图片文件名是 blur_img_test1.jpg。

```
import matplotlib.pyplot as plt
from matplotlib import image
def plot_img(filepath):
    fig, ax = plt.subplots()
    fig.set_size_inches(10, 10)
    img = image.imread(filepath)
    ax.imshow(img)
    ax.grid(False)
plot_img("blur_img_test1.jpg")
```

对于输出的图片，现在我们进行放大后，图片不会显示像素点，反而更高清化了。

```
python main.py "blur_img_test1.jpg"
```

等半分钟左右处理完毕，新的文件名是在原来的文件名基础上添加后缀生成的，这就方便我们辨认。

```
plot_img("blur_img_test1_intermediate_.jpg")
```

20.3　srez 库的代码分析

我们对 srez 库的主要文件和函数代码进行分析，以便于读者对该库有一个更好的认识。srez 库主要文件有主入口代码文件（srez_main.py），创建模型代码文件（srez_model.py）和训练模型代码文件（srez_train.py）。接下来，我们将一一解释文件中主要函数的作用。

20.3.1　主入口函数代码分析

在 srez_main.py 文件中的_train()代码函数解释是：先获取数据所有图像的路径，其次将其分割成训练数据集和测试数据集，然后创建生成器和鉴别器模型，再创建优化器，最后训练模型。

```
def _train():
    # 初始化全局的 TensorFlow 的 session
    sess, summary_writer = setup_tensorflow()
    # 获取所有图片的文件路径
    all_filenames = prepare_dirs(delete_train_dir=True)
    # 将图片路径分割成训练数据集和测试数据集
    train_filenames = all_filenames[:-FLAGS.test_vectors]
    test_filenames  = all_filenames[-FLAGS.test_vectors:]
    # 对图片数据集的输入队列进行异步处理
    train_features, train_labels = srez_input.setup_inputs(sess, train_filenames)
    test_features,  test_labels  = srez_input.setup_inputs(sess, test_filenames)
    # 在训练时添加一些噪点，降噪自编码器
    noise_level = .03
    noisy_train_features = train_features + \
                           tf.random_normal(train_features.get_shape(), stddev=noise_level)
    # 创建模型
    [gene_minput, gene_moutput,
     gene_output, gene_var_list,
     disc_real_output, disc_fake_output, disc_var_list] = \
            srez_model.create_model(sess, noisy_train_features, train_labels)
    # 创建生成器的 loss
    gene_loss = srez_model.create_generator_loss(disc_fake_output, gene_output,
    train_features)
    # 创建鉴别器的 loss
    disc_real_loss, disc_fake_loss = \
                         srez_model.create_discriminator_loss(disc_real_output, disc_fake_output)
    disc_loss = tf.add(disc_real_loss, disc_fake_loss, name='disc_loss')
    # 创建优化器
    (global_step, learning_rate, gene_minimize, disc_minimize) = \
```

```
            srez_model.create_optimizers(gene_loss, gene_var_list,
                                         disc_loss, disc_var_list)
    # 训练模型
    train_data = TrainData(locals())
    srez_train.train_model(train_data)
```

20.3.2 创建模型代码分析

_train()函数里要执行的创建模型代码,也就是在 srez_model.py 文件里的 create_model()函数,该函数中用已定义的_generator_model()函数来创建生成器模型,用_discriminator_model()函数来创建鉴别器模型。

```
def create_model(sess, features, labels):
    # 获取图片的 shape
    rows      = int(features.get_shape()[1])
    cols      = int(features.get_shape()[2])
    channels  = int(features.get_shape()[3])
    # 创建生成器输入图像
    gene_minput = tf.placeholder(tf.float32, shape=[FLAGS.batch_size, rows, cols, channels])
    # 创建生成器模型,定义变量的作用范围,该范围下的每个层的名称都有前缀 gene
    with tf.variable_scope('gene') as scope:
        gene_output, gene_var_list = \
                    _generator_model(sess, features, labels, channels)
        scope.reuse_variables()
        gene_moutput, _ = _generator_model(sess, gene_minput, labels, channels)
    # 创建一个真实数据的鉴别器 Tensor
    disc_real_input = tf.identity(labels, name='disc_real_input')
    # 创建鉴别器模型,定义变量的作用范围,该范围下的每个层的名称都有前缀 disc
    with tf.variable_scope('disc') as scope:
        disc_real_output, disc_var_list = \
                 _discriminator_model(sess, features, disc_real_input)
        scope.reuse_variables()
        disc_fake_output, _ = _discriminator_model(sess, features, gene_output)
    return [gene_minput,      gene_moutput,
            gene_output,      gene_var_list,
            disc_real_output, disc_fake_output, disc_var_list]
```

20.3.3 训练模型代码分析

最后要执行_train()函数里的训练模型代码,也就是在 srez_train.py 文件里的 train_model()函数,该函数用来训练模型,保存检查点则用_save_checkpoint()函数。训练过程中,保存图像训练的状态效果图等信息用_summarize_progress()函数。

```
def train_model(train_data):
    td = train_data
    # 初始化全局变量
    td.sess.run(tf.global_variables_initializer())
    # 设置基本参数
    lrval      = FLAGS.learning_rate_start
    start_time = time.time()
```

```python
        done = False
        batch = 0
        # 缓存测试数据集的 features 和 labels
        test_feature, test_label = td.sess.run([td.test_features, td.test_labels])
        # 通过开始循环来训练模型,
        while not done:
            batch += 1
            # 随机抽取一个 loss 初始值
            gene_loss = disc_real_loss = disc_fake_loss = -1.234
            feed_dict = {td.learning_rate : lrval}
            # 训练模型
            ops = [td.gene_minimize, td.disc_minimize, td.gene_loss, td.disc_real_loss,
            td.disc_fake_loss]
            _, _, gene_loss, disc_real_loss, disc_fake_loss = td.sess.run(ops, feed_dict=
            feed_dict)
            # 每训练10 个批次,打印一次统计日志信息
            if batch % 10 == 0:
                elapsed = int(time.time() - start_time)/60
                print('进度[%3d%%], ETA[%4dm], Batch [%4d], G_Loss[%3.3f], D_Real_Loss
                [%3.3f], D_Fake_Loss[%3.3f]' %
                    (int(100*elapsed/FLAGS.train_time), FLAGS.train_time - elapsed,
                     batch, gene_loss, disc_real_loss, disc_fake_loss))
                # 判断是否要结束训练
                current_progress = elapsed / FLAGS.train_time
                if current_progress >= 1.0:
                    done = True
                # 更新学习率
                if batch % FLAGS.learning_rate_half_life == 0:
                    lrval *= .5
            # 每训练到指定的次数时,保存一次状态效果图。这里设置的是10,也就是每训练10 次保存一张
            if batch % FLAGS.summary_period == 0:
                # 通过测试数据集的表现来保存状态效果图
                feed_dict = {td.gene_minput: test_feature}
                gene_output = td.sess.run(td.gene_moutput, feed_dict=feed_dict)
                _summarize_progress(td, test_feature, test_label, gene_output, batch, 'out')
            # 保存检查点文件,对于参数 checkpoint_period 我们设置的是100,也就是每训练100 次,保存一次检查点文件状态
            if batch % FLAGS.checkpoint_period == 0:
                _save_checkpoint(td, batch)
        _save_checkpoint(td, batch)
        print('训练完毕!')
```

20.4 小结

我们继续使用 LFW 人脸图像数据集,通过使用 DCGAN 实现的 srez 开源库来训练 LFW 数据集,生成新的高清图像。最后我们对 srez 代码库的 3 个主代码块,分别是入口函数、创建模型和训练模型代码块,进行了代码分析。

第 21 章

移花接木

移花接木，简单地从名字上看，就是将 A 对象的属性嫁接到 B 对象上。其原理确实如此，达到这一效果需使用到的技术就是 CycleGAN，也就是循环一致的对抗网络（Cycle-Consistent Adversarial Networks）。它是一种复杂的生成对抗网络模型，不需要图像一对一地配对就可以实现训练模型和生成模型。本章将讲解 3 种 CycleGAN 模型，第一种是根据苹果生成橘子，第二种是根据马生成斑马，第三种是根据男性面貌生成女性面貌。这 3 种模型反过来也是可以生成的。

21.1 基本信息

先看代码运行后的效果，这可能会让读者更加有兴趣探究 CycleGAN 模型。所以我们在训练模型前，可以先来预览这 3 种模型生成的效果图，当然了，也可以基于此技术生成更多、更好玩的例子。这 3 种深度学习模型，除了可以实现从 A 到 B 的生成，也可以实现从 B 到 A 的生成。因为它在训练模型时，就已经训练好了两个模型，分别是顺序的和逆序的。

21.1.1 3 种模型效果预览

第一种 CycleGAN 模型是根据橘子生成苹果（orange2apple）的模型，根据橘子的原始图片生成苹果图片后的效果图，如图 21.1 所示。

第二种 CycleGAN 模型是根据马生成到斑马（horse2zebra）的模型，根据马的原始图片生成斑马图片后的效果图，如图 21.2 所示。

第三种是根据男性人脸面貌生成女性人脸面貌（male2female）的模型，根据男性人脸的原始图片生成女性人脸图片后的效果图，如图 21.3 所示。

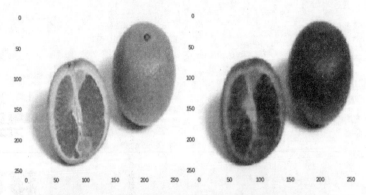

图 21.1　左图是橘子，右图是生成的苹果图片效果

图 21.2　左图是马，右图是生成的斑马

图 21.3　左图是原始男性人脸面貌，中间是原始女性人脸面貌，右图是生成的女性人脸面貌

21.1.2　环境准备

tensorflow = 1.12.0。

21.1.3　图片数据集准备

3 种图片数据集都可以通过前言中提到的链接下载。其中，有两种图片的数据集是在 Github 上公开的，也可以通过访问链接进行下载。这个页面有十几种数据集，本章我们采用 apple2orange 和 horse2zebra 的图片数据集，读者有兴趣也可以尝试用其他的图片数据集训练有趣的模型。数据集在本代码仓库中可查找。

21.1.4 CycleGAN 网络模型架构

CycleGAN 引入了循环一致性的约束,它从源分布转换到目标分布,再转换到源分布,如图 21.4 所示。图 21.4 清晰地描述了这个网络模型的架构。该架构有两个鉴别器,一个是对应的源图像,一个是对应的生成的图像。生成器也有两个,一个用来从源图像生成目标图像,我们称之为 A2B;另一个是反转,即从目标图像生成源图像,我们称之为 B2A。

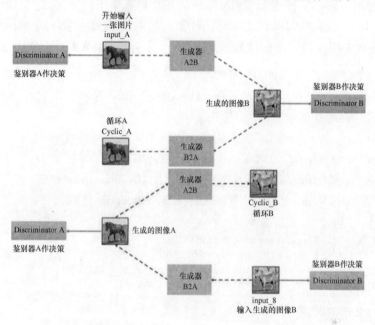

图 21.4 CycleGAN 网络模型的架构

21.2 基于 CycleGAN 根据苹果生成橘子

我们有两个生成器网络(Generator Network)和两个鉴别器网络(Discriminator Network),它们之间互相对抗竞赛。生成器从所需分布中生成图片样本,鉴别器尝试去鉴定该样本来自真实分布样本还是由生成器产生,生成器和鉴别器互相训练。生成器先从 A 到 B,再从 B 到 A,以这种循环周期的方式来训练。

21.2.1 下载代码库

通过 git 命令克隆该仓库下来,它包含了几个文件,分别解释如下。
- download_dataset.sh:下载图片数据集。
- build_data.py:构建图片数据,使其创建 tfrecords 格式的文件。
- train.py:训练模型。
- export_graph.py:导出计算图,以便后续使用。

- inference.py:推理生成图像。

```
git clone 请指定本CycleGAN-TensorFlow-1代码仓库地址
```

下载完毕后,进入 CycleGAN-TensorFlow-1 目录。

21.2.2 图片数据处理

我们先通过以下脚本文件将 apple2orange 的图片数据集下载下来,它的训练集包含了 995 张苹果图片和 1019 张橘子图片。该脚本将图片数据集下载下来后,会自动解压到 data 目录下,然后以 4 个文件夹来代表训练集图片和测试集图片,分别是 trainA 和 trainB,以及 testA 和 testB。

```
bash download_dataset.sh apple2orange
```

然后,将数据集里的图片转换成 TensorFlow 的 TFRecords 格式的文件,执行该脚本文件需要的 4 个参数分别解释如下。

- 参数 X_input_dir:训练时输入的 A,也就是苹果图片集目录。
- 参数 Y_input_dir:训练时输入的 B,也就是橘子图片集目录。
- 参数 X_output_file:苹果图片数据集输出的 tfrecords 格式文件。
- 参数 Y_output_file:橘子图片数据集输出的 tfrecords 格式文件。

```
python3 build_data.py \
--X_input_dir "data/apple2orange/trainA" \
--Y_input_dir "data/apple2orange/trainB" \
--X_output_file "data/apple2orange/apple.tfrecords" \
--Y_output_file "data/apple2orange/orange.tfrecords"
```

21.2.3 训练模型

通过刚才生成的文件来训练模型,单 GPU 的情况下,这个训练过程需要 6～10 小时才能训练完毕,训练的迭代次数取决于模型的效果。也就是说,比如我们可以在训练 1 万次的时候停止训练,看看模型的效果,如果效果不好,我们再接着训练。参数 X 和 Y 分别解释如下。

- 参数 X:苹果图片集的 tfrecords 文件。
- 参数 Y:橘子图片集的 tfrecords 文件。

正常训练模型的代码如下。

```
python3 train.py \
--X "data/apple2orange/apple.tfrecords" \
--Y "data/apple2orange/orange.tfrecords"
```

训练时输出如下。

```
INFO:root:-----------Step 0:-------------
INFO:root:  G_loss   : 5.2771946907043469
INFO:root:  D_Y_loss : 0.8465369701385498
INFO:root:  F_loss   : 4.4309234356254654
INFO:root:  D_X_loss : 0.7143246003433243
……
```

```
INFO:root:-----------Step 10200:-------------
INFO:root:   G_loss    : 3.5771946907043467
INFO:root:   D_Y_loss  : 0.5465438943385490
INFO:root:   F_loss    : 2.7265782356262207
INFO:root:   D_X_loss  : 0.2124302342423074
```

假如我们在 10200 次时停止了训练，此时模型会自动保存最后一次的训练状态和检查点文件。然后我们可以尝试去看看模型的效果，如果不太好，我们就继续训练。继续训练的代码如下。

```
python3 train.py \
--X "data/apple2orange/apple.tfrecords" \
--Y "data/apple2orange/orange.tfrecords" \
--load_model "checkpoints/20190118-0255"
```

继续训练该模型，需要加一个参数 load_model，它的值是检查点文件的目录路径。因为是继续训练模型，所以该参数会先把最后一次检查点文件读取到 TensorFlow 计算图中，然后继续训练模型，训练时输出如下。

```
……
INFO:root:-----------Step 21700:-------------
INFO:root:   G_loss    : 3.296196699142456
INFO:root:   D_Y_loss  : 0.2065090537071228
INFO:root:   F_loss    : 3.176140546798706
INFO:root:   D_X_loss  : 0.2601272165775299
```

读者可以根据自己的需要来停止或继续训练模型，直到得到想要的最佳模型。这里训练到 21700 次就终止了训练。

21.2.4 导出模型

训练完模型后，会在 checkpoints 目录下生成检查点文件，对此我们可以将其导出为扩展名为 pb 的文件，以方便之后生成图片时直接使用模型。导出的模型可以是 A2B 或 B2A 的模型，该脚本文件的参数分别解释如下。

- 参数 checkpoint_dir：检查点文件的目录路径。
- 参数 XtoY_model：导出 A2B 的模型的模型文件名，这里是 apple2orange.pb。
- 参数 YtoX_model：导出 B2A 的模型的模型文件名，这里是 orange2apple.pb。
- 参数 image_size：图片大小。

```
python export_graph.py \
--checkpoint_dir checkpoints/20190118-0255 \
--XtoY_model apple2orange.pb \
--YtoX_model orange2apple.pb \
--image_size 256
```

执行完毕后，会在当前目录创建一个叫作 pretrained 的新文件夹，它的目录下包含两个文件，就是 apple2orange.pb 和 orange2apple.pb 文件。

21.2.5 测试图片

通过使用 inference.py 文件来推理生成新图片，执行此脚本文件的参数分别解释如下。
- 参数 model：训练模型的模型文件。
- 参数 input：输入的图片文件的路径，即我们要通过这张图片来生成新的图片。
- 参数 output：新生成图片的输出文件名。
- 参数 image_size：图片大小。

```
python inference.py \
--model pretrained/apple2orange.pb \
--input data/apple2orange/testA/n07740461_41.jpg \
--output output_sample.jpg \
--image_size 256
```

推理生成完毕后，我们想让这张图片显示出来，就定义 plot_image()函数，传入一个图片地址即可。

```
import os
from PIL import Image
import matplotlib.pyplot as plt
def plot_image(img_path):
    img = Image.open(img_path)
    plt.figure()
    plt.imshow(img)
    plt.grid(False)
    plt.show()
```

输出源苹果图片。

```
plot_image("data/apple2orange/testA/n07740461_41.jpg")
```

输出结果如图 21.5 所示。

图 21.5 源苹果图片

输出生成的橘子图片。

```
plot_image("output_sample.jpg")
```

输出结果如图 21.6 所示。

图 21.6　生成的橘子图片

然后我们再测试一张从橘子到苹果的图片，对于 plot_image() 函数，把它的输出图片参数名称修改为源图片名称加后缀的形式，显示橘子图片的代码如下。

```
plot_image("data/apple2orange/testB/n07749192_401.jpg")
```

输出结果如图 21.7 所示。

图 21.7　源橘子图片

显示生成的苹果图片，它的文件名是以源图片名称来作为后缀的。

```
plot_image("output_sample_n07749192_401.jpg")
```

输出结果如图 21.8 所示。

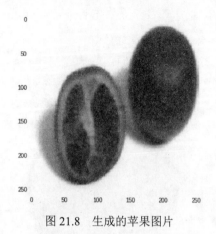

图 21.8 生成的苹果图片

21.3 基于 CycleGAN 根据马生成斑马

上一节已经下载了代码库，脚本文件和命令也都使用了一遍。本节还是使用这些脚本文件，命令也不变，变的只是参数。这次使用到的数据集是 horse2zebra.zip（马到斑马）的图片数据集。

21.3.1 图片数据处理

通过以下脚本文件将图片数据集下载下来，它的训练集包含了 1067 张马的图片和 1334 张斑马的图片。该脚本将图片数据集下载下来后，会自动解压到 data 目录下，然后以 4 个文件夹来代表训练集和测试集，分别是 trainA 和 trainB，以及 testA 和 testB。

```
bash download_dataset.sh horse2zebra
```

然后，将数据集里的图片转换成 TensorFlow 的 TFRecords 格式的文件，执行该脚本文件。

```
python3 build_data.py \
--X_input_dir "data/horse2zebra/trainA" \
--Y_input_dir "data/horse2zebra/trainB" \
--X_output_file "data/horse2zebra/horse.tfrecords" \
--Y_output_file "data/horse2zebra/zebra.tfrecords"
```

21.3.2 训练模型

通过刚才生成的文件来训练模型，可能需要几个小时才能训练完，这取决于读者的计算机的性能。

```
python3 train.py \
--X "data/horse2zebra/horse.tfrecords" \
--Y "data/horse2zebra/zebra.tfrecords"
```

在训练过程中，我们可以通过 tensorboard 工具来实时查看训练的情况。

```
tensorboard --logdir checkpoints/20190118-0326
```

参数 logdir 的值就是训练时日志存储的目录路径，这里就是在 checkpoints 目录下。该目录下有检查点文件，它们是以 checkpoint 开头的；有训练日志文件，以 events 开头。然后在浏览器中打开输出的提示地址，来浏览训练时的损失值走势图。这里截取了 tensorboard 两张图，分别是生成器的损失值走势图和判别器的损失值走势图，输出结果如图 21.9 和图 21.10 所示。

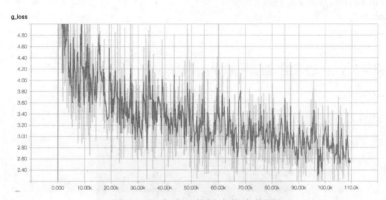

图 21.9　生成器的损失值走势图

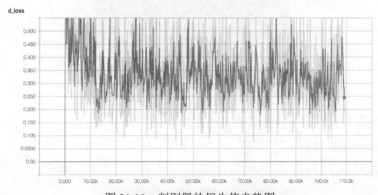

图 21.10　判别器的损失值走势图

21.3.3　导出模型

训练完模型后，会在 checkpoints 目录下生成检查点文件，对此我们可以将其导出为扩展名为 pb 的文件，以方便之后生成图片时直接使用该模型。导出模型包含 A2B 模型和 B2A 模型。

```
python export_graph.py \
--checkpoint_dir checkpoints/20190118-0326 \
--XtoY_model horse2zebra.pb \
--YtoX_model zebra2horse.pb \
--image_size 256
```

执行完毕后，在当前目录下的 pretrained 文件夹里会有两个 pb 文件，分别是 horse2zebra.pb 和 zebra2horse.pb 文件。

21.3.4 测试图片

通过使用 inference.py 文件来推理生成新的斑马图片，这里我们以两匹马的原始图片作为例子。

```
python inference.py \
--model pretrained/horse2zebra.pb \
--input horse.jpg \
--output zebra.jpg \
--image_size 256
```

推理生成完毕后，我们来看这张马的原始图片。

```
plot_image("horse.jpg")
```

输出结果如图 21.11 所示。

图 21.11　马的原始图片

再来看由这张马的原始图片生成的斑马图片。

```
plot_image("zebra.jpg")
```

输出结果如图 21.12 所示。

图 21.12　生成的斑马图片

再次推理生成一张斑马的图片，我们来看这张马的原始图片。

`plot_image("horse2.jpg")`

输出结果如图 21.13 所示。

图 21.13　马的原始图片

再来看由这张马的原始图片生成的斑马图片。

`plot_image("zebra2.jpg")`

输出结果如图 21.14 所示。

图 21.14　生成的斑马图片

21.4　男性和女性的人脸面貌互换

男性人脸面貌和女性人脸面貌互换可以使用上面的 CycleGAN 技术来实现，但是本节我们将用另一种方式，这种方式的表现效果在特定条件下是非常好的。虽然它的泛化能力没有 CycleGAN 训练的模型好，但足以做到使男性和女性的人脸面貌互换。这次我们使用的人脸图片数据来自 LFW。

21.4.1　环境准备

- numpy = 1.15.3。
- dlib = 19.16.0。

- cv2 = 3.4.3。

克隆本次的 git 代码仓库。

`git clone 请指定本 faceswap.git 代码仓库地址`

该代码库里 faceswap.py 文件中用到了 dlib 库，它是一个 C++ 的工具包，涵盖了大量复杂的机器学习算法，已经被学术界和工业界广泛使用。安装方法是先下载该安装包的压缩文件，下载 dlib-19.16.tar.bz2 压缩文件，地址在本代码仓库。

下载压缩文件后，解压到当前目录，然后进入 dlib-19.16 目录进行安装。

`python setup.py install`

安装完后，再去下载一个已训练过的模型，该模型是用于 face_landmarks 的，读者可以通过前言中提到的地址下载 dlib 和模型。这里最重要的代码文件就是 faceswap.py 脚本文件，它用来读取模型和计算处理输入的图像的 Python 脚本文件。

21.4.2　计算和生成模型

人脸图片也可以通过前言中提到的地址下载，这次我们先将女性人脸面貌转换成男性人脸面貌，女性人脸面貌图片的文件名是 77_female_2.jpg，男性人脸面貌的图片文件名则是 Victor_2.jpg。

`python faceswap.py Victor_2.jpg 77_female_2.jpg`

输出结果如图 21.15 所示。

图 21.15　左图为原女性人脸面貌图片，中间为原男性人脸面貌图片，右图是生成的图片

我们再来看由男性人脸面貌图片到女性人脸面貌图片的生成过程，女性人脸面貌图片的文件名是 99_female_4.jpg，男性人脸面貌图片的文件名是 Victor_3.jpg。

`python faceswap.py 99_female_4.jpg Victor_3.jpg`

输出结果如图 21.16 所示。

图 21.16　左图为原男性人脸面貌图片，中间为原女性人脸面貌图片，右图是生成的图片

21.4.3 代码分析

从刚才我们使用的 faceswap.py 脚本文件以及后面跟着的两个图片文件名地址可以看出，这就是将其中一张图片的人脸面部特征融合到另一张带有人脸的图片上。这种操作一般分为 4 个步骤，分别如下。

（1）检测面部标志。
（2）旋转、缩放和平移第一个图片以适应第二个图片。
（3）调整第一个图片中的人脸特征点，使其与第二张图片的人脸特征点匹配。
（4）将第一张图片的特征混合在第二张图片的顶部，并使色彩适配和平衡。

我们以男性人脸面貌生成女性人脸面貌来举例，通过 faceswap.py 文件中的 get_landmarks() 的相关函数来获取人脸面部特征值，演示如图 21.17 所示。

图 21.17　步骤 1：检测人脸的面部特征值

此时两张图片的人脸特征都被检测出来了，然后我们通过平移、缩放和旋转将第一张图片上的人脸跟第二张图片的人脸对齐。例如，鼻子对鼻子，眼睛对眼睛，嘴巴对嘴巴等。演示如图 21.18 所示。

第一张输入图片　　　　第二张输入图片　　　将两张人脸的位置和角度对齐

图 21.18　步骤 2：对齐两张人脸的位置和角度，第三张覆盖图片是 0.5 的透明度（为方便观看显示）

然后我们调整第一张图片的人脸特征点，使其与第二张图片的人脸特征点匹配，再把除了人脸以外的多余图像删除，演示如图 21.19 所示。

图 21.19　步骤 3：删除多余的图像并使两张人脸叠加

最后混合两张人脸为一体，使第一张人脸的色彩均衡并与第二张匹配，演示如图 21.20 所示。

图 21.20　最终演示图

21.5　小结

本章采用的是加州大学伯克利分校的一名学生收集并开放给大家研究使用的 apple2orange 和 horse2zebra 图像数据集。本章基于 TensorFlow 的 CycleGAN 生成技术，根据苹果生成橘子，根据马生成斑马，反之也可以。最后我们使用 faceswap 开源库实现了男女人脸互换。